'This volume is the first of its kind to collect viewpoints on formal modelling in transition studies. It provides details on modelling philosophy as well as how to apply basic techniques, such as system dynamics or agent-based modelling. A limited team of authors guarantees coherence and complementarity of the contributions.'

Jeroen van den Bergh, ICTA, Universitat Autònoma de Barcelona, ICREA, Autonomous University of Barcelona and VU University Amsterdam

'Modelling is the language that underpins policy and industrial decision-making. This book makes a wide-reaching contribution to boost global efforts to build and apply formal models of transitions research, so that this exciting multi-disciplinary field can fulfil its huge potential to address critical problems such as the global energy transition!'

Professor Neil Strachan; Director, University College London Energy Institute

'Contemporary society is dominated by ever more complex transitions and transformations ranging from new technologies to human value systems. The contributors to this book succeed in showing how simulations, particularly those built around on systems dynamics and agent-based models, can provide new insights into change and difference. Essential reading for analysts and policy makers involved in our social future.'

Michael Batty, University College London

Modelling Transitions

Modelling Transitions shows what computational, formal and data-driven approaches can and could mean for sustainability transitions research, presenting the state-of-the-art and exploring what lies beyond.

Featuring contributions from many well-known authors, this book presents the various benefits of modelling for transitions research. More than just taking stock, it also critically examines what modelling of transformative change means and could mean for transitions research and for other disciplines that study societal changes. This includes identifying a variety of approaches currently not part of the portfolios of transitions modellers. Far from only singing praise, critical methodological and philosophical introspection are key aspects of this important book.

This book speaks to modellers and non-modellers alike who value the development of robust knowledge on transitions to sustainability, including colleagues in congenial fields. Be they students, researchers or practitioners, everyone interested in transitions should find this book relevant as reference, resource and guide.

Enayat A. Moallemi is a Research Fellow at School of Life and Environmental Sciences, Deakin University, Melbourne. His research is focused on computational and participatory approaches for modelling socio-ecological systems under the uncertainties of future global change. Enayat's research is applied to a range of areas, such as renewable energies, sustainable mobility, and the Sustainable Development Goals, aiming to advance robust decision-making and adaptive planning. Enayat obtained his PhD from the University of Melbourne, where he worked on model-based energy policy analysis. In his PhD, Enayat developed a theoretical transition framework and an exploratory system dynamics model for investigating future energy transition pathways under uncertainty. Enayat was a visiting researcher at the Faculty of Technology, Policy and Management, TU Delft (The Netherlands) in 2016 and at the Fraunhofer Institute for Systems and Innovation Research ISI in Karlsruhe (Germany) in 2018.

Fjalar de Haan is a theoretician, developing computational and mathematical approaches for a scientific understanding of transitions to sustainability. Fjalar has an MSc in theoretical physics (Leiden University) and did his PhD on transitions (Erasmus University). He has been exploring the fringe of transitions theory and modelling in a variety of sectoral contexts including health care, urban water management and energy, as part of international, interdisciplinary teams, project based with industry, and in curiosity-driven solo projects. Fjalar currently is Lecturer on Sustainability Transitions at the Melbourne School of Design, The University of Melbourne.

Routledge Studies in Sustainability Transitions
Series editors: Johan Schot, John Grin and Jan Rotmans

Transitions to Sustainable Development
New Directions in the Study of Long Term Transformative Change
John Grin, Jan Rotmans and Johan Schot
In collaboration with Frank Geels and Derk Loorbach

Automobility in Transition?
A Socio-Technical Analysis of Sustainable Transport
Edited by Frank W. Geels, René Kemp, Geoff Dudley and Glenn Lyons

Food Practices in Transition
Changing Food Consumption, Retail and Production in the Age of Reflexive Modernity
Edited by Gert Spaargaren, Peter Oosterveer and Anne Loeber

Governing the Energy Transition
Reality, Illusion or Necessity?
Edited by Geert Verbong and Derk Loorbach

Urban Sustainability Transitions
Edited by Niki Frantzeskaki, Vanesa Castán Broto, Lars Coenen and Derk Loorbach

Modelling Transitions
Virtues, Vices, Visions of the Future
Edited by Enayat A. Moallemi and Fjalar J. de Haan

Modelling Transitions

Virtues, Vices, Visions of the Future

Edited by Enayat A. Moallemi and Fjalar J. de Haan

Routledge
Taylor & Francis Group
LONDON AND NEW YORK

earthscan
from Routledge

First published 2020
by Routledge
2 Park Square, Milton Park, Abingdon, Oxon OX14 4RN

and by Routledge
52 Vanderbilt Avenue, New York, NY 10017

Routledge is an imprint of the Taylor & Francis Group, an informa business

British Library Cataloguing-in-Publication Data
A catalogue record for this book is available from the British Library

Library of Congress Cataloging-in-Publication Data
A catalog record for this book has been requested

ISBN: 978-0-367-17406-4 (hbk)
ISBN: 978-0-429-05657-4 (ebk)

Typeset in Bembo
by Apex CoVantage, LLC

MIX
Paper from responsible sources
FSC® C013056

Printed and bound in Great Britain by
TJ International Ltd, Padstow, Cornwall

Contents

1 Foreword

The importance of transitions modelling

Jan Rotmans

We do not live in an era of change, but in a change of era. The world around us is changing drastically at a rapid pace: the shifting world order, the new digital industrial revolution, rising populism, unprecedented environmental problems such as climate change, accelerated sustainability, etc. In this change of era, we need interpreters as the seismographers of the radically shifting spirit of the times, just as during the last great transition period, the second half of the nineteenth century.

But interpreters need explanatory tools to interpret the Zeitgeist. And among the most important tools to describe and explain transitions are computer simulation models. In particular transitions models, a relatively young shoot at the modelling tree.

Regular models, dominated by economic models, struggle with the representation of multiple agents, scales, uncertainty and pluralism, and transition dynamics (tipping points). Regular economic models are general equilibrium or optimisation models, fully price driven, with one type of actor that behaves in a (bounded) rational manner. That means that they are not able to represent the non-linear dynamics of a transition, which makes them unsuitable tools for modelling transitions. That was a major reason for developing transitions models, around 15 years ago.

Transitions models should do better, and be able to capture the intricate dynamics of transitions, encapsulating phenomena as emergence, co-evolution and self-organisation, and tipping points. This means at the conceptual level the involvement of niche and regime agents, and the dynamics between these agents, also influenced by landscape signals. It also means the incorporation of multiple scale levels with different kinds of dynamics. This means the representation of social/institutional factors in conjunction with emerging and existing technologies. And finally, the model design and implementation need to be done in the context of stakeholder participation, based on co-creation and co-production.

This is a daunting challenge, and over the last 15 years we have produced quite a few of interesting examples of transitions models. However, the current examples are highly simplified and primitive, and still far away from the required complexity level we need to address transition phenomena adequately.

Nevertheless these highly simplified transitions model examples already produce interesting results. There is definitely an urgent need for many more examples of transitions models, mimicking transformative change.

While regular traditional models fail in simulating transition dynamics, the new generation of transitions models is still in its infancy. That creates a model vacuum, with an increasing gap between the complexity of reality and the simplicity of our models. In a model-vacuum era, we can only experiment with new types of models and produce experimental results rather than reliable projections. Meanwhile we need to work hard to build up a mature transitions modelling community, and generate more financial means and institutional power to scale up the current transitions modelling niche and make it a modelling regime itself.

More than ever in these transitional times, there is a need for explorative transitions models that provide insights and guidance in highly uncertain and unstable times. I wholeheartedly recommend this interesting book, which gives the state of the art of the rising field of transitions modelling, as well as a plethora of interesting transitions modelling examples that might surprise you by their counter-intuitive results. I am grateful to the authors that they took responsibility and produced this challenging book on theoretical and practical examples of transitions models. These models are an emerging phenomenon that might become quite important for the future of our society and economy in transition: a child on its way to maturity.

2 Prologue

Enayat A. Moallemi and Fjalar J. de Haan

Transitions modelling is a growing niche within the sustainability transitions community. Transitions models are defined as the application of modelling techniques to investigate the profound and pervasive transformation of societal systems on generational timescales. Transitions modelling has been part of the transitions research portfolio since its inception. A more coordinated interest in this niche started mainly from 2007 and after the week-long workshop at Leiden University's Lorentz Centre on Computational and Mathematical Approaches to Societal Transitions and the special issue with the same topic in *Computational and Mathematical Organization Theory* the following year (Timmermans and de Haan 2008). Based on recent publications and increased attention from modellers and non-modellers in the field, we believe that transitions modelling is becoming ever more important and deserves to be a central part of transitions research. Among the multiple reasons behind this growing importance are the diversity and heterogeneity of this area in terms of adopted modelling techniques, implemented sectoral domains and connections to other fields.

- A unique feature of transitions modelling is its diversity and heterogeneity in terms of adopted modelling techniques, from pure mathematical modelling with partial differential equations (de Haan 2008), to agent-based modelling (ABM) (Bergman et al. 2008; Chappin and Dijkema 2010; Holtz 2010; Köhler et al. 2009), to system dynamics modelling (Moallemi et al. 2017; Papachristos 2018; Valkering et al. 2017; Raven and Walrave 2018), and to complex systems and evolutionary modelling (de Haan et al. 2016; Frenken 2006; Safarzynska 2013).
- Transitions models are also diverse in terms of their sectoral domains, such as energy (Li, Trutnevyte, and Strachan 2015; Moallemi et al. 2017; Trutnevyte et al. 2014), water (de Haan et al. 2016; Rauch et al. 2015), transport (Bergman et al. 2008; Köhler, Turnheim, and Hodson 2018) and urban systems (Valkering et al. 2017).
- Another important feature of transitions modelling is its critical connections with other established fields. Transitions models can adopt techniques from or are capable of being used by other fields of modelling, as Köhler

et al. (2018) argue. Amongst these connections we find: integrated assessment and eco-innovation modelling (Köhler et al. 2006; Alkemade et al. 2015), evolutionary-economics modelling (Safarzynska 2013), complex systems modelling (Frenken 2006; Kauffman 1993; Zeppini, Frenken, and Kupers 2014), socio-ecological systems modelling (Schlueter et al. 2012) and exploratory modelling (de Haan et al. 2016; Moallemi and Malekpour 2018).

For these and other reasons we think a book on modelling transitions is timely. We wanted to critically investigate what modelling of transformative change means and could mean for transitions research and for other disciplines that study societal changes and could potentially benefit from transition concepts. This led us to examine both the virtues and the vices of modelling and to look forward to approaches that are currently not part of the standard toolkit of modellers in transition research. The volume has given due attention to the state of the art of transitions modelling but with the explicit aim of evaluating the contributions to the broader transitions field and the modelling lessons learnt. To reflect these various aims, the volume is organised as follows.

The volume starts with an introductory part, including a foreword and this prologue by the editors. It continues with an overview of transitions modelling an approach to understanding transitions.

1 *Foreword.* J. Rotmans, a pioneer of transitions research and transitions modelling, introduces us to the ongoing and growing importance of modelling in the study of transitions and transformative change.
2 *Prologue.* The editors, E. A. Moallemi and F. J. de Haan, highlight the significance of this volume, provide an overview of chapters and acknowledge the contributions of those who helped to shape this volume.
3 *Transitions modelling: status, challenges and strategies.* J. Köhler and G. Holtz review and summarise some of the approaches to modelling innovation for sustainability. They identify three stylised strategies for developing transition models and compare the past research based on their scope, the emphasis they place on depicting social aspects and their use of widely accepted (formal) theories and assumptions.

Part 1: virtues and vices

Part 1 explores the meanings of modelling for transitions research, what it is, what it could be and what it can do for the field. This part is both looking inward (i.e. critical self-reflection) and looking outward (i.e. regarding the role of modelling in the development of the field).

4 *Making it a science: aspirations and apprehensions of transitions research.* F. J. de Haan discusses the possibility and desirability of a transitions science. The chapter dispels arguments against scientific approaches for transitions

research and provides concrete recommendations, including a central role for modelling in the broadest sense.

5 *Modelling and social science: problems and promises.* F. Bianchi and F. Squazzoni look at the role of model and modelling in social sciences, in particular the role of agent-based models (ABMs). The chapter provides a taxonomy of the use of ABMs in social sciences in different ways and at different levels of abstraction. The chapter critically appraises previous literature of this area and sheds light on challenges and problems towards integration of theory and empirical data via modelling. This chapter is a rich source for readers with an interest in using models in transitions research and other social sciences.

Part 2: state of the art

Part 2 provides an overview of what has been done in transitions modelling, both in the sense of the models out there and in the sense of methods, techniques and applications. The emphasis will be on lessons learnt.

6 *Modelling the multi-level perspective: the MATISSE agent-based model.* J. Köhler revisits the MATISSE model and describes the way the model was developed to implement the multi-level perspective framework. He uses two applications: a transition to household sustainable mobility and a transition to low emissions shipping, involving different interpretations of regime–niche interactions. This gives readers insights into one of the few classics in transitions modelling and helps them to understand how the structure of a model can be adaptable to different cases.

7 *Considering actor behaviour: agent-based modelling of transitions.* G. Holtz and E. J. L. Chappin continue with the role of ABMs as one of the most popular modelling techniques in social sciences. They focus on the conceptualisation of actors and institutions in transitions research and highlight challenges for ABMs of transitions that arise from the multitude of actors and institutions as well as from the diversity of theoretical perspectives to describe them. This chapter provides readers with four modelling strategies to cope with the typical challenges that they face in the use of ABMs in transitions research, each illustrated with examples from the literature, discussed and compared to each other. The chapter offers a timely contribution to the literature at the interface of ABM and transitions research.

8 *System dynamics methodology and research: opportunities for transitions research.* G. Papachristos and J. Struben explore the potential contributions from system dynamics modelling to transitions research, as another established method for modelling transformative change. They explain how foundational concepts of system dynamics can contribute to transition research using multiple applications corresponding to different levels of transitions. The discussion of these applications raises a number of future opportunities for integration of system dynamics into transitions research.

9 *Socio-technical representation of electricity provision across scales.* A. Rojas and F. J. de Haan use the idea of formalising socio-technical systems as a complementary tool for the study of transitions. They introduce an analytical framework to map utility-based socio-technical systems of service provision. The framework emphasises the networked social structure and nested technical structure of these systems. The framework can be used to develop computational models for energy transition modelling as well as aid quantitative and qualitative case analysis. This chapter uses three examples, drawn from the electricity system in Australia at different points in time, to illustrate how the framework represents actors, (technical) components of the system and their relationships.

Part 3: the future of modelling transitions

Part 3 investigates what avenues could be pursued in modelling transitions. Here the emphasis will lie on approaches that are not often used (e.g. exploratory modelling) and those that push the boundaries of what modelling means (e.g. data-driven approaches).

10 *Models as scenario tools for developing robust transformative plans.* S. Malekpour investigates normative explorative planning in transitions research under uncertainty. She explains how transitions models could contribute to processes of long-term planning, to inform policy decisions and to facilitate stakeholder engagement. The chapter offers insights for readers interested in managing transitions about how to envision desired futures and to draw pathways to realisation while remaining robust to uncertain futures.

11 *Participatory modelling in sustainability transitions research.* J. Halbe undertakes an extensive analytical review of participatory modelling in sustainability transitions research. The chapter presents a framework which includes central dimensions of participatory modelling: model use, modes of knowledge capture and exchange, timing of stakeholder involvement and methods for stakeholder participation. The chapter then reviews and compares previous participatory approaches to modelling transitions based on the dimension of this framework to identify patterns, emphases and gaps.

12 *Data-driven transitions research: methodological considerations for event-based analysis.* F. J. de Haan, A. M. Arranz and W. Spekkink develop a methodological argument for 'data-driven' transitions research. They argue that transitions research currently is not, or hardly at all, data driven, and discuss why they think this should be changed. The chapter provides an important contribution to methodological debate in transitions research by discussing trends towards and practices of data-driven research in other areas, and by clearly building on substantial hands-on experience with the methodologies proposed.

13 *Exploratory modelling of transitions: an emerging approach for coping with uncertainties in transitions research.* Enayat A. Moallemi, Fjalar J. de Haan and J. Köhler focus on the systematic treatment of uncertainties in transitions. The

chapter has a look at the discussions around the concept of uncertainty in transitions research and discusses the adoption of exploratory modelling – as an emerging computational approach for coping with uncertainties – to transitions. The chapter defines potential roles for exploratory modelling and analyses challenges and opportunities of using this approach in coping with complexities, facilitating systematic experimentation and advancing policy analysis in transitions research.

14 *Epilogue*. The editors, Fjalar J. de Haan and Enayat A. Moallemi, summarise and conclude the volume. The Epilogue serves as a roadmap for future transitions modelling work.

We hope that this volume with the outlined chapters will become a reference with continuing relevance for multiple audiences:

- *Researchers on the fringes* where the transitions field overlaps with adjacent fields. Prominent examples are the model-based decision-making approaches from the broader field of operations research and decision science where the design of robust policies is investigated for accelerating system transformations.
- *Theorists of transitions* where modelling approaches put concepts to the test in a formal manner and use models to inform theory development.
- *Empirical researchers of transitions* where analysing data and reproducing observed dynamics *in silico* generate empirically oriented insights from case studies through data mining, statistics, formal ontologies and databases.
- *Transition managers, practitioners and policy makers* where transitions models can help to create portfolios of possible futures which can be used to design robust policies.

We believe that transitions modelling is in a stage where it has need of a volume that can be a point of reference. The recent position papers by Holtz et al. (2015) and Köhler et al. (2018) provided a good overview of recent work and a delineation of the research agenda of the nascent sub-field. For this reason, we are delighted that G. Holtz and J. Köhler have written one of our introductory chapters. We acknowledge their insightful comments and discussions on the early preparatory work leading to this volume. We also greatly thank our chapter authors for bringing their best of ideas and work to this volume, for helping us with the internal review of chapters and for bearing with us throughout this journey.

The editors, on behalf of all chapter authors, would like to acknowledge and sincerely thank the time and valuable comments of our chapter reviewers which have helped us to substantially improve the quality of this volume (in alphabetical order): Paul Van Baal (École Polytechnique Fédérale de Lausanne, Switzerland), Sibel Eker (International Institute for Applied Systems Analysis, Austria), Marletto Gerardo Ettore (University of Sassari, Italy), Paula Hansen (Curtin University, Australia), Friedrich Krebs (University of Kassel,

Germany), Judy Lawrence (University of Wellington, New Zealand), Francis Li (University College London, UK), Bonno Pel (Université libre de Bruxelles, Belgium), Neil Strachan (University College London, UK), Jason Thompson (The University of Melbourne, Australia) and Warren Walker (TU Delft, The Netherlands).

Last but not least, the publication of this volume was not possible without the support of the series editors of the Routledge Studies in Sustainability Transitions, the three Js: Johan Schot, John Grin and Jan Rotmans. We sincerely thank them for their trust and guidance. We also greatly acknowledge the hard work and dedicated support of Routledge Editorial and Production team, in particular, Julia Pollacco and Rebecca Brennan.

References

Alkemade, Floortje, Gaston Heimeriks, Antoine Schoen, Lionel Villard, and Patricia Laurens. 2015. "Tracking the internationalization of multinational corporate inventive activity: National and sectoral characteristics." *Research Policy* 44 (9): 1763–72.

Bergman, Noam, Alex Haxeltine, Lorraine Whitmarsh, Jonathan Köhler, Michel Schilperoord, and Jan Rotmans. 2008. "Modelling socio-technical transition patterns and pathways." *Journal of Artificial Societies and Social Simulation* 11 (3): 7.

Chappin, Emile J.L., and Gerard P.J. Dijkema. 2010. "Agent-based modelling of energy infrastructure transitions." *International Journal of Critical Infrastructures* 6 (2): 106–30.

de Haan, Fjalar J. 2008. "The dynamics of functioning investigating societal transitions with partial differential equations." *Computational and Mathematical Organization Theory* 14 (4): 302–19. http://doi.org/10.1007/s10588-008-9034-2.

de Haan, Fjalar J., Briony C. Rogers, Rebekah R. Brown, and Ana Deletic. 2016. "Many roads to Rome: The emergence of pathways from patterns of change through exploratory modelling of sustainability transitions." *Environmental Modelling & Software* 85: 279–92. http://doi.org/10.1016/j.envsoft.2016.05.019.

Frenken, Koen. 2006. "Technological innovation and complexity theory." *Economics of Innovation and New Technology* 15 (2): 137–55.

Holtz, Georg. 2010. *Modelling System Innovations in Coupled Human-Technology-Environment Systems* (PhD diss., Institute of Environmental Systems Research), Department of Mathematics/Computer Science University of Osnabrück.

Holtz, Georg, Floortje Alkemade, Fjalar de Haan, Jonathan Köhler, Evelina Trutnevyte, Tobias Luthe, Johannes Halbe, George Papachristos, Emile Chappin, Jan Kwakkel, and Sampsa Ruutu. 2015. "Prospects of modelling societal transitions: Position paper of an emerging community." *Environmental Innovation and Societal Transitions* 7: 41–58. http://doi.org/10.1016/j.eist.2015.05.006.

Kauffman, Stuart A. 1993. *The Origins of Order: Self-organization and Selection in Evolution.* Oxford: Oxford University Press.

Köhler, Jonathan, Fjalar de Haan, Georg Holtz, Klaus Kubeczko, Enayat A. Moallemi, George Papachristos, and Emile Chappin. 2018. "Modelling sustainability transitions: An assessment of approaches and challenges." *Journal of Artificial Societies and Social Simulation* 21 (1): 8. http://jasss.soc.surrey.ac.uk/21/1/8.html.

Köhler, Jonathan, Michael Grubb, David Popp, and Ottmar Edenhofer. 2006. "The transition to endogenous technical change in climate-economy models: A technical overview to the innovation modeling comparison project." *The Energy Journal*: 17–55.

Köhler, Jonathan, Bruno Turnheim, and Mike Hodson. 2018. "Low carbon transitions pathways in mobility: Applying the MLP in a combined case study and simulation bridging analysis of passenger transport in the Netherlands." *Technological Forecasting and Social Change*. http://doi.org/10.1016/j.techfore.2018.06.003. https://www.sciencedirect.com/science/article/pii/S0040162518308783

Köhler, Jonathan, Lorraine Whitmarsh, Björn Nykvist, Michel Schilperoord, Noam Bergman, and Alex Haxeltine. 2009. "A transitions model for sustainable mobility." *Ecological Economics* 68 (12): 2985–95. http://doi.org/10.1016/j.ecolecon.2009.06.027.

Li, Francis G.N., Evelina Trutnevyte, and Neil Strachan. 2015. "A review of Socio-Technical Energy Transition (STET) models." *Technological Forecasting and Social Change* 100: 290–305. http://doi.org/10.1016/j.techfore.2015.07.017.

Moallemi, Enayat A., Lu Aye, Fjalar J. de Haan, and John M. Webb. 2017. "A dual narrative-modelling approach for evaluating socio-technical transitions in electricity sectors." *Journal of Cleaner Production* 162: 1210–24. http://doi.org/10.1016/j.jclepro.2017.06.118.

Moallemi, Enayat A., and Shirin Malekpour. 2018. "A participatory exploratory modelling approach for long-term planning in energy transitions." *Energy Research & Social Science* 35: 205–16. http://doi.org/10.1016/j.erss.2017.10.022.

Papachristos, George. 2018. "System dynamics modelling and simulation for sociotechnical transitions research." *Environmental Innovation and Societal Transitions* 31: 248–261. http://doi.org/10.1016/j.eist.2018.10.001.

Rauch, W., P. Bach, R. Brown, B. Rogers, F. de Haan, D. McCarthy, M. Kleidorfer, M. Mair, R. Sitzenfrei, C. Urich, A. Deletic, L. Sharp, E. Westling, S. Tait, R. Ashley, and M. Rychlewski. 2015. "Enabling change: Institutional adaptation." In *Climate Change, Water Supply and Sanitation: Risk Assessment, Management, Mitigation and Reduction*, 408. London: IWA Publishing.

Raven, Rob, and Bob Walrave. 2018. "Overcoming transformational failures through policy mixes in the dynamics of technological innovation systems." *Technological Forecasting and Social Change*. http://doi.org/10.1016/j.techfore.2018.05.008.

Safarzynska, Karolina. 2013. "An evolutionary model of energy transitions with interactive innovation-selection dynamics." *Journal of Evolutionary Economics* 23 (2): 271.

Schlueter, Maja, Ryan R.J. McAllister, Robert Arlinghaus, Nils Bunnefeld, Klaus Eisenack, Frank Hoelker, Eleanor J. Milner-Gulland, Birgit Müller, Emily Nicholson, and Martin Quaas. 2012. "New horizons for managing the environment: A review of coupled social-ecological systems modeling." *Natural Resource Modeling* 25 (1): 219–72.

Timmermans, Jos, and Hans de Haan. 2008. "Special issue on computational and mathematical approaches to societal transitions." *Computational and Mathematical Organization Theory* 14 (4): 263–5. http://doi.org/10.1007/s10588-008-9037-z.

Trutnevyte, Evelina, John Barton, Áine O'Grady, Damiete Ogunkunle, Danny Pudjianto, and Elizabeth Robertson. 2014. "Linking a storyline with multiple models: A cross-scale study of the UK power system transition." *Technological Forecasting and Social Change* 89: 26–42. http://doi.org/10.1016/j.techfore.2014.08.018.

Valkering, Pieter, Gönenç Yücel, Ernst Gebetsroither-Geringer, Karin Markvica, Erika Meynaerts, and Niki Frantzeskaki. 2017. "Accelerating transition dynamics in city regions: A qualitative modeling perspective." *Sustainability* 9 (7): 1254.

Zeppini, Paolo, Koen Frenken, and Roland Kupers. 2014. "Thresholds models of technological transitions." *Environmental Innovation and Societal Transitions* 11: 54–70.

3 Transitions modelling

Status, challenges and strategies

Jonathan Köhler and Georg Holtz

1 Introduction

The relatively new field of sustainability transitions research has opened up new questions and opportunities for modelling of change in socio-technical-environmental systems.[1] The premise of transitions research is that:

> environmental problems, such as climate change, loss of biodiversity, resource depletion (clean water, oil, forests, fish stocks) . . . require shifts to new kinds of (socio-technical) systems, shifts which are called 'sustainability transitions' (Markard, Raven, and Truffer 2012).
>
> Therefore, a central aim of transitions research is to conceptualise and explain how radical changes come about in the way societal functions are fulfilled.
>
> (STRN 2017, p. 1)

The ambition of sustainability transitions modelling is to show possible pathways of change and contribute to the understanding of these radical changes in socio-technical-environmental systems. This ambition is rooted in the perspective that quantitative modelling methods can deliver insights that other methodologies cannot. Holtz et al. (2015) identify potential benefits of simulation modelling for transitions research: a systematic and explicit structure, the ability to undertake systematic experiments on the modelled structure and the ability to examine the dynamics of complex socio-technical systems.

However, because of the high level of complexity of transitions, modelling them comes with several challenges. Transitions are processes involving multiple social, technical, political, market and environmental subsystems. While transitions take up to several decades, their non-linear dynamics emerge from co-evolution of these sub-systems on multiple temporal and spatial scales (from local to global). As a consequence, the problem of which aspects to model and which to leave out to enable a model of appropriate complexity is particularly difficult. Furthermore, transitions modelling has the ambition of representing qualitative system change (i.e. a change to a different system structure with different elements and feedbacks) in complex socio-technical systems, which poses an additional core challenge (Köhler et al. 2018). Another fundamental difficulty is that the combination of multiple sub-systems and scales makes it necessary to decide what is 'in' a model and what is 'out'. Since the number of studies is still small, there are few examples of what can be studied and how. Another consequence of this is that there are no commonly used datasets or data structures designed to reflect transitions theories or patterns. As with the models themselves, data has to be collected or combined from sources that were not necessarily intended to examine the characteristics of transitions. There is, once again fundamentally, no agreed indicator for when a transition has happened or how close to a transition a socio-technical-environmental system is.

Capturing the complex interactions and the resulting non-linear dynamics inherent to transitions in simulation models implies the need to move beyond the established approaches to modelling innovation for sustainability, which have been dominated by techno-economic models. These are most often based on microeconomic theory and optimisation to develop scenarios of least cost technological change to meet emissions goals, especially for greenhouse gas (GHG) mitigation (Köhler 2019a). These do not take full account of the changes in socio-technical-environmental systems required for large-scale changes for sustainability. They are also limited in their representation of behavioural change, although some recent work with integrated assessment models does address this issue (Köhler 2019a). Driven by the ambition to make the benefits of quantitative modelling fruitful for transitions research, the (sub-)community of transition modellers has explored different innovative approaches to tackle these challenges.

The use of simulation modelling methods in research on (sustainability) transitions is still in an early stage of development. A first review of the field was conducted by Timmermans and Haan (2008), who found almost no modelling research and proposed some mathematical and computational approaches. Safarzyńska, Frenken and van den Bergh (2012) reviewed evolutionary approaches and Zeppini, Frenken, and Kupers (2014) discussed threshold models with regard to their use in sustainability transition research. Halbe et al. (2015) explored parallels of transition modelling to and lessons to be learned from integrated assessment models, environmental modelling and socio-ecological modelling, and reviewed different roles of modelling in transitions research. Li, Trutnevyte, and Strachan (2015) identify a new category of models: socio-technical energy transitions (STET) models. They review these models and consider the extent to which they capture the characteristics of transitions. Köhler et al. (2018) provide the most recent review and summarise six different modelling approaches for sustainability transitions. Overall, the reviews show that there is no generally accepted set of approaches or even theories for modelling sustainability transitions. These limitations require simulation approaches to be developed and combined in new ways to further the understanding of transitions and to support policy and stakeholder processes. This will contribute to the development of a transitions science (Chapter 4 by de Haan in this volume).

This chapter discusses characteristics of transitions and the particular challenges that they pose for simulation modelling. It summarises some of the different research objectives and approaches used in modelling transitions and discusses their strengths and weaknesses. The differing approaches are compared by positioning them along three dimensions: their coverage (or scale), the emphasis they place on depicting socio-technical processes and also their use of widely accepted (formal) theories and assumptions. Finally, we reflect on the state of the art and discuss promising research directions.

2 The characteristics of transitions and the implications for transitions models

The idea of sustainability transitions arose from ideas of evolutionary economics as adapted to the idea of strategic niche management (Rip and Kemp 1998). The idea of radical innovations struggling against a dominant regime in a slowly changing landscape reflects the idea of selection in a competitive environment with fitness criteria. These ideas complement neo-Schumpeterian ideas of radical technological change leading to Kondratiev waves (or long waves) of growth (Köhler 2012). These were most clearly discussed in Freeman and Louçã (2002), who developed a theory of Kondratiev waves to explain industrial revolutions and long-term patterns of economic growth. In doing this, they adopted the idea of revolutionary new technologies and industries growing and replacing the pre-existing regime. Köhler (2012) shows that the multi-level perspective on transitions (Grin, Rotmans, and Schot 2010), which has developed to form a new line of research into the potential for radical socio-technical change to a sustainable system, closely follows the structure of Freeman and Louçã (2002). The pathways of change in such systems are understood to involve co-evolution between the political, scientific, economic, technology and culture sub-systems. Geels (2002) adopted a similar approach.

These ideas have important implications for the understanding of transition processes. Transitions involve radical, not marginal change. They involve highly non-linear processes, in which a small early change can create a niche industry and market which then has unstable but accelerating growth. Sometimes this growth continues and the new technology and industry matures and stabilises to become the new dominant design and regime. Köhler et al. (2018) argue that these processes are polycentric, involving multiple actors and subject to behaviour/social practice and expectations, cultural changes, technology and economy trends, institutional change, environmental changes and policy. The different social, geographical and economic scales all have the possibility of playing a decisive role, from grassroots movements to international and global policy. The consequence of this is that transitions are complex open systems with pathways of transitions being determined by both exogenous (landscape) and endogenous (niche innovation) changes.

Köhler et al. (2018) identify the following characteristics that models of transitions should ideally be able to represent (their incorporation in modelling approaches is reviewed in Section 3 below):

2.1 Non-linear behaviour

A typical pattern of transition dynamics is that of an S-curve, in which the rate of change is initially slow, then accelerates and then slows down again as the new technologies and industries mature and stabilise. A transitions model should be able to reproduce such variations of the rate of change and other dynamics through which the end state of the transition is not proportional to changes in the initial state. An important feature of transition processes is path dependence. 'Lock-in', understood in transitions ideas as a restriction on the ability of a system to change in response to internal or external pressures, is a form of path dependence often considered to apply to the regime.

2.2 Qualitatively different system states

The technologies, institutions, culture and markets fulfilling a particular societal function all change; that is, new elements are included, old ones are dropped, elements might adapt and the interactions between elements are reconfigured.

2.3 Changes in social values and norms

Transitions involve changes in the value system of society and bring about new cultural norms. There are changes in the decision rules and more fundamentally in the nature of the decisions being taken.

2.4 Representation of diversity and heterogeneity

Transitions depend upon diversity for change. In the case of sustainability transitions, this implies the involvement of diverse actor groups (producers, consumers, politicians, NGOs, etc.) and actors within these groups being heterogeneous (e.g. producers following different strategies, consumers having different preferences). These differences are critical to the transition process;[2] the process cannot be understood as a change in a single average behaviour.

2.5 Representation of dynamics across different scales and levels

There is also a consensus in the literature that transitions bridge different scales and levels. The multi-level perspective recognises three different levels: niche, regime and landscape. Geels and Schot (2010) see this as an application of Giddens' theory of structuration (1984), in which agents act within a set of social structures. Agents' actions can change these structures, which means that there are feedbacks between the micro and macro levels in societal systems. Furthermore, developments on various spatial scales (local to global) and temporal scales (days to decades) interact during transition processes.

2.6 Open processes and uncertainties or contingencies

Because transitions are open, co-evolutionary processes, they imply structural uncertainty. The future configuration of socio-technical-environmental systems is not known and ex-ante displays a bewildering array of possibilities. Because the systems are non-linear and may respond to small differences with large changes (or to large efforts in policies with small responses), future uncertain contingencies may have a decisive impact on the transition pathway and final outcome. This is especially true in the early stages of a transition, before the new technology, practices, markets and institutions mature to a new 'dominant design'. Models therefore need to be able to generate a wide range of outcomes and be able to produce different scenarios from differing input changes. The challenge here is to design the level of the abstraction of the model such that changes in the socio-technical-environmental structure become part of the dynamics of the model (Holtz et al. 2015).

This a very challenging set of characteristics for transitions modelling. Developing an 'ideal' transitions model might not even be possible. This implies that a model needs to be designed to address a suitable combination of these characteristics for the particular issue to be addressed.

3 Modelling objectives and approaches

Holtz et al. (2015) follow the structure proposed by Halbe et al. (2015) for the classification of modelling and discuss different uses of models in transitions studies. Models can be used for understanding transitions, for developing policy insights and also as a part of stakeholder processes. Transitions research has developed a particular approach – transitions management – for the practical implementation of transformative ideas, practices and measures. However, specific transition models for supporting such stakeholder processes have not yet been extensively developed (Chapter 11 by Halbe in this volume).

When using models for understanding transitions, models can help to make the structure of the system being investigated explicit. Dynamic models of transitions can also help to show and explain the links between emergent phenomena and the overall system dynamics to the underlying structure and processes. Simulation models can also be used to run structured sensitivity analyses to investigate the uncertainty space and to test hypotheses about the system and phenomena being studied, an approach called exploratory modelling (Chapter 13 by Moallemi, de Haan, and Köhler in this volume; Chapter 10 by Malekpour in this volume; Moallemi and Malekpour 2018). They can be used to simulate pathways that show the possibilities for transitions.

Modelling for case-specific policy analysis is a very important use of models of sustainable innovation. Holtz et al. (2015) argue that the models need to use theories and assumptions that stakeholders find acceptable/plausible. In the case of transitions models, the uncertainty of model structures and outcomes poses a particular problem in providing results that stakeholders can understand and find useful. Again, exploratory modelling is an approach that is being developed to address this uncertainty. Through making the causal factors which determine the system pathway clear, models can give insights into the conditions under which different transitions pathways occur.

Köhler et al. (2018) review and classify six different approaches that have been used to analyse transitions, which come from different schools of thought and use different methods.

Energy-economy models and integrated assessment models (IAMs) have used management and/or economics methods to represent innovation (Köhler et al. 2006; Clarke et al. 2014). They include top-down (macroeconomic) models, bottom-up models of economic sectors with a range of technologies represented mainly by cost and emissions characteristics and hybrid models (combining some top-down and bottom-up features).

Evolutionary economics models represent processes of change and competition in a population of decision makers or agents using three core concepts from evolutionary biology (Safarzyńska, Frenken, and van den Bergh 2012): variation, selection and differential replication. Concepts applied in evolutionary economics models include: bounded rationality, path dependence and lock-in, group selection and co-evolutionary dynamics (van den Bergh and Gowdy 2009; Gazheli, Antal, and van den Bergh 2015). These ideas mean that evolutionary economics models can be used to study changes in consumer preferences, social structure and institutions, in addition to technological innovation. An important area of application has been demand-and-supply dynamics, which have generated insights directly relevant to transitions. Some of the most prominent models in the literature are Silverberg, Dosi, and Orsenigo (1988); Windrum and Birchenhall (2005); Windrum, Ciarli, and Birchenhall (2009); and Safarzyńska and van den Bergh 2010).

Complex systems research has developed frameworks and themes such as 'complex, adaptive systems' (Holland 1992), 'self-organisation' (e.g. Kauffman 1993), 'self-organised criticality' (e.g. Bak 1996) and a variety of others. The unifying joint characteristic of these approaches is that system representations of typically simple elements and local non-linear interactions lead to emergent patterns and non-linear dynamics at the system level. While transitions studies use concepts from complexity studies, in particular non-linearity and co-evolution, direct applications of complexity models in transitions modelling are limited. Frenken (2007) reviews complexity models in three areas: NK models,[3] complex network models and percolation models. This category also includes diffusion models, reviewed by Geroski (2000) for S-curves of innovation diffusion, including epidemic models, probit models and density-dependence models. Finally, we would also subsume here simple threshold models from physics, biology, economics and sociology as reviewed by Zeppini, Frenken, and Kupers (2014).

Computational social science (CSS) often uses agent-based modelling (ABM) to analyse behaviour of complex social systems from a bottom-up perspective (Chapter 7 by Holtz and Chappin in this volume). ABMs have become an important approach to simulation modelling in the social sciences (Chapter 5 by Bianchi and Squazzoni in this volume). ABM allows systemwide structures and dynamics to emerge from an underlying dynamic system of interacting (heterogeneous) agents. Agents can be defined on different levels of aggregation, from individuals to larger social entities (e.g. organisations, regimes). ABM allows model behaviours and decisions of agents with various levels of sophistication, from simple rules to elaborate representations of processes of cognition or internal organisation. Another main feature is the representation of agents' interactions, not only including (strategic) behaviours, but also social influences such as norms. The MATISSE model (Köhler et al. 2009) is an example of an ABM applied to sustainability transitions (Chapter 6 by Köhler in this volume).

System dynamics (SD) models (Chapter 8 by Papachristos and Struben in this volume) combine stock-and-flow system variables with feedback relationships to describe dynamic interactions between multiple system components. Different states of a system are captured with the values of the stock variables. Their quantitative values change through the accumulation of flow variables which are the results of interactions between several internal and external variables. Stock accumulation processes create path-dependent system behaviours, and the interaction of multiple feedback loops of various strength and on different temporal scales creates complex non-linear dynamics. SD is a very general concept which can be applied to many different systems on different temporal and spatial scales. Typically, SD modelling applications take an aggregated view of the system components and do not deal with actor-level changes.

Socio-ecological systems (SES) models combine dynamics of social and ecological systems. SES modelling is based on literatures dealing with natural resources modelling in ecology, economics and conservation. Schlüter et al. (2012) noted a similarity to transitions research in the way in which SES adopts approaches and methods from other fields to consider change in socio-technical-environmental systems. SES are addressed as complex, co-evolutionary adaptive systems, with a more central consideration of co-evolution between social systems and ecosystems than is usual in other transitions research. Strategies for resilience in natural resource management are a central area of research. SES models incorporate multi-domain and multi-level feedbacks, path dependency and involvement of multiple actors (Halbe et al. 2015).

Köhler et al. (2018) conclude that these approaches do in principle provide the modelling tools to address the main characteristics of transitions to sustainability in socio-economic-environmental systems outlined above. They include non-linear behaviour and path dependency. CSS ABMs and evolutionary models can represent heterogeneity well. The complexity science/evolutionary approaches have mostly been used for micro-level analyses, while macro-level analyses are still mostly the domain of economics models or IAMs, with a few SD models. The approaches all require further development to address uncertainties and contingencies. The SES models, and for climate change the global IAMs, are the only models that try to fully represent feedbacks between socio-technical systems and ecosystems, The existing literature is still developing from a small base, so there are many open questions. The limited ability of most of the models in these literatures to represent qualitatively different system states, and especially the normative and cultural aspects of sustainability transitions (which are also not addressed in other modelling literatures on environmental modelling) is an important area to be addressed. Furthermore, these model types are not a comprehensive review; approaches such as innovation diffusion models and the models from mathematical sociology are examples of other areas which could also be applied to transitions problems.

4 A comparison of strategies to develop transition models

The discussion so far has revealed three fundamental aspects of transitions and transitions research that come with challenges for modelling transitions:

- Scope: transitions have an inherently large scope in terms of aspects covered, time and geography. They emerge over long time-scales (decades) from interacting developments in multiple areas (technology, institutions, culture, social practices, policy, environment) across different levels (from local to global), all of which potentially play a decisive role in the process.
- Established theories, conceptual frameworks and assumptions: the need to go beyond current eco-innovation modelling approaches has been clearly identified, but no generally accepted set of approaches or even theories for modelling sustainability transitions exists.
- Inclusion of social aspects: qualitative research on transitions has clearly shown that institutional, cultural, socio-economic and socio-political aspects including changes in social values and norms as well as diversity and heterogeneity of the involved actors are fundamental aspects of transitions. The explicit consideration of these kind of characteristics and processes has been identified as a main gap of the existing modelling approaches (Köhler et al. 2018).

These three aspects have a strong influence on the complexity and uncertainty[4] associated with modelling transitions, which we briefly elaborate on. Covering a large scope in a model comes at a price – either of high model complexity in terms of the number of variables and processes to be included in the model, or in terms of a need for bold abstraction and simplification. The alternative of a 'partial' model – only considering a restricted set of aspects and/or scope enables the restricted set to be more closely analysed, while acknowledging that system feedbacks from outside the chosen set are being (artificially) exogenised.

The lack of established formalised theories necessitates either drawing on existing formalisations that may not fully match the problem, or the development of new concepts and formalisations. Further, it brings an additional challenge for defining indicators and collecting data. STRN (2017) identify indicators for quantitative analysis of transitions as an area which requires further research. However, in order to define indicators and collect data, a precise definition of the objects to be modelled and their quantification is required. The corollary of this is that since there is a lack of established theories, there is inevitably a lack of established indicators and data. This can be partly overcome by using data collected for other purposes, such as data on investment in renewable energy technologies or the uptake of low carbon powered vehicles, but such descriptive data can only partially address questions of whether a niche or technological innovation system has the potential to grow, or what the weaknesses of an established system (e.g. power generation and supply or mobility) are.

Developing a model including institutional, cultural, socio-economic and socio-political aspects increases model complexity and the uncertainty about the validity of the model structure (cf. Holtz et al. 2015). Social science insights often exist in qualitative terms only, and their formalisation involves inevitably some interpretation by the modeler (see also Chapter 5 by Bianchi and Squazzoni in this volume). Furthermore, different schools of thought may propose different concepts and theories (e.g. concerning the representation of bounded-rational actors and institutions; Chapter 7 by Holtz and Chappin in this volume). A model necessarily employs a certain conceptual frame and leaves others unexplored. Finally, socio-political processes are decisive at particular stages of transitions, but at the same time contingent on potentially very specific circumstances of the actors involved and the institutional setting, and thus difficult to include in models.

Challenges that arise from high model complexity and uncertainty for calibration, validation and interpretation of models have been discussed elsewhere (Holtz et al. 2015; Andersson, Törnberg, and Törnberg 2014; Brugnach et al. 2008; Jackson 2014). Here, we want to emphasise that taking the above considerations together, it follows that developing a transition model that at the same time (1) covers a transition in its full scope, (2) uses (only) widely accepted theories and assumptions and (3) covers also the multiple social aspects of transitions is not practically feasible – at least not with current knowledge about transitions and the current level of formalisation of social science theories. To cover a transition in its full scope with a model of manageable complexity requires the definition of model elements and processes on a level of abstraction and aggregation at which currently no widely accepted formalised theories exist.

As a consequence, in developing a transition model, researchers need to decide where they are going to locate their model along the above three dimensions in a way that fits their purpose: how comprehensively they cover a transition, the extent to which they use accepted theories and formalisations and to what extent they try to include the social dimensions of transitions.

We identify three main streams of model development that we relate to the three dimensions in Figure 3.1. A first stream, which we name 'customised models' puts emphasis on covering a transition in its full scope while addressing

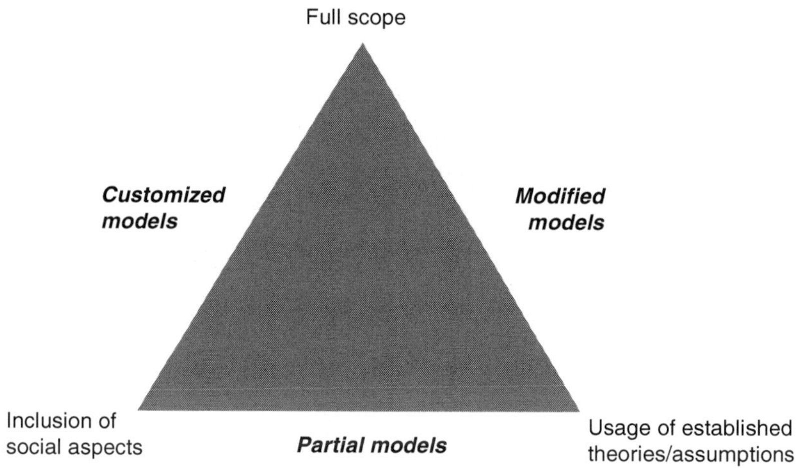

Figure 3.1 Illustration of strengths and limitations of three stylised model classes.

the social aspects of transitions discussed in Section 2 and by Köhler et al. (2018) through incorporating the 'grand lines' of insights gained by qualitative transition studies, for example, as expressed in the multi-level perspective or the technological innovation systems (TIS) framework. The objective and main benefit of these modelling exercises is to enhance the understanding of transitions through making use of the particular strengths of modelling methods, while remaining 'compatible' with qualitative conceptual and theoretical developments in transitions research through starting from the same set of basic concepts and perspectives. The most prominent example for this type of model is the MATISSE model (Chapter 6 by Köhler in this volume). Other examples include Papachristos (2011), Walrave and Raven (2016), Walz et al. (in review) and de Haan et al. (2016). The particular challenge these models face is that they have limited possibilities to draw on widely accepted formalisations and have to develop their own concepts and assumptions based on a limited conceptual knowledge base in the literatures.

A second stream, which we term 'modified models', starts from established modelling traditions that feature a large scope and make use of accepted theories. It applies such models to analysing transitions, thereby potentially refining the representation of the social science aspects in the models. In one sub-stream are the models from complexity studies (Frenken 2007; Geroski 2000) discussed earlier. Another sub-stream are modifications of techno-economic and integrated assessment models. We observe the latter type of models to be developed or existing models being modified for studying long-term system change in particular for the transport sector (McCollum et al. 2017; Pettifor et al. 2017) and for the energy sector (Mantzos et al. 2016; Mantzos et al.

2017; Li and Strachan 2017; Daly et al. 2014; Ramea et al. 2018; Tattini et al. 2018; Venturini et al. 2018). An example is the POTEnCIA model developed by the EU Joint Research Centre (Mantzos et al. 2016; Mantzos et al. 2017). POTEnCIA is an EU-wide energy sector model that represents a detailed capital stock for the industry, residential, services, agriculture and transport sectors for the EU28 member states. It simulates the investment in and use of this capital stock in yearly time-steps for a typical projection timeline to 2050. POTEnCIA deviates from the established energy sector optimisation models in several ways. For example, the typically used fully rational social planner for the whole energy sector is split into representative agents for each sector, year and vintage,[5] and each of those agents performs economic optimisation under constraints. The 'social' is included via altering technical and economic aspects of the model such as, for example, discount rates for households being higher than nominal capital cost annuities to reflect subjective financing capabilities; increased energy needs of units to represent non-rational use of energy; multiplication of price with a 'market-acceptance' factor to cover a range of issues related to lack or congestion of infrastructure and to consumer behaviour.

The benefit of starting from established (complexity/energy/transport) models or modelling traditions is that the underlying formalisations and assumptions are accepted in the respective communities and by stakeholders. Those type of models therefore have a lower hurdle of acceptance by the respective stakeholders when using them to underpin policy advice. The techno-economic and integrated assessment models typically feature a high level of techno-economic detail and a large scope in terms of the economic sectors and technological alternatives covered. However, the established formalisms are typically only able to incorporate certain social and institutional aspects in simplified ways. Moreover, these aspects and their dynamics over time (if any) are typically defined externally to the model and do not evolve as part of or in response to the overall transition dynamics emerging in the simulation run. As a consequence, these models have a limited ability to simulate transitions as open, co-evolutionary and contingent processes.

A third stream of 'partial' models refrains from covering a transition in its full scope for the sake of increased coverage of social science aspects and the possibility to draw on established theories. In principle, a large number and huge diversity of models could be related to this stream. Focusing on models that feature important characteristics identified for transition models – such as the ability to represent non-linear dynamics, co-evolution and the representation of diversity and heterogeneity – the evolutionary economics models and models from the CSS tradition should be mentioned here in particular. The lower level of aggregation of these models allows them to draw on empirical knowledge about and theories of particular actor groups (e.g. consumers, firms), the details of their (strategic) interactions and the institutional context in which they act. To achieve this, those models have to focus on some but 'cut off' other areas and related actor groups and interactions. For example, models from evolutionary economics typically focus on interactions of (heterogeneous) consumers and firms in the marketplace. Complementary work streams in the CSS field, for example, zoom in on opinion dynamics or norms and conventions (Meyer, Lorscheid, and Troitzsch 2009). A side effect of this kind of focus is that the models do not explicitly model the co-evolution of their focal system with other systems, though these co-evolutionary interactions are crucial for the understanding of transitions in their full scope and duration. As a consequence, the insights gained are limited to more short-term effects at the (spatial) scale analysed. The challenge is then to relate the specific insights to the larger transitions picture. Holtz (2012) provides a discussion of and suggestion for the bridging of aggregation levels – from particular insights to the full scope of a socio-technical transition.

This classification of modelling strategies into three main streams is neither all-encompassing nor does it allow a clear allocation of all of the existing models into one of the categories. For example, Yücel (2010) developed models that cover transition in their full scope, but neither start from existing models (as would be the case for 'modified models') nor build on established transitions theories (as would be the case for 'customised models'). Instead, they use insights from qualitative research. Moallemi and Malekpour (2018) relate their model approach to some transition concepts (Haan et al. 2014), but also make use of sector-specific and stakeholder knowledge. The purpose of the proposed scheme is rather to highlight some of the main trade-offs in transition modelling and to provide a framework to discuss and compare different modelling strategies.

5 Reflections and research directions

The discussion of the three dimensions discussed in Section 4 – scope, the application of established theories and the inclusion of social aspects – shows that different 'inroads' have been made in the modelling of transitions. It also shows that capturing core features of transitions processes – combining economic, technical, social and institutional sub-systems in a co-evolutionary structure, in particular – remains a challenge.

This challenge includes the representation of the potentials for qualitatively different system states (Köhler et al. 2018). A fundamental issue thereby is moving beyond the representation of system change as a (severe) shift in pre-defined technical or behavioural options, but also to allow for the potential emergence of completely new system states or system elements (e.g. new roles of actors such as 'prosumers' in the electricity system; new market designs). There are some modelling streams that have developed concepts and approaches for this, while taking a smaller scope. The evolutionary economics and complex systems models are intended to allow for emergent system properties or phase changes (Safarzyńska and van den Bergh 2010; Zeppini, Frenken, and Kupers 2014; Haan 2008). CSS ABMs (Epstein and Axtell 1996; Chapter 5 by Bianchi and Squazzoni in this volume; Chapter 7 by Holtz and Chappin in this volume) and evolutionary economics models, including recombinant innovation

(Zeppini and van den Bergh 2011; Frenken, Izquierdo, and Zeppini 2012), indicate approaches for generating new institutions and social/economic structures. The representation of cognitive and regulative institutions, together with changes in cultural norms and values (Elsenbroich and Gilbert 2014) is also an area of investigation. Recent transition modelling efforts which have tried to represent such qualitative changes while taking a larger scope have needed to limit the level of detail that is represented, that being the case for the MATISSE model (Köhler et al. 2009; Köhler, Turnheim, and Hodson 2018; Chapter 6 by Köhler in this volume). Achieving a deeper understanding of this core aspect of transition dynamics, however, needs to be a joint effort of modellers together with qualitative empirical and theoretical/conceptual work. One important role for simulation modelling is to contribute to the development of new precise theories that live up to the task of understanding co-evolutionary change across different domains.

Another core challenge is rooted in the fact that concepts such as niche and regime from the transitions/multi-level perspective or innovation functions in the technological innovation systems framework do not have agreed definitions that are precise enough to be directly translated into quantitative structures and software. While this is a challenge for transitions modellers, it is also an opportunity to contribute to theorising and analysis in the fields of sustainability transitions. More generally, a further research task in transition studies, to which modelling can contribute, is the development of indicators for sustainability transitions. There are no generally agreed indicators or datasets for transitions. Simulation modelling exercises therefore have to interpret such concepts for any given analysis and develop new indicators and data for assessing transition dynamics (Köhler 2019b; STRN 2017; see also Chapter 12 by de Haan, Martinez Arranz and Spekkink in this volume). Current models use indicators derived from previous modelling literatures – emissions quantities, economic outcomes, activity such as energy production or transport volumes. However, since transitions are multi-dimensional processes, there are no indicators that directly provide a measure of whether a transition has occurred – or not. In the current sustainability transitions literature, these questions are answered descriptively or qualitatively. Developing clearer indicators is of relevance for transitions studies beyond the modelling sub-field.

The relevance of the different challenges outlined in this chapter and how best to deal with them is also a question of how models are embedded in a broader setting. There are a few more general methodological concepts that have been developed for using modelling in transitions research. Exploratory modelling has already been mentioned as a way to address the very high degrees of uncertainty in transitions models (this theme is explored in Chapter 13 by Moallemi, de Haan, and Köhler in this volume). Given the importance placed on stakeholder processes, especially in the transitions management area, the use of simulation modelling in participatory stakeholder processes could open up a new field of application for transitions modelling. Halbe et al. (2015) review these possibilities and the limited number of applications so far (this theme is explored in Chapter 11 by Halbe in this volume). A related area is the relationship between transitions modelling and qualitative analysis in transitions studies. The potential for combining qualitative and quantitative methods for compensating the limitations of both types of methods has been addressed in various proposals for 'bridging' (Turnheim et al. 2015; Köhler, Turnheim, and Hodson 2018; Trutnevyte et al. 2014; McDowall 2014; Holtz et al. 2015); 'linking' (Trutnevyte et al. 2014), 'hybrid approaches' (McDowall 2014) and 'integration' (Holtz et al. 2015). Better understanding these potentials and also the pitfalls of linking qualitative and quantitative approaches is a vibrant and important field of research.

This brief review has shown that modelling of sustainability transitions is a field of research that is in its infancy. This brings both opportunities and challenges. It is necessary to 'start from first principles' in building transitions models, because there are no commonly accepted ways of modelling, no commonly used theories, indicators or data. Building a transitions model therefore involves developing a theory of what to model as well as writing a model structure, collecting data and running simulations. This is also the opportunity: the model-based analysis of transitions to sustainability provides methodological guidance for exploring new conceptual and policy insights into these processes.

Notes

1 Because sustainability transitions research is directed to the interaction between socio-technical systems and the environment, we refer to the objects of study as socio-technical-environmental systems.

2 For example, consumers in particular market niches provide the opportunity for technological learning that is required before entering mass markets is possible.

3 In an NK model, each individual in the set of *N* elements interacts with *K* other elements within a fitness landscape.

4 Uncertainty in the meaning adopted here covers uncertain parameter values, but more importantly the uncertainty associated with the choice of a model's structure; that is, drawing boundaries, selecting a level of abstraction and aggregation, selection and representation of model components and processes.

5 A 'vintage' being a number of installations with particular characteristics from a particular time step (e.g. particular type of boiler, car or industry motor drive).

References

Andersson, Claes, Anton Törnberg, and Petter Törnberg. 2014. "Societal systems – Complex or worse?" *Futures* 63: 145–57.

Bak, Per. 1996. *How Nature Works: The Science of Self-organized Criticality*. New York: Springer New York. http://doi.org/10.1007/978-1-4757-5426-1.

Bianchi, Frederico, and Flaminio Squazzoni. 2019. "Modelling and social science: Problems and promises." In *Modelling Transitions – Virtues, Vices, Visions of the Future*, edited by Enayat Moallemi and Fjalar de Haan. This volume: Routledge, Abingdon.

Brugnach, Marcela, Claudia Pahl-Wostl, Karl-Erich Lindenschmidt, Judith A.E.B. Janssen, Tatiana Filatova, Annaleen Mouton, Georg Holtz, Peter van der Keur, and Nadia Gaber.

2008. "Chapter four complexity and uncertainty: Rethinking the modelling activity." *Developments in Integrated Environmental Assessment* 3: 49–68.

Clarke, Leon, Kejun Jiang, Kiego Akimoto, Mustafa Babiker, Geoffrey Blanford, Karen Fisher-Vanden, Jean-Charles Hourcade et al. 2014. "Assessing transformation pathways." In *Climate Change 2014: Mitigation of Climate Change. Contribution of Working Group III to the Fifth Assessment Report of the Intergovernmental Panel on Climate Change*, edited by O. Edenhofer, R. Pichs-Madruga, Y. Sokona, E. Farahani, S. Kadner, K. Seyboth, and A. Adler. Cambridge: Cambridge University Press.

Daly, Hannah E., Kalai Ramea, Alessandro Chiodi, Sonia Yeh, Maurizio Gargiulo, and Brian Ó. Gallachóir. 2014. "Incorporating travel behaviour and travel time into TIMES energy system models." *Applied Energy* 135: 429–39.

Elsenbroich, Corinna, and Nigel Gilbert. 2014. *Modelling Norms*. Dordrecht: Springer. http://gbv.eblib.com/patron/FullRecord.aspx?p=1317791.

Epstein, Joshua M., and Robert Axtell. 1996. *Growing Artificial Societies: Social Science From the Bottom Up*. Complex adaptive systems. Washington, DC: Brookings Institution Press. http://search.ebscohost.com/login.aspx?direct=true&scope=site&db=nlebk&db=nlabk&AN=1813.

Freeman, Chris, and Francisco Louçã. 2002. *As Time Goes by: From the Industrial Revolutions to the Information Revolution*. Oxford: Oxford University Press.

Frenken, Koen. 2007. "Technological innovation and complexity theory." *Economics of Innovation and New Technology* 15 (2): 137–55. http://doi.org/10.1080/10438590500141453.

Frenken, Koen, Luis R. Izquierdo, and Paolo Zeppini. 2012. "Branching innovation, recombinant innovation, and endogenous technological transitions." *Environmental Innovation and Societal Transitions* 4: 25–35.

Gazheli, Ardjan, Miklós Antal, and Jeroen van den Bergh. 2015. "The behavioral basis of policies fostering long-run transitions: Stakeholders, limited rationality and social context." *Futures* 69: 14–30.

Geels, Frank W. 2002. "Technological transitions as evolutionary reconfiguration processes: A multi-level perspective and a case-study." *Research Policy* 31 (8–9): 1257–74. http://doi.org/10.1016/S0048-7333(02)00062-8.

Geels, Frank W., and Johan Schot. 2010. "The dynamics of socio-technical transitions: A socio-technical perspective." In *Transitions to Sustainable Development: New Directions in the Study of Long Term Transformative Change*, edited by John Grin, Jan Rotmans, and J.W. Schot. Routledge studies in sustainability transitions. New York: Routledge.

Geroski, Paul A. 2000. "Models of technology diffusion." *Research Policy* 29 (4–5): 603–25. http://doi.org/10.1016/S0048-7333(99)00092-X.

Grin, John, Jan Rotmans, and J.W. Schot, eds. 2010. *Transitions to Sustainable Development: New Directions in the Study of Long Term Transformative Change*. Routledge studies in sustainability transitions. New York: Routledge. http://site.ebrary.com/lib/alltitles/docDetail.action?docID=10370143.

Haan, Fjalar J. de. 2008. "The dynamics of functioning: Investigating societal transitions with partial differential equations." *Computational and Mathematical Organization Theory* 14: 302–19.

Haan, Fjalar J. de. 2019. "Making it a science – Aspirations and hesitations of transitions research." In *Modelling Transitions – Virtues, Vices, Visions of the Future*, edited by Enayat Moallemi and Fjalar de Haan. This volume. Abingdon: Routledge.

Haan, Fjalar J. de, Briony C. Ferguson, Rachelle C. Adamowicz, Phillip Johnstone, Rebekah R. Brown, and Tony H.F. Wong. 2014. "The needs of society: A new understanding of transitions, sustainability and liveability." *Technological Forecasting and Social Change* 85: 121–32.

Haan, Fjalar J. de, Briony C. Rogers, Rebekah R. Brown, and Ana Deletic. 2016. "Many roads to Rome: The emergence of pathways from patterns of change through exploratory modelling of sustainability transitions." *Environmental Modelling & Software* 85: 279–92.

Halbe, Johannes. 2019. "Participatory modelling in sustainability transitions research." In *Modelling Transitions – Virtues, Vices, Visions of the Future*, edited by Enayat Moallemi and Fjalar de Haan. This volume. Abingdon: Routledge.

Halbe, Johannes, Dominic Reusser, Georg Holtz, Marjolijn Haasnoot, Annette Stosius, Wibke Avenhaus, and Jan Kwakkel. 2015. "Lessons for model use in transition research: A survey and comparison with other research areas." *Environmental Innovation and Societal Transitions* 15: 194–210.

Holland, John Henry. 1992. "Complex adaptive systems." *Deadalus* 121: 17–30.

Holtz, Georg. 2012. "The PSM approach to transitions: Bridging the gap between abstract frameworks and tangible entities." *Technological Forecasting and Social Change* 79 (4): 734–43. http://doi.org/10.1016/j.techfore.2011.10.005.

Holtz, Georg, Floortje Alkemade, Fjalar de Haan, Jonathan Köhler, Evelina Trutnevyte, Tobias Luthe, Johannes Halbe, et al. 2015. "Prospects of modelling societal transitions: Position paper of an emerging community." *Environmental Innovation and Societal Transitions* 17: 41–58.

Holtz, Georg, and Emile J.L. Chappin. 2019. "Considering actor behaviour: Agent-based modelling of transitions." In *Modelling Transitions – Virtues, Vices, Visions of the Future*, edited by Enayat Moallemi and Fjalar de Haan. This volume. Abingdon: Routledge.

Jackson, Piper J. 2014. "Software solutions for computational modelling in the social sciences." In *Theories and Simulations of Complex Social Systems*. Vol. 52, edited by Vahid Dabbaghian and Vijay K. Mago, 5–17. Intelligent systems reference library. Berlin, Heidelberg: Springer.

Kauffman, Stuart A. 1993. *The Origins of Order: Self-Organization and Selection in Evolution*. Oxford: Oxford University Press. http://gbv.eblib.com/patron/FullRecord.aspx?p=648038.

Köhler, Jonathan. 2012. "A comparison of the neo-Schumpeterian theory of Kondratiev waves and the multi-level perspective on transitions." *Environmental Innovation and Societal Transitions* 3: 1–15.

Köhler, Jonathan. 2019a. "Advances in modelling sustainable innovation: From technology bias to system theories and behavioural dynamics." In *Handbook of Sustainable Innovation*, edited by F. Boons and A. McMeekin. Routledge, Abingdon.

Köhler, Jonathan. 2019b. "Modelling the multi-level perspective: The MATISSE agent-based model." In *Modelling Transitions – Virtues, Vices, Visions of the Future*, edited by Enayat Moallemi and Fjalar de Haan. This volume: Routledge, Abingdon.

Köhler, Jonathan, Fjalar de Haan, Georg Holtz, Klaus Kubeczko, Enayat Moallemi, George Papachristos, and Emile Chappin. 2018. "Modelling sustainability transitions: An assessment of approaches and challenges." *Journal of Artificial Societies and Social Simulation* 21 (1). http://doi.org/10.18564/jasss.3629.

Köhler, Jonathan, Michael Grubb, David Popp, and Ottmar Edenhofer. 2006. "The transition to endogenous technical change in climate-economy models: A technical overview to the innovation modeling comparison project." *The Energy Journal* 27 (SI Endogenous Technological Change and the Economics of Atmospheric Stabilisation): 17–55.

Köhler, Jonathan, Bruno Turnheim, and Mike Hodson. 2018. "Low carbon transitions pathways in mobility: Applying the MLP in a combined case study and simulation bridging analysis of passenger transport in the Netherlands." *Technological Forecasting and Social Change*. http://doi.org/10.1016/j.techfore.2018.06.003.

Köhler, Jonathan, Lorraine Whitmarsh, Björn Nykvist, Michel Schilperoord, Noam Bergman, and Alex Haxeltine. 2009. "A transitions model for sustainable mobility." *Ecological Economics* 68 (12): 2985–95. http://doi.org/10.1016/j.ecolecon.2009.06.027.

Li, Francis G.N., and Neil Strachan. 2017. "Modelling energy transitions for climate targets under landscape and actor inertia." *Environmental Innovation and Societal Transitions* 24: 106–29.

Li, Francis G.N., Evelina Trutnevyte, and Neil Strachan. 2015. "A review of Socio-Technical Energy Transition (STET) models." *Technological Forecasting and Social Change* 100: 290–305.

Malekpour, Shirin. 2019. "Models as scenario tools for developing robust transformative plan." In *Modelling transitions – Virtues, vices, visions of the future*, edited by Enayat Moallemi and Fjalar de Haan. This volume. Abingdon: Routledge.

Mantzos, Leonidas, Nicoleta A. Matei, Máté Rózsai, Peter Russ, and Antonio S. Ramirez, eds. 2017. *POTEnCIA: A New EU-wide Energy Sector Model*. New York: IEEE.

Mantzos, Leonidas, Tobias Wiesenthal, Nicoleta A. Matei, Mate Rozsai, Elena N. Cawood, Ioanna Kourti, Anastasios Papafragkou, Peter Russ, and Antonio S. Ramirez. 2016. *POTEnCIA Model Description-Version 0.9*. New York: IEEE.

Markard, Jochen, Rob Raven, and Bernhard Truffer. 2012. "Sustainability transitions: An emerging field of research and its prospects." *Research Policy* 41 (6): 955–67. http://doi.org/10.1016/j.respol.2012.02.013.

McCollum, David L., Charlie Wilson, Hazel Pettifor, Kalai Ramea, Volker Krey, Keywan Riahi, Christoph Bertram, Zhenhong Lin, Oreane Y. Edelenbosch, and Sei Fujisawa. 2017. "Improving the behavioral realism of global integrated assessment models: An application to consumers' vehicle choices." *Transportation Research Part D: Transport and Environment* 55: 322–42.

McDowall, Will. 2014. "Exploring possible transition pathways for hydrogen energy: A hybrid approach using socio-technical scenarios and energy system modelling." *Futures* 63: 1–14.

Meyer, Matthias, Iris Lorscheid, and Klaus G. Troitzsch. 2009. "The development of social simulation as reflected in the first ten years of JASSS: A citation and co-citation analysis." *Journal of Artificial Societies and Social Simulation* 12 (4): 12.

Moallemi, Enayat A., Fjalar J. de Haan, and Jonathan Köhler. 2019. "Exploratory modelling of transitions: An emerging approach for coping with uncertainties in transitions research." In *Modelling Transitions – Virtues, Vices, Visions of the Future*, edited by Enayat Moallemi and Fjalar de Haan. This volume: Routledge, Abingdon.

Moallemi, Enayat A., and Shirin Malekpour. 2018. "A participatory exploratory modelling approach for long-term planning in energy transitions." *Energy Research & Social Science* 35: 205–16.

Papachristos, Georg. 2011. "A system dynamics model of socio-technical regime transitions." *Environmental Innovation and Societal Transitions* 1 (2): 202–33.

Papachristos, Georg, and Jeroen Struben. 2019. "System dynamics methodology and research: Opportunities for transitions research." In *Modelling Transitions – Virtues, Vices, Visions of the Future*, edited by Enayat Moallemi and Fjalar de Haan. This volume. Abingdon: Routledge.

Pettifor, Hazel, Charlie Wilson, David McCollum, and Oreane Y. Edelenbosch. 2017. "Modelling social influence and cultural variation in global low-carbon vehicle transitions." *Global Environmental Change* 47: 76–87.

Ramea, Kalai, David S. Bunch, Christopher Yang, Sonia Yeh, and Joan M. Ogden. 2018. "Integration of behavioral effects from vehicle choice models into long-term energy systems optimization models." *Energy Economics* 74: 663–76.

Rip, Arie., and Rene. Kemp. 1998. "Technological change." In *Human Choice and Climate Change*, edited by S. Rayner and E.L. Malone, 327–992. Columbus, OH: Battelle Press.

Safarzyńska, Karolina, Koen Frenken, and Jeroen C.J.M. van den Bergh. 2012. "Evolutionary theorizing and modeling of sustainability transitions." *Research Policy* 41 (6): 1011–24. http://doi.org/10.1016/j.respol.2011.10.014.

Safarzyńska, Karolina, and Jeroen C.J.M. van den Bergh. 2010. "Demand-supply coevolution with multiple increasing returns: Policy analysis for unlocking and system transitions." *Technological Forecasting and Social Change* 77 (2): 297–317. http://doi.org/10.1016/j.techfore.2009.07.001.

Schlüter, Maja, Ryan R.J. Mcallister, Robert Arlinghaus, Nils Bunnefeld, Klaus Eisenack, Franz Hölker, et al. 2012. "New horizons for managing the environment: A review of coupled social-ecological systems modeling." *Natural Resource Modeling* 25 (1): 219–72. http://doi.org/10.1111/j.1939-7445.2011.00108.x.

Silverberg, Gerald, Giovanni Dosi, and Luigi Orsenigo. 1988. "Innovation, diversity and diffusion: A self-organisation model." *The Economic Journal* 98 (393): 1032. http://doi.org/10.2307/2233718.

STRN. 2017. *Research Agenda for the Sustainability Transitions Research Network*. https://transitionsnetwork.org/about-strn/research_agenda/.

Tattini, Jacopo, Kalai Ramea, Maurizio Gargiulo, Christopher Yang, Eamonn Mulholland, Sonia Yeh, and Kenneth Karlsson. 2018. "Improving the representation of modal choice into bottom-up optimization energy system models – The MoCho-TIMES model." *Applied Energy* 212: 265–82.

Timmermans, Jos, and Hans de Haan. 2008. "Special issue on computational and mathematical approaches to societal transitions." *Computational and Mathematical Organization Theory* 14 (4): 263–5. http://doi.org/10.1007/s10588-008-9037-z.

Trutnevyte, Evelina, John Barton, Áine O'Grady, Damiete Ogunkunle, Danny Pudjianto, and Elizabeth Robertson. 2014. "Linking a storyline with multiple models: A cross-scale study of the UK power system transition." *Technological Forecasting and Social Change* 89: 26–42.

Turnheim, Bruno, Frans Berkhout, Frank Geels, Andries Hof, Andy McMeekin, Björn Nykvist, and Detlef van Vuuren. 2015. "Evaluating sustainability transitions pathways: Bridging analytical approaches to address governance challenges." *Global Environmental Change* 35: 239–53.

van den Bergh, Jeroen C.J.M., and John M. Gowdy. 2009. "A group selection perspective on economic behavior, institutions and organizations." *Journal of Economic Behavior & Organization* 72 (1): 1–20. http://doi.org/10.1016/j.jebo.2009.04.017.

Venturini, Giada, Jacopo Tattini, Eamonn Mulholland, and Brian Ó. Gallachóir. 2018. "Improvements in the representation of behavior in integrated energy and transport models." *International Journal of Sustainable Transportation*: 1–20.

Walrave, Bob, and Rob Raven. 2016. "Modelling the dynamics of technological innovation systems." *Research Policy* 45 (9): 1833–44.

Windrum, Paul, and Chris Birchenhall. 2005. "Structural change in the presence of network externalities: A co-evolutionary model of technological successions." *Journal of Evolutionary Economics* 15 (2): 123–48. http://doi.org/10.1007/s00191-004-0226-8.

Windrum, Paul, Tommaso Ciarli, and Chris Birchenhall. 2009. "Environmental impact, quality, and price: Consumer trade-offs and the development of environmentally friendly technologies." *Technological Forecasting and Social Change* 76 (4): 552–66. http://doi.org/10.1016/j.techfore.2008.04.012.

Yücel, Gönenç. 2010. *Analyzing Transition Dynamics: The Actor-option Framework for Modelling Socio-technical Systems*. Delft: Delft University of Technology.

Zeppini, Paolo, Koen Frenken, and Roland Kupers. 2014. "Thresholds models of technological transitions." *Environmental Innovation and Societal Transitions* 11: 54–70.

Zeppini, Paolo, and Jeroen C.J.M. van den Bergh. 2011. "Competing recombinant technologies for environmental innovation: Extending Arthur's model of lock-in." *Industry and Innovation* 18 (3): 317–34. http://doi.org/10.1080/13662716.2011.561031.

Part 1

Virtues and vices

4 Making it a science

Aspirations and apprehensions of transitions research

Fjalar J. de Haan

1 Introduction

Making transitions research a science should be, in my opinion, the highest priority on the agenda of the field. Higher than, for instance, striving for policy relevance and influence. Indeed, one could argue (as I do) that it is questionable to strive for such things while transitions research is not yet scientific. Note that I am not only saying that the body of knowledge to have come from transitions research is not yet mature enough to be called science. I am saying that transitions research at the time of writing is itself, by and large, not scientific or even unscientific.

I expect there to be many transitions researchers who wholeheartedly disagree with the paragraph above. Moreover, I expect them to divide neatly into two groups who disagree in two complementary ways. One group would argue that transitions research does not need to become a science because transitions are not necessarily a scientific subject and therefore cannot be treated scientifically. Another group would argue that transitions research is scientific research already, that it is social science and while this entails a different suite of methods and approaches it is science nonetheless.

To the first group, transitions must be something like art – and I do not say this to ridicule – and transitions research a bit like art critique. Their thinking seems to be along the lines of: while art can be studied scientifically, the appreciation and pursuit of art are at best marginally aided by this. If you identify with this group, I disagree with you and most likely you with me, but our disagreement is one of general principle and I do not expect it to be amenable to argument.

To the second group, transitions (and likely various other subjects) belong to a special class of phenomena: the human phenomena. These researchers would see a range of reasons why the standards and practices of science (which they would call 'hard science') do not apply in their field. Amongst their reasons are things like the unpredictability of human action, normativity, (inter)subjectivity and so on. If you identify with this group, I disagree with you and most likely you with me, but our disagreement is more subtle and I will try to make an argument for my case in Section 2.

1.1 The importance of transitions

That transitions are important is perhaps the one notion that unites transitions researchers in their endeavours. Transitions, to me, are important in several ways, but the unifying importance is *practical*. I usually summarise this importance with reference to *persistent problems* and *pervasive change*.

Persistent problems are, in a way, the negative side effects of the systems we rely upon to meet our human needs on societal scales. Rotmans (2005) sees persistent problems as symptoms of an unsustainable society. Schuitmaker (2012) points out that persistent problems are not so much malfunctions of the systems but undesired consequences of the way they work. From this perspective, transitions are ways to overcome persistent problems. If the systems are intrinsically problematic, then the systems need to structurally change.

Pervasive change seems to come closer to the way transitions were described in the pioneering literature. The classic *More Evolution than Revolution* (Rotmans et al., 2001), for example, refers to transitions as the confluence of developments in several domains, leaving open whether these developments are purposefully pursued or 'emergent'. Pervasive change is thus the transitions that happen to us whether we want it or not. From this perspective, transitions are something to deal with. Positively speaking they present opportunities. Negatively speaking they need to be navigated to mitigate potential harm.

Whether we 'merely' want to navigate a transition or purposefully pursue one, it seems obvious that we need to *understand* them, and I think this is the unifying importance that is a major motivation of most, if not all, transitions researchers.

Even if one is not interested in the possibility to navigate or pursue transitions, or if one is deeply skeptical about the possibilities to do so, then still there is the noble quest to understand the movements of the societies we are part of. The changing systems we have created to sustain us and of which we are a part, determine a large extent the modern human condition. Transitions are therefore eminently worthy of study as a contribution to human culture at large. Even if one does not appreciate these lofty connotations, then there is still curiosity. One wants to know because one does not yet. Because of the pleasure of finding things out, as Feynman (1999) would have it. Perhaps because of the beauty of what we will find. Perhaps because we cannot help but want to know. Transitions are a Grand Intellectual Challenge.

1.2 The desirability of a transitions science

Curiosity and the Grand Intellectual Challenge are enough to make me desire a transitions science, regardless of its utility. To paraphrase a quote generally attributed to[1], again, Feynman: science is like sex, it has practical applications but that's not why we do it. But for those of you who also want transitions research to matter to actual transitions (likely most researchers in the field, as per my assessment in Section 1.1), I ask: why is there so precious little scientific research on transitions?

Leaving aside, for just this moment, whether the dominant mode of transitions research produces knowledge that *can* matter, one really wonders why a field so concerned with real-world issues like sustainability does not embrace more scientific modes of operating. Is it a perverse irony of this field? That its 'regime' of narrative case study based research is so strong it discourages budding transition scientists? Or worse? It is not difficult to recognise as downright hostility the not-even-thinly-veiled patronising tone of regime advocates like McDowall and Geels (2017). While that article was a response to Holtz et al.'s (2015) position paper on modelling, my estimate is that their response to anything even more explicitly in favour of a 'hard' science approach would invoke even stronger corrective responses and scorn – especially given their antipathy to anything reeking of 'positivism'.

I can think of no other field that matters, in the sense that transitions research could, putting its trust so narrowly in an approach not known for yielding reliable, accurate knowledge. Medicine and psychology may share the ubiquity of case research with transitions, but these fields also employ laboratory experiments, modelling and randomised trials while the case research is routinely meta-analysed and a single case report is likely to be viewed as mere anecdotal evidence. I mention these two fields also because they share many of the complexities, ethical implications and empirical difficulties of transitions research[2]. There is surely a role for case-based research – narrative or otherwise – in a science of transitions, which I will discuss extensively in Section 3.6. The problem is that at this point, the transitions research play seems written as a monologue.

It is of no relevance whether or not the objects of study in those fields are more amenable to 'hard' scientific methods. The point is that if we do not know anything scientifically about such issues, we know very little at all and whatever we opine or may think they mean is no basis for policy or otherwise intervention. Imagine public health services preparing for a pandemic solely on the basis of detailed historical accounts of previous outbreaks. Not that such accounts are wholly inappropriate for future strategy and policy – if only to learn from obvious mistakes – but what is needed, what the public rightfully expect, is that strategy and policy are based on a thorough, tested, mechanistic understanding of how disease is contracted, propagated and treated. It is ethically objectionable to even suggest important decisions be informed by transitions research if not scientific.

1.3 Science as aspiration

I have pointed out several times already that most transitions research is at the moment not scientific. Precisely what attitudes and practices are problematic in this sense, I will elaborate extensively in Section 3. As a matter of course, since this seems to be a strictly negative assessment, I will outline concrete steps that can be taken right now by transitions researchers to make it a science. This I will do in the concluding part of this chapter (Section 5), based on my — philosophically uncontroversial – interpretation of the intimate relation between theorising, models and empiricism (Section 4), taking practical cues from other fields that face similar challenges. Before dismantling the arguments against the possibility of a transitions science (Section 2), I will now sketch what I – and I believe most scientists – consider the nature and features of scientific research.

Despite the household notion of the 'scientific method', science is *not* a matter of method. Advanced sciences have advanced methods of course, but this is a consequence of their progress, not the reason for it. As their theories become more accurate, they need more precision[3] to know whether observations match those theories or not. If science is not a matter of method, then what *is* it a matter of? It is a matter of honesty. An intellectual honesty going far beyond not wanting to lie. It is the acceptance of the possibility of being wrong, moreover it is actively setting oneself up to be proved wrong constantly. Feynman (1974; 1985; 1999) calls it 'a kind of leaning over backwards' and urges publications of 'negative' results as well as 'positive'. Dennett (2013, p. 27) says[4]: 'In science you make your mistakes in public'.

Proving a hypothesis wrong also progresses knowledge, perhaps even more than finding confirming evidence does. So, rather than cushioning conjectures in vague terms so no-one is ever quite sure what is stated or denied, we should acknowledge that we are conjecturing and phrase ourselves refutably. This prescription is usually attributed to (Popper, 1957) though it just summarises the scientific attitude. One of the very worst things is to be 'not even wrong', which Pauli reputedly liked to point out. Theoretical speculation, conjecture and conceptual design are of central importance in scientific research, the above is certainly not meant to decry speculation as such. In fact, one sees its importance acknowledged in the explicit division of labour in some advanced sciences, where theoretical research is represented by specialised branches of otherwise empirical fields. The art of conjecture is to say something that is not trivial, that is, not just a label for what is empirically known, and that could be false.

It is important to appreciate the role of method correctly, because wellintended scientific aspirations can easily lead to naïve imitation of methods common in the natural sciences. Minsky (1998) referred to this tendency as 'physics envy'. This may lead to what Feynman (op. cit.) has described as 'cargo-cult science[5]', in which methods of the 'hard' sciences, like quantification and statistics are applied while missing the point of being scientific, and therefore the rationale of those methods, completely. It is not that they apply those methods badly, nor is it that they do not apply at all, that may or may not be the case. The point is that using such methods does not *make* the research scientific.

Science seeks understanding of classes of phenomena and this understanding is considered to be brought about by *explanation*. Hence philosophy of science has since long had a strong focus on the nature of explanation. Theories, from this perspective, can be thought of as bodies of meta explanations, generic formulae that, when filled-out with the specifics of the situation at hand, explain the phenomenon of concern. Hempel's Covering-Law Model (see e.g. Hempel and Oppenheim, 1948) is perhaps the best known example, but later models based on causal mechanisms and functional explanation (see e.g. Salmon, 1990) are in this sense quite similar.

This perspective makes generalisation a key part of any scientific explanation. To explain something first requires one to identify it as an instance of a class of phenomena. One then invokes the law or mechanism about that class of phenomena with the parameters pertinent to the specific phenomenon under consideration. If the observed now logically follows, it is explained.

In other words, *no explanation without generalisation*. This seems in stark contrast with historical explanations where there are only particulars and contingencies. Such explanations usually rely on the course of events to provide the explanation. Every next event is expected given the previous ones in the sequence. Implicitly, every step involves an elliptical invocation of a law or mechanism considered so obvious it can go without saying (e.g. Roberts, 1996, for an explication of this view of historical explanation). If done well, this may satisfy as an explanation of a particular event or case (but see Section 3.4 for common pitfals), it does not typically contribute to understanding the class of phenomena the sequence of events was a particular instance of.

This perspective also makes the capacity of *prediction* a central desideratum of any scientific explanation. Amongst natural scientists, this would not cause much of a stir, but I expect researchers of social phenomena may object that this is too much to ask – or that it is not of central importance. Is it really? 'Any rationally acceptable answer to the question "why did X occur?" must offer information which shows that X was to be *expected* – if not definitely … then at least with reasonable probability', said Hempel (quoted in Rosenberg, 1993, my emphasis). I feel this is a fair, if not modest, demand. This capacity of prediction is also precisely what makes theories testable. Moreover, it is this predictive aspect that makes science *matter* in the way transitions research wants to matter.

That science is empirical is one of its defining characteristics. Yet there are common misunderstandings about the role of empiricism in science. It does patently not mean, for instance, that scientific hypotheses or theory are always obtained inductively. Spotting a pattern inductively is definitely a way to come to a hypothesis, but there are infinitely many hypotheses for any one pattern observed so to say induction *leads* to a hypotheses seems to strong. The practice of analytic theorising is much more a form of inductioninspired speculation. Moreover, it is often the noticing of something *not* fitting a pattern that gets the hypothesising going.

How science is empirical and the *practical* relation of theory and experiment in sound science have rarely been more accurately summed up, likely never with less jargon, than by Feynman. In his lectures on the character of physical law (1965), he said: 'In general, we look for a new law by the following process. First, we guess it. [LAUGHTER] Then, we compute – well, don't laugh, that's really true. Then we compute the consequences of the guess [...] we see what it would imply. And then we [...] compare to experiment or experience. [...] If it disagrees with experiment, it's wrong. And that simple statement is the key to science. [...] That's all there is to it'.

The empirical test provides science with an ultimate arbiter. Clearly, the guessing is not arbitrary, it is educated and may take cues from established theory or other considerations, like elegance. Clearly also, where Feynman says 'compute', one can safely substitute 'infer' – it is the logical implications of our hypotheses that matter, they endow hypotheses with the predictive powers that make them testable. Though Feynman was a physicist and the quoted lectures were on the character of physical law, his summary of the practice of good science is noticably devoid of any reference to a particular empirical domain. In fact, he literally speaks '[i]n general' about 'science' and I do not see any reason to restrict his lessons to physics or even the natural sciences (see also Section 2). Feynman's lesson also does not refer to any specific method. Not for the experimenting, not for the guessing, nor for the way one should go about comparing guesses to experiments. It is not a matter of method.

2 The case against a transitions science

I mentioned in the introduction that there are numerous transitions researchers who believe that the standards and practices of what they would refer to as 'hard' science do not apply to their field – the people I referred to as 'the second group'. Defending this belief they would bring to bear various technical arguments against the possibility of a true transitions science. I do not disagree with all the issues they raise, but I think their conclusion does not follow. I will try to distill the common kernels of their arguments and show how I think they are *non sequitur*. Apart from these technical arguments there is a common non-argument based on misguided ideas about philosophy. I will discuss this also.

Whatever may be said in favour or against these arguments, I feel that there are 'deeper' reasons for people to oppose a scientific approach to transitions. These reasons, I think are based on a misguided perspective on the social sciences vis-á-vis the natural sciences – namely that there would be a separation between the two and, moreover, that this supposed separation would warrant different standards and methods. I will point out what I think are the origins of this perspective and urge to move beyond it.

2.1 Technical arguments

The supposed reasons against a 'hard' scientific treatment of transitions can be grouped into the following three categories:

Too complex – According to this argument, societies, or at any rate the dynamics of transitions, are just too much for the crude methods of science. The generalisations that the scientific approach is after, require simplifications that make a caricature out of the rich fabric of historically unfolding transitions. This latter point is no doubt true, but out of many (empirically tried and tested) caricatures we build scientific theories: robust generalisations across a class of phenomena. The argument from complexity did not stop biology and medicine from charting and understanding organisms and human bodies, which are stupefyingly complex systems composed of tens of *trillions* of cells, which are themselves very complex systems. Organisms have functional structures and dynamics on and across many scales, covering about 20 orders of magnitude (see e.g. West et al., 2002). Certainly, this presents a massive challenge, but the continuing progress of chemistry, biology, genetics and medicine shows that complexity of arguably larger[6] proportions can be understood with scientific methods. Such progress should inspire transitions researchers.

No lessons from the past – This argument roughly says that whatever hope one has of generalisations with predictive power, such generalisations would be intimately tied up with the historically contingent structures of the societies they would be found in. These structures themselves are prone to change (transitions!) and then those generalisations no longer apply. In other words, whatever we learn about transitions in the 20th and 21st Centuries will be void when a new socio-politico-economic era dawns, perhaps heralded in by the very transitions we pursue. Karl Popper (1957) has put forward such an argument and Rosenberg (1993) explores it to identify limits to prediction in social science. The argument seems to me not flawed and Martínez Arranz (2017) raised it in the context of transitions.

The in-principle impossibility Popper raises, which Rosenberg discusses, I find not very intimidating – at least not for the prospect of a transitions science. The notion that we can never predict e.g. future technological advances (because it would imply the possibility to extrapolate future knowledge from current knowledge, which implies we *already* have the knowledge required to attain it) is amusingly paradoxical. However, such a science of technological progress would meet limits of prediction much earlier. Even in Newtonian physics where such *inprinciple* impossibilities are absent (remember Laplace's demon), practicalities like non-linearity, measurement uncertainties and many-body problems lard the ointment with flies. The ambitions of a transitions science, though storming the heavens, would be much more modest than that.

Impossibility of mind-reading – This last category of objections against a scientific approach to transitions, argues in the following fashion: Transitions are the product of human actions, and therefore we have to deal with the intricacies of the human mind, like issues of agency, (inter) subjectivity and meaning. Something the 'hard' sciences do not have to, or refuse to deal with. This argument is flawed in two separate ways. Firstly, issues of agency, subjectivity and meaning *are* amenable to scientific inquiry. The interdiscipline of cognitive science has made serious headway in these areas and again this should inspire transitions researchers. The second way in which the argument is flawed is this: the (supposed) fact that human action is contingent on subjective perceptions and meanings attributed, does patently *not* mean that they do not display regularities that can be captured in 'hard' science-like covering laws – which can even be in terms of subjective perceptions or meanings if appropriate. Behavioural science is certainly finding that human actions are quite predictable. In fact, everyone routinely 'reads mind' all the time in social situations and deviations from our 'predictions' give rise to humor or affront. Moreover, these issues may matter much less than we are inclined to think – I will return to this shortly.

2.2 Philosophy non-argument

Some object that a 'hard' scientific approach in the vein of what I am suggesting is based on a different philosophy of science, one that may not be appropriate to social sciences. This is of course just another form of the separatist argument of 'the second group'. In this specific form, however, it also reveals a thorough misunderstanding of the role of philosophy of science. Many social scientists seem to be under the impression that one has to *choose* a particular philosophy of science, like if it were methodology and one simply has to find the one best suited to the topic. This is clearly mistaken.

Philosophy of science does not aim to be a foundation of science, it aims to clarify what science is. One of the ironies of philosophy of science is that it is acutely aware of the fact that science has been getting by without philosophy very well thank you. Moreover, the successes of empirical science 'sans philosophy' inspired early analytic philosophers and philosophers of science, like Russell and the Vienna Circle to remodel philosophy in the image of science – *not* the other way around.

This misunderstanding leads to the absurd situation of graduate students having to write perfunctory thesis chapters where they declare their allegiance to e.g. pragmatism as a supposed fundament and justification for their mixed methods approach. Clearly, whoever uses statistics or tries to measure something operates in a positivist tradition, while qualitative researchers are shepherded towards interpretivism. I think this is a real and entrenched problem in academic research training, perpetuated by popular 'research design' textbooks (e.g. Creswell (2018) – five editions and over 100,000 citations on Google Scholar) that artificially distinguish philosophical 'worldviews' and identify them with certain methods.

Another way in which philosophy is commonly misconstrued in transitions research and other social sciences is through terminology. The terms 'ontology' and 'epistemology' are routinely used to mean pretty much 'concepts' and 'methodology'. Ontology is a branch of metaphysics[7] interested in what the world ultimately consists of, e.g. matter or mind, and what that may mean for the existential status of such things as universals. Epistemology is theory of knowledge. What constitutes knowledge? How can we be sure that something is actually knowledge?

To refer to one's conceptual framework as an 'ontology' suggests a metaphysical commitment to those concepts being 'really out there'. The scientific modus operandi is to not care what the metaphysical status of a concept is. What is the metaphysical status of centres of gravity? Do species really exist[8]? What shall it matter? To refer to one's methodological approach as an epistemology is also confused, albeit in a slightly different manner. It is mistaking the means to acquire something (here, knowledge) with the nature of the something. Like saying, work is the essence of money. Discussing ways of obtaining money is simply distinct from what money is and is not – unless one wants to get *really* philosophical about it. This is not to say there are no epistemological concerns about transitions worthy of philosophical inquiry, just that when transitions researchers use the term, typically they mean something else.

2.3 Deeper reasons

I think that underneath the technical objections raised against the possibility of a transitions science lurks something deeper, something rooted in fear, a twofold fear.

The first is psychological, it is fear of inadequacy. This would apply to those who feel a scientific approach would actually be superior, but lack the requisite knowledge and skills – or the motivation to obtain them. The resulting cognitive dissonance is alleviated by adjusting attitudes, towards supposing 'hard' science approaches to not apply to their topic. Perhaps the prevalence of this fear is not that great. At any rate, it is unnecessary, see my comments about 'no matter of method'.

The second is philosophical, it is the pre-Copernican fear of us humans losing our special place in the universe. Especially the third genre of technical objections, with its deference to mind as the great separator, is a child of this fear. I believe the idea of a sharp separation between human phenomena and the rest of the world, and its concomitant necessity of special modes of inquiry, to be an echo of antiquated ideas with a shared core of human self-importance: The sharp separation between earthly and celestial mechanics – an idea no longer tenable after Newton. Humans as distinct, and placed above, animals and the rest of nature – ridiculous after Darwin.

'Life and all its glories are thus united under a single perspective, but some people find this vision hateful, barren, odious. They want to cry out against it, and above all, they want to be magnificent exceptions to it', said Dennett (1995, p. 144). But we *are* part of the animal kingdom, part of the great Tree of Life and part of the fascinating physical universe in which it sprouted. We should take this at face value in our quest to understand transitions.

Even if we come to learn that certain species of birds have a rich inner life and complex cultural interactions based on the meanings they attribute to aspects of their world, I doubt we would then conclude that scientific methods and explanations no longer apply. Nor would we conclude they then transcend biology, that their behaviour and social structures are then somehow of a totally different order. Fields like sociobiology (Wilson, 1975) and subsequently evolutionary psychology (Symons, 1979; Tooby and Cosmides, 1992;

Barkow et al., 1992) have taken Dennett's single perspective at face value, and indeed attracted the expected controversy and scorn – so much for the intrinsic unpredictability of human behaviour.

'The social sciences need to take seriously their status as divisions of biology' and they therefore 'must proceed in accordance with the explanatory and evidential strictures of biology', says Rosenberg (2017, p. 341–342) in a most lucid recent piece of philosophy of (social) science. He makes abundantly clear how this is not a reductionist thesis in neither of the common meanings of the term 'reductionist'. It is the practical observation that the methods of the social sciences should be no different than those of biology, and moreover, that the 'findings, theories, and models of human behavior and of human groups [would be] no different in kind from those biology is familiar with'.

That this is not a singular voice advocating an isolated scientistic daydream can be gleaned from recent developments. Fields that traditionally looked at human phenomena in isolation from the rest of nature or did consider other animals as part of their jurisdiction at all, begin to adopt the single perspective. An example of the former, is the research on the evolution of leadership looking at the human manifestations of the phenomenon on equal footing with those of other animals (e.g. King et al., 2009; Van Vugt, 2006). An example of the latter is how some archaeologists have broadened the scope of their discipline to study the material cultural history of non-humans (Haslam et al., 2009, 2016; Proffitt et al., 2016; Wong, 2017) —this should be especially inspiring to transitions researchers as it entails the disappearance of the artificial distinction between human and non-human technological innovation. These developments routinely find their way to major publications of the natural (Nature, Current Biology) and 'human-centred' sciences (Personality and Social Psychology Review, Journal of Human Evolution).

3 What is wrong with transitions research?

The practices I will raise in this section are not necessarily problematic in principle. As I mentioned in the introduction, science is not a matter of method. That, however, does not mean that every method is as good as the next. Some are certainly worse than others. Some are so because they invite a way of thinking that almost precludes a scientific approach. Whenever this is the case, I will attempt to clarify how this happens.

In the following, I will discuss practices that I think are pervasive in the field to the point of being typical. I am certainly not suggesting that all researchers employ them, nor that each researcher employing them is necessarily being unscientific. The point is more subtle. Because of the prevalence of these practices in the field, people do not question their legitimacy from a scientific point of view or otherwise. A whole community cannot be wrong, right? Therefore, the dissenters are wrong. Their approach must not apply.

3.1 Fetish of the peculiar

Plainly put, much research on transitions tries to explain what is *special* about a particular transition or part of it, whereas the aim should be to find out what is *common* across transitions. This is not to say that an explanation of what seems to be an exception cannot be enlightening. Indeed, often scientific theories are illustrated by how they subsume supposed exceptions under a general principle.

This is, however, not the kind of explanation-of-the-peculiar common in transitions research. There, general principles have often not yet been found and thus explanations of apparent exceptions will necessarily be accounts of the particular contingencies involved. At best these will actually constitute explanations, but it would be a failure of any scientific theory of transitions if many such exceptions remain to be explained in this manner.

This lure of peculiarism is great as what stands out cries for explanation. It is amplified by the large reliance on case studies as the source of empirical material. The virtue of case studies is that they provide in-depth, in-detail accounts of phenomena but this virtue tempts to forego the search for generalisation. Even amongst the comparative case studies around, the tendency is to focus on what sets them apart.

The fetish of the peculiar is self reinforcing. The more the focus is on what is special, the less we search for general principles, the more we lack scientific explanations for apparent exceptions, the greater the reliance on peculiarist explanation.

3.2 Fuzzy concepts that move

There is nothing problematic about concepts being fuzzy or their interpretation shifting over time. Indeed, it is quite common for the central concepts of a science or discipline to be vaguely defined. Biology is not troubled much by the lack of a precise definition of 'life' or 'species' and physicists do not bat an eyelid about shifting back and forth between an interpretation of space as 'empty expanse' or 'stuff' that can bend and curve. So why worry about concept fuzziness and motion?

Sometimes it seems that more precise definitions are a sign of maturity of the field (e.g. the concrete definition of 'genes' in terms of DNA molecules), but this precision may just as well be a consequence of the progress as the cause. Moreover, I think that pedantically emulating the conceptual precision of the exact sciences will lead to a theoretical variant of the cargo-cult science I discussed in the introduction. The point is that conceptual precision has to be earnt – empirically.

When we are still in the explanatory dark, our concepts in fact likely *need* to be a bit fuzzy, though obviously not as fuzzy as to mean anything and everything. The testing of hypotheses, however, necessitates quite precise operational definitions of concepts to unambiguously know whether something is an instance of the class of phenomena we are trying to explain. In this way we find out which definitions do work for us and it is quite possible that, as we learn, several related concepts emerge with subtle technical distinctions where once we had just one. In biology, 'heredity' was made more precise with the term 'gene', which became even more precise when its relation to the nucleic acids was discovered, subsequently giving rise to even subtler concepts like 'cistron' and 'operon'. The study of mechanics led to several related but distinct and well-defined concepts like velocity, speed, momentum and kinetic energy, when natural philosopy essentially had one big, fuzzy concept of 'motion'.

The point I made above is related to the confusion between precision and accuracy. Precision is about how much margin we leave in our description, the more fuzzy, the less precise. One is accurate when one's predictions, or otherwise claims are within the margins set by our precision. Science aims for, and progresses towards being accurate with increasing precision. When we have no way to be accurate yet, because our science is nascent at best, we should not formulate our concepts with – effectively arbitrary – precision. If we cannot see the target because we are in the dark, our estimate of its location should not be precise. However, when we send our arrows probingly, we should do so with precision so we can ascertain what aim was most accurate.

3.3 Nuance and sophistication

When one presents a piece of social theory at a conference, a typical question runs something like: 'Very interesting, but how does it address *X*? How can your theory leave out *X* while it is so obvious that *X* is highly relevant in transitions?' For *X* can be substituted the asker's favourite concept, e.g. 'power', 'love' or 'networks'. Clearly, the pressure is on for the theorist or analyst to weave in *X* and 'address' it. This pressure is unfortunate because the question suggests a manifestly unscientific approach. The question should be whether the theory *works*, whether it explains things. If it does, without *X*, then this is an indication that *X* may not have been that important after all.

Even without such pressure there seems to be an innate tendency towards more nuance and sophistication in social theorising and this is a problem. One could object that transitions are multi-faceted processes in which not only *X*, but also the rest of the alphabet is important. That may well be, but that does not mean that a theory needs to 'address' all of them in one fell swoop. By conceptually overloading theory in advance, it simply becomes more difficult to figure out what really matters in what situation. It is referred to as 'nuance' or 'sophistication' but it really is vagueness, it *decreases* resolution. More of the phenomenon will fit the theory in the same way that a vague picture of a face will match more people[9]. There is less information – literally, see Shannon (1948) – in a nuanced picture of a phenomenon.

Simplification, by contrast, has become somewhat of a dirty word. But simplification is good. In fact, oversimplification is better. If complexity is a barrier to understanding what is going on, then simplification is part of the solution. I certainly do not mean that all aspects of transitions should be reduced to a simple equation with concept *X*. If the mechanics of transitions truly are multifaceted – I think they are, but it is an open research question – then we should end up with many bits of theory providing piecemeal explanations based on few concepts. The grand task of future transition theoreticians is then to unify these and explain their interrelations.

One particularly problematic aspect of 'nuance' and 'sophistication' is its progressive nature, especially when combined with frameworkism (see Section 3.5). When a theoretical framework does not quite fit, the tendency is not to question the validity of the assumptions of the frame, rather, the framework is 'improved' by adding to it. I think this dynamic is extremely prevalent in transitions-theoretical work, see for example the response to the supposed lack of a range of letters of the alphabet in the Multi-Level Perspective in Geels (2011, p. 29–31). I am not original in my observation of the scourge of 'nuance' and how it thwarts progress in social theorising, see Healy's (2017) brilliant exposé of the problem in sociological theory broadly, upon which much of the above few paragraphs is based.

3.4 Not-even just-so stories

Many of what are supposed to be explanations in transitions research really are not. Transitional phenomena are routinely 'explained' by means of an account of the circumstances, like a sequence of events, that lead up to them or of which they are part. While this may sound reasonable, these are often not explanations. Explanation (see Section 1.3) requires a deductive step from a general principle. If the right conditions obtain, the general principle (covering law, mechanism) holds and the phenomenon is logically entailed. Even if it concerns a proper historical explanation, where every next event is elliptically explained (i.e. by implicitly assuming a general principle) by the preceding ones, one does not necessarily obtain an explanation of the phenomenon at large.

This is perhaps best seen as a level issue. The event sequence resides on one level of explanation, while the phenomenon it describes lives on another. The set of explanations (elliptic or otherwise) of how or why one event followed another may constitute an explanation of the sequence but not necessarily of the phenomenon. The song is not explained by the string of notes. Ironically, this suffers from precisely the kind of reductionism detractors acuse scientific approaches of. Yet a narrative presenting a historical process as the logical unfolding of a sequence of events can be extremely compelling, leaving us feeling we truly understand what happened.

The reason we fall for these historicing pseudo-explanations is psychological and the effect has been known for ages. A recent warning against it was issued by Taleb (2007) in his book Black Swan. He points out that we humans have a limited ability to hear a story with a sequence of events and not conclude the whole process to be explained by this. He calls it the *narrative fallacy*. It can be a proper logical fallacy, a kind of *post hoc / cum hoc ergo propter hoc* or *non causa pro causa* fallacy or the abovementioned level confusion.

A possibly nefarious consequence of this fallacy is that analysts are tempted to include only those circumstances they think 'explain' the outcome, that is, those that help build up a consistent, plausible narrative. Not only may they thereby be thoroughly locking in their – possibly unconscious – biases, they may prevent others to inductively form rival hypotheses since they are presented with a subset of the circumstances. In this sense, the problem is not limited to narrative explanation, though it seems extra susceptible. One straightforward way to avoid it is to present relevant hypotheses explicitly, in a testable (predictive) form, so others can at least search themselves for evidence in favour and against.

Dewey identified a related fallacy as far back as 1896 in the context of psychological research[10]. He called it the psychological or historical fallacy. It arises when one thinks something to be a cause or crucial factor of some state of affairs that is really an outcome of that state of affairs. For example, compare (1 – fallacy) *these things happened and* this *is why*, with (2 – fact) *because these things happened,* this *is the case*. The constitution guarantees regular elections, that is why we live in a democracy. The historical fallacy is certainly not unique to narratives, but I believe they are especially prone.

In transitions theorising it may take the following form, consider a statement like: niche innovations have difficulty to break through to the mainstream, unless the regime is destabilised by internal or external stressors[11]. A narrative of the events 'explained' by this would, for example, recount the struggle and demise of one or several innovations in the presence of a stable regime. The point is that the stability of the regime follows from the absence of niche innovations breaking through, not the other way around.

The reason I refer to all this under the header of 'Not-Even Just-So Stories' is that just-so stories, when used appropriately (like in adaptationist evolutionary explanations) at least explicitly present the hypothesis that is purported to explain the phenomenon in question (e.g. some adaptive trait).

3.5 Frameworkism

It is rare for disciplinary scientists to not work within some theoretical frame or other and this is not necessarily problematic. Theoretical frameworks provide (1) a system for the *organisation* of data under investigation (the framework part) and (2) a *template* for explanations of phenomena to which those data pertain (the theoretical part). The merit of the first is not having to take into account everything in the world. The merit of the second is having building blocks of explanation. In physics, statistical mechanics is a framework for studying the macroscopic behaviour of gasses (amongst other things) with devices like partition functions as explanatory building blocks. In biology, evolutionary theory is framework for studying the dynamics of adaptive traits, with mechanisms of heredity and selection as explanatory building blocks. Sometimes multiple frameworks can be applied, not necessarily implying rival explanations, for example classical thermodynamics is another framework to study gasses.

There are potential downsides of any theoretical framework also. Frameworkism, where the framework part dominates the empirical research, may lead to the exclusion of data or whole cases that would provide interesting insight, for instance proving a hypothesis wrong. Ironically, such a selection bias especially shields off from empirical testing precisely the hypotheses of the theoretical framework itself. If the hypotheses underlying the theoretical frame are not questioned, for example because they are considered obvious, there is a real danger that the framework grows effectively immune to empirical testing. As a body of 'evidence' supporting it is amassed, there is less tendency to question it.

The dominant theoretical framework in transitions research is the Multi-Level Perspective (MLP), in particular in the form and way of application popularised by Geels since the early Noughties. Part of the allure of the MLP, according to Smith et al. (2010), is that '[i]ts terminology of niche, regime and landscape provides a language for organising a diverse array of considerations into narrative accounts of transitions'. More than just the language, the MLP provides a basic storyline, or in Geels' (2011) own words, it 'provides such a plot for the study of transitions'. The foundational assumptions of the MLP have been imported from other fields like evolutionary economics and remain basically unquestioned. For example, the assumption that the technologies and practices replacing or changing the regime in a transition always grow from niches is effectively a dogma[12], whereas it should be put forward as an explicit conjecture, as a hypothesis to be tested.

3.6 Narrative case studies

In fact, there are two problems with narrative case studies: (1) they are narrative, and (2) they are case studies. Dawkins (1976, p. 6) concisely captures a key issue by saying: 'I am not trying to make a point by telling stories. Chosen examples are never serious evidence for any worthwhile generalization'. But there are other issues also and, this all notwithstanding, there is surely a role for narrative case studies in transition science. I will discuss considerations about the narrative and the case study aspects sequentially.

Narratives

I already discussed some of the issues with narrative research, like an increased proneness to certain fallacies. The root of the issues with narrative research is simply its reliance on ordinary language. Avoiding the ambiguities and imprecision that come with ordinary language requires extraordinary efforts. Communication in ordinary language depends crucially on context and the other party's ability to infer what the speaker means, usually from her own experience as a language user. This also means that the information in verbal communication is compressed (in the computer science sense of the term) relying on the possibly, likely, imperfect decompression facilities of the other end of the line.

Anyone who has ever had to make something truly unambiguously clear, for example as part of a rigorous philosophical argument, or better still a mathematical proof, knows how incredibly hard this can be. For this reason, philosophy and mathematics take recourse to formalism and technical, precisely-defined terms. The sciences also do this and it helps, see also my discussion about 'fuzzy concepts that move' in Section 3.2. The more precise a science needs to be in its statements, and as a rule this necessity grows as a field matures, the greater the reliance on arcane terminology and formalism and this is a good thing. Precision is important so we can be proved wrong. Ordinary language often is an enemy of precision.

A much deeper problem of the use of ordinary language became clear in the early 20th Century when philosophers realised that the structure of reality is not necessarily reflected in the structures of language. This linguistic turn was probably initiated by Russell, who used predicate logic to analyse 'problematic' sentences, in his 'On Denoting' (1905) which still stands as a point of reference in this area of philosophy. The application of predicate logic in Russell's theory of denoting evaporated the perceived problems, inspiring Wittgenstein, Carnap and many others to investigate logic and the logical structure of the world. In other writings, Russell demonstrates how whole philosophical schools (German idealism, existentialism) are effectively based on artifacts of natural language. I am not sure whether this deeper problem has any practical consequences for transitions theorising.

I suppose we first have to deal with the 'shallow' problems of using ordinary language before we can even appreciate the deeper one. At any rate, we have every reason to be suspicious of narrative. Nevertheless, there seems to be a prevailing sense amongst many transitions researchers that narrative approaches are not only all right, but in fact superior. Clearly, someone has to be wrong here.

In favour of narrative, I hear such arguments as that they are better suited to deal with complexity. How this should be the case I can for the life of me not understand. If one wants to understand something complex, one should pick it apart to figure out how it all fits and works together. In other words, one should *analyse*. The more complex, the greater the need for precision and piecemeal analysis and narratives are eminently unsuited to meet these needs, as per all of the previous. If the whole could be understood as such, without breaking it down, the connotations of 'complexity' would not apply. Perhaps narratives are better at conveying *that* something is complex, but what of it? I do not think that is true either.

A prime example of a pro-narrative argument of this genre can be found in Andersson et al. (2014), eagerly cited by McDowall and Geels (2017). They assess the prospects of applying formal science methods to societal systems and conclude there is not much hope: 'formal science in general would be incapable of dealing with these systems' (p. 148). Narratives are the way to go because 'the strength of narrative is [...] that it is the only mode of theorizing that is not obviously married to assumptions of a lack of either complicatedness or complexity[13] (p. 154).

This is a combination of logical fallacy and naïveté. The fallacy is this: Just because formal theorising employs strident abstractions and simplifications, does *not* mean it is married to any assumptions. Abstractions and simplifications are based on assumptions, it is true, but if the obtained formalised hypotheses explain the phenomena we have learnt something – perhaps that the simplification was warranted, and not everything mattered after all. This is not marriage, this is dating to find the true one. The naïveté is of course that narrative is much more deeply marred by assumptions, that moreover are not tested or even made explicit like they are in formalisations.

Another common genre of arguments in favour of narrative uses some sort of appeal to metaphysics. More specifically, this goes through a supposed distinction between 'process' and 'variance' theories. Variance theories explain by looking at the changes in key variables that describe the phenomenon of interest, whereas process theories explain by looking at sequences of events. How events in such sequences are described, if not as changes in key variables is unclear. Also, descriptions of a phenomenon over time in terms of a sequence of changes (events) or sequence of states (key variables) are equivalent and can always be converted into each other in principle[14].

Apart from not thinking much of this distinction, I also cannot find any argument as to why one would want to use one or the other – it seems just another us-and-them distinction (we are not unscientific, we are just different). Nevertheless, the literature seems decided that process theories are the best match for transitions (e.g. Geels, 2011; McDowall and Geels, 2017). As said, sometimes the motivation is metaphysical, referring to 'processbased ontology, which sees things as temporary instances of processes-in-motion[15]' (McDowall and Geels, 2017, p. 5).

Process metaphysics goes back as far as Heraclitus obviously, but the modern exponents apparently base themselves on Whitehead, whose philosophy of process seems to draw on that of Bergson[16]. The Bergsonian view of change was summarised by Russell (1946, p. 763) as '[t]rue change can only be explained by true duration; it involves an interpenetration of past and present, not a mathematical succession of static states. This is what is called a "dynamic" instead of a "static" view of the world'. Or 'process' instead of 'variance' apparently. Bergson's philosophy of change and time is demonstrated by Russell to be based on an error in relation to the separation between past and present, but this does not mean process metaphysics is impossible in principle.

I suppose there is not much virtue in arguing against someone's metaphysics. It is per definition out of reach of empirical testing – otherwise it would be plain physics. And just so long as it is consistent with fact and not selfcontradictory there is no way to settle such an argument. What I *can* argue regarding process theories and metaphysics is that no-one seems able to truly analyse phenomena purely in terms of change, that is, without implicitly employing a state-based description.

Narrative, in other words, ordinary language, is supposed to be the mode for process-theoretical analyses *par excellence* but this is naïve. Narrative is just as ill-equipped for process theory as mathematics. Always, when one verbally describes an event, some change, one describes it in terms of agents of change and objects of change: 'Newton changed the way we think'. Newton is the agent, the way we think the object of change. Further analysis would describe how the way we think changed by describing how various aspects of it changed. Thus we are back at what essentially is a variance-theoretical description.

No matter how one tries, one describes change in terms of changing states. Ordinary language is simply unable to do it otherwise, see also my earlier remarks about the 'deeper' problem of ordinary language use. *All* languages have nouns and verbs and verbs take subjects (the agents of change) and sometimes objects (the objects of change). When there is no object in a declarative sentence ('I walk', 'it rains'), the change would still be analysed in terms of states ('I was there, now I am here', 'it was dry, now it is wet'). Perhaps a special language can be constructed in which change can be described in terms of change. My guess is that it is fundamentally impossible – perhaps someone in formal linguistics will be able to prove it as a theorem.

Case studies

That concludes my thoughts on narratives for now. What about case studies? Case studies and narratives come as a package deal in the transitions field, it seems. Indeed, the 'gold standard' in empirical transitions research remains the Geelsian, $N = 1$, narrative case study. That there are intrinsic problems with relying so much on this approach does not seem widely acknowledged. In fact, it seems to be the accepted way to go as the field's official research agenda points out: 'The overall direction for future transitions research involves a continued dedication to in-depth single-case research designs, also as new topics and transitions contexts are being explored' (Köhler et al., 2019, p. 19).

Whence this dedication? And is it appropriate? What are case studies good for and good at? Although I suspect that the dedication to narrative case studies does not stem from any supposed methodological merit – it may just be another manifestation of the fetish of the peculiar – I do want to highlight some of the merits of case studies.

I think case studies (narrative or otherwise) are appropriate, if not preferred, in the roles described below. There are certainly other legitimate roles for case studies, for example with an aim of sense-making, inspiration, advocacy or providing precedent. Though these are valuable, I do not consider them to be scientific ends and therefore I shall not discuss them further. In the following, case or case study may refer to historical, narrative or any other relevant variety.

Illustration – This refers to situations where a case is used to clarify the meaning and application of concepts or theory. In these situations, the case is not meant as empirical evidence but as a guide to application or as demonstration of the kind of results that can be obtained. For illustrative purposes, it does not matter in principle whether the case is factual, dramatised or hypothetical.

Paradox resolution – This may be a special case of illustration, where a puzzling phenomenon is shown to be a particular instance of a general principle. This is very common practice in science education. If not for illustrative purposes, paradox resolution is usually a form of inquest (see below), where a case presents a counter-intuitive phenomenon and the analyst uses *known* principles, i.e. a solid theory to explain the phenomenon.

Inquest – Here the aim is to learn about the specific case, to identify how circumstances or a certain course of events led to some outcome. Public inquiries, royal commissions and trials effectively conduct case studies in this vein with an emphasis on fact finding. Investigative journalism also operates in this manner. It is important to realise that the facts thus found are case specific and do not warrant generalisation on their own.

Speculation – Cases can suggest generalisations and in this fashion be a source for inductive hypothesis building. Especially if the case is typical for some class of phenomena or if there are numerous cases in which the same patterns are observed, this can make a strong case for a hypothesis. The caveat is of course the status of the hypotheses thus obtained. They are *conjectures*, and should be explicitly designated as such. One cannot test hypotheses with the same data that suggested them, fitting the existing data is but a *minimum* criterion for a hypothesis.

The only other role I see for the kind of narrative case studies so prevalent in transitions research is as part of a systematic effort towards the building of a large empirical dataset covering very many cases. As long as the concepts are still fuzzy and the mechanisms are unclear, cataloguing cases in great detail is not only appropriate but also desirable, but it comes at a cost. Systematically using large amounts of qualitative, especially narrative, data is hard. Just as it is hard to systematically catalogue such data in order to make them available programmatically, i.e. using software. But just because it is hard should not mean we should settle for an unscientific approach. I will discuss this further in the concluding section. See also my Chapter 12 with Martínez Arranz and Spekkink in this volume.

4 Theory, models and empiricism

In this section, I will argue that central to scientific practice is an interplay between theory, models and empirical data. Science, roughly speaking, aims at understanding by means of constructing *theories* of classes of phenomena. Science also, again roughly speaking, is *empirical*, meaning that observation and experiment provide the inspiration and touchstone for scientific knowledge. Experiment is the conscience of science, theory her soul. Models, I claim, perform a *mediating* role between theory, empirical data and human understanding. And they do so in several different ways.

I should clarify upfront that my notion of 'model' is much broader than the one the rest of this book employs, which refers mainly to simulation models. The sort of model that I am particularly interested in here is what could be called 'scientific models', by which I mean models that are instruments in the context of scientific investigation. This notion ranges from Bohr's model of the atom, through to Iterated Prisoner Dilemmas as models for the evolution of altruism and Lotka-Volterra style equations for population dynamics, all the way to computer models like in agent-based social simulation.

Given that I claimed models to mediate between three things, there are in fact three distinct mediating roles for models. I will treat them sequentially, but before I do so, I will need to digress briefly on the character of theory.

4.1 Theory, models and meta-models

I am assuming, perhaps foolishly, that the notions of 'empirical data' and 'understanding' are sufficiently clear and that the examples I gave of the kind of models I mean will suffice for now. To assume such things about 'theory' is dangerous, so an explication for the purposes of this chapter is appropriate.

My views on what theory is and does are an amalgam of several that are all in the mainstream of philosophy of science. For the interested, these are views ranging from classics, like Carnap (1956) and Hempel (1966), to more contemporary central figures, like Salmon (1990) and van Fraassen (1980) and several others. The issues on which these and other authors disagree, or where I disagree with them, are either not relevant here or about technical philosophical matters, e.g. whether one should be realist or anti-realist about theoretical concepts. It is in this sense I stated in the introduction that my views are philosophically uncontroversial.

Theories are both the repositories of knowledge acquired and the key to knowledge new. As a repository of knowledge, a theory contains a collection of what could be called 'templates of explanation' about a certain class of phenomena. These templates, depending on the theory, may be formal or fuzzy but at any rate they are generic. Applying theory, then, consists of taking a template and substituting the particulars of the case at hand for the generic terms in the template. This application is precisely what modelling is.

For example, Maxwell's theory of electrodynamics explains the class of phenomena related to electricity, magnetism and optics. The theory provides an explanatory template in the form of four partial-differential equations (the Maxwell Equations) and one additional equation (the Lorentz Force Law). For any particular case (hypothetical or actual), one provides the charge distributions, which determine the fields (or the other way around) and then explains the phenomenon by mathematically deriving it as a prediction. Describing the case context (charge distributions) with the explanatory template (Maxwell's equations) of the theory constitutes a model.

In the above case, the model is fully particular but this need not be the case at all. Clearly, explanatory templates can be set-up for commonly recurring empirical situations by equipping them with only those particulars these situations have in common. Or similarly, for situations that are variations on a theme, the particulars can be parameterised. The models thus obtained are of course closer to the common understanding (if there is such a thing) of the term 'model', though the term 'meta-model' would perhaps be more apt. Such meta-models apply to a class of phenomena that is a subclass of the phenomena the theory is about. This does not exclude the possibility that the (meta-) model applies to various phenomena outside of the scope of the theory also. This is so exceedingly common that it is in itself a motivation for model-based research.

Taking the above view to its logical conclusion, one obtains that theories themselves are meta-models, or patchworks of meta-models. Here also, one sees the aspirational nature of formal theory. While our understanding is dim our theory may be formulated in fuzzy terms, allowing various operational definitions and corresponding models. As our understanding gains accuracy, the precision of our concepts can gain concurrently.

4.2 Between theory and understanding

This is perhaps the most abstract of the mediating roles. I suspect many theoreticians in the sciences employ it intuitively without questioning it or even be aware of it being at work. Working with theory always entails working with more or less specific models based on that theory. This holds whether the theory is established or in development.

Using an established theory to inform one's general understanding of phenomena is based on creating and manipulating *ad hoc* (meta-) models of various situations the theory is supposed to cover. In a very literal sense, this is how students in the mathematical sciences learn theories, working through scores of problems with pencil and paper. In a less explicit, but still literal, sense this is how everyone learns theory, also when it is not-so-formal. Having ingested the raw material of a theory, one mentally tries it on familiar cases and tests it on apparent exceptions. Developing theory makes use of the very same creating and manipulating of *ad hoc* (meta-) models.

The more explicit (tentative) models are formulated, the more they lend themselves for active exploration. Exploration can mean thought experiments, where one tries to take one's assumptions to their logical extremes, either to find the limits of applicability or simply see what would happen. Using simulation, a model can be 'run' to produce patterns over time, showing how one's nascent theory behaves in ways that the naked mind is unable to. Another virtue of explicit models, especially when formalised is that they enable checking consistency, both internally and with existing, established theory.

It should be noted that models can, often do and perhaps even should, involve deliberate simplifications and distortions of what one knows to be the case in 'reality'. Common examples are 'course graining', where one ignores dynamics on smaller scales (e.g. by looking at whole-of-industry dynamics) or idealisations, where certain phenomena are ignored (e.g. frictionless pendula). Even when such ignoring seems utterly unjustified (friction often does matter) such models are still useful. In fact, they can be better than models taking many such factors into account because they look at the *pure* phenomenon, hence they are called idealisations. Holland (2006) emphasises the use of toy models, which are deliberately oversimplified models, to learn about complex, adaptive systems.

4.3 Between empirical data and understanding

Making sense of empirical data entails viewing them as particular instances of generic categories. This holds both in quantitative and qualitative contexts. An example of the former is counting adoptions of a certain technology. An example of the latter is identifying something as a type of event like a government intervention, or a type of actor like a frontrunner. Understanding comes from finding patterns among the data and then capture those patterns in terms of hypotheses. Again, both when capturing data in generic categories and in the capturing the patterns in hypotheses, one is making models.

Close-to-data models like these may be truly *ad hoc*, serving only to see whether there is any pattern at all, or to figure out whether a certain theory or other model can fruitfully be applied. Such models may be evanescent mental constructs, never even written up, while 'eye-balling' an interview transcript, say, or they may be based on formal ontologies[17] and employ statistical techniques.

It is not always necessary to provide the categories or concepts *a priori*. Standard methods in statistics like factor analysis or principal component analysis enable the identification of relevant categories on the basis of the data themselves, this is commonly used in psychological research.

4.4 Between theory and empirical data

This is perhaps the most straightforward role of models and in a sense the short-circuiting of the previous two roles. This role works in two directions, namely: (1) models enable empirical data to be explained by framing them in the terms of the theory, and (2) they enable the derivation of predictions from the theory which can be tested against empirical data.

5 Conclusion – concrete steps towards transitions science

I would like to conclude this chapter by suggesting several, relatively concrete, things that could done to move towards a science of transitions. Some of these are being done already but none seem to be anywhere near the norm in the field. For the sake of overview, I will group my suggestions under a theory and data related header, despite their interrelatedness.

5.1 Theory development as modelling and vice versa

Taking my view of theory as meta-model at face value (see Section 4.1), I think theory development can progress more, and more systematically if it is approached as modelling. I do not mean that everyone should write computer models, nor that such modelling is all we should do. I mean an approach to theorising where hypotheses are explicitly framed, preferably in a case a-specific format, in such a way that consequences can be derived from them. At the very least, that will enable straightforward hypothesis testing against data. More ambitiously speaking, it makes them amenable to formal modelling.

Such an approach has been advocated for the social sciences by people like Hedström and Swedberg (1998, see also Hedström 2005). They refer to hypotheses framed in this fashion as 'social mechanisms' and see a significant role for (computer) modelling. In Chapter 5 of this volume, Bianchi & Squazzoni also discuss this view. Ideally, any candidate mechanism is deployed in several model forms, for example based on different operational definitions of the concepts involved and using various modelling approaches, ranging from verbal descriptions with rigorous definitions, through to mathematical formulae or simulation models. All of these can then be explored as toy models, as part of other models and, of course, tested against empirical material. In a slogan: *one mechanism, multiple models, many data.*

Though not a strict necessity to do so, many hypotheses actually can be framed in one or more semi-quantitative forms. I say 'semi' because I am not referring to metrics or indicators. What I mean is that a phrase like 'frontrunners play an important role in early phases of transitions' can be phrased as 'more frontrunners advocating practice X, lead to an increased early uptake of that practice in comparison to cases where they are absent'. In a format like the latter version, correlations can explicitly be investigated. Moreover, most hypotheses will have several of these semi-quantitative versions.

I have already discussed some of the virtues of formalisation in Section 4. The trajectory of the natural sciences has invariably been towards more formalisation as they progressed. Like with the argument about precision, I cannot infer that formalisation is behind that progress. My guess is that it did help and one of the factors would have been that it adds deductive argument and mathematical proof to the toolbox of the theoretician. Formalisation should not be confused with quantification, in other words concepts do not have to correspond to quantities like in physics. Linguistics, as a science, essentially begins with Chomsky's formalisation effort, which was a formalisation of the structure of language.

5.2 Dealing with data differently

Though narrative case studies have a role in scientific transitions research, different ways of engaging with empirical material will be necessary. Moreover, altogether different sources of data need to be harnessed. This, in combination with the model-based theorising I discussed above, will – so I argue – greatly strengthen the approach.

There is really no strong reason to single out the case as the preferred unit of analysis. Ideally, the kind of empirical data to use should be informed by the hypotheses we are trying to test. In this regard, it can be instrumental to think in terms of data *sets* rather than case data. The epilogue of this volume discusses in some detail the kinds of data sets that would be relevant for scientific transitions research. To give an idea, one thinks of financial flows and how they change, text mining of news articles or twitter feeds, energy use of industries, morphological patterns in built environments, organisational networks, statistics of organisational life cycles (founding, merging, death), social networks of key transition actors and so on and so forth. There are many kinds of data from which to learn about transitions that we are not even considering.

One way in which case-oriented research may have continued relevance is through meta analysis. Fields like medicine and psychology, similar in complexity of subject, routinely use meta analysis to check findings across many cases. This helps resolve conflicting hypotheses based on different cases and find anomalies. Using the considerable volume of transitions case studies for meta analysis would allow testing key assumptions and new hypotheses. Ideally, existing case studies would be cast into a form in which they can be queried and then made available as a database for the entire research community. With Martínez Arranz and Spekkink (Chapter 12 of this volume), I discuss meta-analysis, case databases and other data-driven methods for transitions research in more detail.

Letting go of the case-orientation, dealing with data differently and more effectively is not going to be easy for two reasons. The first of which is that it simply is not easy. Transitions as a research topic are not special in this sense, constructing good data sets is difficult in any science and careers are built on devising better ways to obtain data. I am always affronted by remarks to the effect that the natural sciences have it easy because their data are quantitative. Or because their topic is 'not complex'. It is in fact incredibly difficult to design experiments that isolate a phenomenon from the complexities it is entangled in. Saying that this is impossible for transitions is just giving up in advance.

The second reason that it is going to be difficult to deal differently with data is because few transitions researchers are currently conversant with the methods and techniques this will require. And a prevailing, but misplaced, attitude of superiority regarding the dominant research practice is not going to do the field any favours in this respect. Not everyone needs to become a modeller, but basic mathematical and programming skills may become indispensable. At any rate, reading secondary sources and synthesising them into a multi-level narrative will no longer suffice to make progress. But there is a reward, the systematic availability of new, preferably large, data sets, will allow the use of tools now mostly out of reach. Notably, a rich suite of statistical and visualisation methods will become available. Not to mention the improvements in the robustness of the knowledge produced.

In conclusion, we should acknowledge that despite much empirical work, we do not know much about transitions scientifically yet. I think there is no shame in that and I am optimistic about the prospect of a transitions science. In the meanwhile, when we find ourselves in a position to influence decisions based on our knowledge of transitions, we should ask ourselves what gives that knowledge the authority and legitimacy to be such a basis. If it is not science, then what is it?

6 Acknowledgements

Without making them complicit in anything this chapter says, I thank the following people: Alfonso Martínez Arranz for our countless and ongoing discussions about empiricism and transitions, as well as various other topics. Wouter Spekkink for discussions about process versus state-based metaphysics and for providing such a rich, in depth review of a draft of this chapter. The text has been improved in many places because of it and I hope I have done his comments justice. Thanks to Angela Rojas for reviewing a draft of this chapter which reduced the instances where it may offend anyone by at least one. And, to Daizee Boucher for sanity checks of some of the arguments as they were taking shape.

Note

1 Wikiquote says it is not in any of the Feynman books. I have a relatively clear but quite probably false memory of reading the quote as an undergrad.
2 See also my Chapter 12 with Martínez Arranz and Spekkink (this volume) on this comparison in relation to meta analysis.
3 See Section 3.2 about the important distinction between accuracy and precision.
4 People who are irked by my repeatedly citing Feynman, will see this as yet another sign of my narrow, physics-based views. Perhaps they are right. Either way it does not invalidate those views. I cite Feynman because of his talent to state in simple but not simplistic terms the common sense of science. To my delight, Dennett, on this topic of publishing 'negative' results, cites the very same Cargo-Cult lecture by Feynman.
5 Cargo cults are a Melanesian phenomenon that has attracted considerable anthropological attention. Melanesian people observed aeroplanes landing with cargo bringing material wealth. When the aeroplanes no longer came, they began imitating the practices associated with the arrival of cargo, for example by building ritual landing strips. Most cargo cults seem to have disappeared since the 1950s though a few persist. Wikipedia has several lemmas on cargo cults, see https://en.wikipedia.org/wiki/Cargo_cult, and see Otto (2009) for an overview article of the anthropology.
6 Using back-of-the-envelope calculations of scale-ranges for mass, surface area or the numbers of individuals in societies, I estimate the systems of society to cover less than 10 orders of magnitude.
7 There is another, very specific, meaning of 'ontology' in computer science that is related to, but different from the usual philosophical meaning.
8 See Dennett (2013) for an interesting discussion about such commonly used but metaphysically difficult notions.
9 See Russell (1923) on vagueness in this sense, including an example much like the picture one here.
10 See also the American Psychological Association's online dictionary at https://dictionary.apa.org/historical-fallacy.
11 An example of this reasoning (not so much the fallacy) can be found in Geels (2002, p. 1272): 'Radical innovations, which are pioneered in niches, have a hard time to break out of the niche-level. If the regime is confronted with problems and tensions emerge,

the linkages in the configuration "loosen up". The configuration becomes "warm" (Callon, 1998). This creates opportunities for radical innovations to escape the niche-level and be incorporated in the ST-configuration'.

12 As can be judged from phrases like 'niches as the locus where radical variety is generated, and regimes as selection and retention mechanism' (Geels, 2002). The retort in Geels (2011) is that there may have been a bias towards bottom-up dynamics and that more attention needs to go to regime and landscape processes. I do not care if it is a bias, it may well be true. The point is it needs to be tested. As regards the bias, in Geels et al. (2016) all the pathways are still described in terms of niche innovations replacing or altering the regime.

13 They do add that this 'does not mean that it is particularly good at dealing with their presence'.

14 See my detailed argument for this in my Chapter 12 with Martínez Arranz and Spekkink in this volume.

15 In reference to a distinction between 'strong' and 'weak' process-based ontologies made by Langley (2009)

16 Russell wrote: 'His philosophy in later years was essentially that of Bergson' (Feinberg and Kasrils, 1969, p. 160).

17 These are formalisations of knowledge domains or 'universes of discourse', see for example Uschold and Gruninger (1996) and Gruber (1993). Though related, this usage of the term 'ontology' is distinct from its use in philosophy. See also Footnote 7.

References

Andersson, Claes, Anton Törnberg, and Petter Törnberg (2014). Societal systems–complex or worse? *Futures 63*, 145–157.

Barkow, Jerome H, Leda Cosmides, and John Tooby (Eds.) (1992). *The Adapted Mind: Evolutionary Psychology and the Generation of Culture*. Oxford, England, UK: Oxford University Press.

Callon, Michel (Ed.) (1998). *The Laws of the Markets*. Oxford, England, UK and Malden, MA, USA: Blackwell Publishers/The Sociological Review.

Carnap, Rudolf (1956). The methodological character of theoretical concepts.

Creswell, John W and J David Creswell (2018). *Research Design: Qualitative, Quantitative, and Mixed Methods Approaches* (5th ed.). Thousand Oaks, CA, USA: SAGE Publications, Inc.

Dawkins, Richard (1976). *The Selfish Gene* (30th Anniversary Edition 2006 ed.). Oxford, UK: Oxford University Press.

Dennett, Daniel C. (1995). *Darwin's Dangerous Idea: Evolution and the Meanings of Life*. New York, NY, USA: Touchstone / Simon & Schuster.

Dennett, Daniel C (2013). *Intuition Pumps: And Other Tools for Thinking*. London, England, UK: Penguin Books Ltd.

Dewey, John (1896). The Reflex Arc Concept in Psychology. *Psychological review 3*(4), 357–370.

Feinberg, Barry and Ronald Kasrils (1969). *Dear Bertrand Russell... A Selection of His Correspondence with the General Public 1950-1968*. London, England, UK: George Allen & Unwin Ltd.

Feynman, Carl and Michelle Feynman (1999). *The Pleasure of Finding Things Out*. New York, NY, USA: Perseus Books.

Feynman, Richard P (1965). *The Character of Physical Law*. London, England, UK: BBC.

Feynman, Richard P (1974, June). Cargo Cult Science. *Engineering and Science 37*(7), 10–13.

van Fraassen, Bas (1980). *The Scientific Image*. Oxford, England, UK: Oxford University Press.

Geels, Frank W. (2002). Technological transitions as evolutionary reconfiguration processes: A multi-level perspective and a case-study. *Research Policy 31*(8–9), 1257–1274.

Geels, Frank W. (2011, June). The multi-level perspective on sustainability transitions: Responses to seven criticisms. *Environmental Innovation and Societal Transitions 1*(1), 24–40.

Geels, Frank W., Florian Kern, Gerhard Fuchs, Nele Hinderer, Gregor Kungl, Josephine Mylan, Mario Neukirch, and Sandra Wassermann (2016). The enactment of socio-technical transition pathways: A reformulated typology and a comparative multi-level analysis of the German and UK low-carbon electricity transitions (1990–2014). *Research Policy 45*(4), 896–913.

Gruber, Thomas R. (1993, June). A translation approach to portable ontology specifications. *Knowledge Acquisition 5*(2), 199–220.

Haslam, Michael, Adriana Hernandez-Aguilar, Victoria Ling, Susana Carvalho, Ignacio de la Torre, April DeStefano, Andrew Du, Bruce Hardy, Jack Harris, Linda Marchant, Tetsuro Matsuzawa, William McGrew, Julio Mercader, Rafael Mora, Michael Petraglia, Hélène Roche, Elisabetta Visalberghi, and Rebecca Warren (2009, July). Primate archaeology. *Nature 460*, 339.

Haslam, Michael, Lydia Luncz, Alejandra Pascual-Garrido, Tiago Falótico, Suchinda Malaivijitnond, and Michael Gumert (2016, July). Archaeological excavation of wild macaque stone tools. *Journal of Human Evolution 96*, 134–138.

Healy, Kieran (2017, June). Fuck Nuance. *Sociological Theory 35*(2), 118–127.

Hedström, Peter (2005). *Dissecting the Social*. Cambridge, New York, Melbourne: Cambridge University Press.

Hedström, Peter and Richard Swedberg (Eds.) (1998). *Social Mechanisms: An Analytical Approach to Social Theory*. Cambridge University Press.

Hempel, Carl Gustav (1966). *Philosophy of Natural Science*. Englewood Cliffs, NJ, USA: Prentice Hall.

Hempel, Carl Gustav and Paul Oppenheim (1948, April). Studies in the Logic of Explanation. *Philosophy of Science 15*(2), 135–175.

Holland, John H. (2006, March). Studying Complex Adaptive Systems. *Journal of Systems Science and Complexity 19*(1), 1–8.

Holtz, Georg, Floortje Alkemade, Fjalar de Haan, Jonathan Köhler, Evelina Trutnevyte, Tobias Luthe, Johannes Halbe, George Papachristos, Emile Chappin, Jan Kwakkel, and Sampsa Ruutu (2015, December). Prospects of modelling societal transitions: Position paper of an emerging community. *Environmental Innovation and Societal Transitions 17*, 41–58.

King, Andrew J., Dominic D.P. Johnson, and Mark Van Vugt (2009, October). The Origins and Evolution of Leadership. *Current Biology 19*(19), R911–R916.

Köhler, Jonathan, Frank W. Geels, Florian Kern, Jochen Markard, Elsie Onsongo, Anna Wieczorek, Floortje Alkemade, Flor Avelino, Anna Bergek, Frank Boons, Lea Fünfschilling, David Hess, Georg Holtz, Sampsa Hyysalo, Kirsten Jenkins, Paula Kivimaa, Mari Martiskainen, Andrew McMeekin, Marie Susan Mühlemeier, Bjorn Nykvist, Bonno Pel, Rob Raven, Harald Rohracher, Björn Sandén, Johan Schot, Benjamin Sovacool, Bruno Turnheim, Dan Welch, and Peter Wells (2019). An agenda for sustainability transitions research: State of the art and future directions. *Environmental Innovation and Societal Transitions 31*, 1–32.

Langley, Ann (2009). Studying processes in and around organizations. *The Sage handbook of organizational research methods 409*, 429.

Leighton, Ralph and Richard P Feynman (1985). *"Surely You're Joking, Mr. Feynman!": Adventures of a Curious Character*. New York, NY, USA: W.W. Norton.

Martínez Arranz, Alfonso (2017, May). Lessons from the past for sustainability transitions? A meta-analysis of socio-technical studies. *Global Environmental Change 44*, 125–143.

McDowall, Will and Frank W. Geels (2017, March). Ten challenges for computer models in transitions research: Commentary on Holtz et al. *Environmental Innovation and Societal Transitions 22*, 41–49.

Minsky, Marvin (1998, February). Consciousness is a Big Suitcase: A Talk With Marvin Minsky. *Edge*.

Otto, Ton (2009). What Happened to Cargo Cults? Material Religions in Melanesia and the West. *Social Analysis: The International Journal of Social and Cultural Practice 53*(1), 82–102.

Popper, Karl R. (1957). *The Poverty of Historicism*. London: Routledge and Kegan Paul.

Proffitt, Tomos, Lydia V. Luncz, Tiago Falótico, Eduardo B. Ottoni, Ignacio de la Torre, and Michael Haslam (2016, October). Wild monkeys flake stone tools. *Nature 539*, 85.

Roberts, Clayton (1996). *The Logic of Historical Explanation*. University Park, Pennsylvania: The Pennsylvania University Press.

Rosenberg, Alex (1993, November). Scientific innovation and the limits of social scientific prediction. *Synthese 97*(2), 161–181.

Rosenberg, Alex (2017, September). Why Social Science is Biological Science. *Journal for General Philosophy of Science 48*(3), 341–369.

Rotmans, J. (2005). *Societal Innovation: Between Dream and Reality Lies Complexity*. Inaugural Addresses Research in Management Series. Rotterdam: ERIM, Erasmus Research Institute of Management. Inaugural address.

Rotmans, J., R. Kemp, and M. van Asselt (2001). More evolution than revolution: Transition management in public policy. *Foresight 3*(1), 15–31.

Russell, Bertrand (1905). On Denoting. *Mind 14*(56), 479–493.

Russell, Bertrand (1923, June). Vagueness. *Australasian Journal of Psychology and Philosophy 1*(2), 84–92.

Russell, Bertrand (1946). *History of Western Philosophy: And Its Connection with Political and Social Circumstances from the Earliest of Times to the Present Day* (Second ed.). London, England, UK: George Allen & Unwin.

Salmon, Wesley C (1990). *Four Decades of Scientific Explanation*. Pittsburgh, PA, USA: University of Pittsburgh Press.

Schuitmaker, Tjerk Jan (2012). Identifying and unravelling persistent problems. *Technological Forecasting and Social Change 79*(6), 1021–1031.

Shannon, C. E. (1948). A Mathematical Theory of Communication. *The Bell System Technical Journal 27*, 379–423, 623–656.

Smith, Adrian, Jan-Peter Voß, and John Grin (2010). Innovation studies and sustainability transitions: The allure of the multi-level perspective and its challenges. *Research Policy 39*(4), 435–448.

Symons, David (1979). *The Evolution of Human Sexuality*. Oxford, England, UK: Oxford University Press.

Taleb, Nassim Nicholas (2007). *The Black Swan: The Impact of the Highly Improbable*. New York, NY, USA: Random house.

Tooby, John and Leda Cosmides (1992). The Psychological Foundations of Culture. In *The Adapted Mind: Evolutionary Psychology and the Generation of Culture*, pp. 19–136. Oxford, England, UK: Oxford University Press.

Uschold, Mike and Michael Gruninger (1996, June). Ontologies: Principles, methods and applications. *The Knowledge Engineering Review 11*(2), 93–136.

Van Vugt, Mark (2006, November). Evolutionary Origins of Leadership and Followership. *Personality and Social Psychology Review 10*(4), 354–371.

West, Geoffrey B., William H. Woodruff, and James H. Brown (2002, February). Allometric scaling of metabolic rate from molecules and mitochondria to cells and mammals. *Proceedings of the National Academy of Sciences 99*(suppl 1), 2473.

Wilson, Edward O (1975). *Sociobiology: The New Synthesis*. Cambridge, MA, USA: Harvard University Press.

Wong, Kate (2017, May). The New Origins of Technology. (Cover story). *Scientific American 316*(5), 28–35.

5 Modelling and social science

Problems and promises

Federico Bianchi and Flaminio Squazzoni

1 Modelling in the social sciences

A model is a simplified representation of an empirical target or phenomenon; that is, smaller scale, fewer details, either circumscribed or isolated from the context, or all of these together (Gilbert and Troitzsch 2005; Squazzoni 2012). It can be expressed in various forms and languages, graphically, by mathematical equations or computational codes. After being translated from our mind (often via paper and pencil) to a formalised implementation, the model becomes an artefact that supports the target's indirect examination by direct manipulation. This is key when direct access to the target is impossible, due to the target scale and the lack of appropriate observational instruments or difficult, due to economic or ethical constraints. The beauty of any formalised model is to make any complex target something that is at least analytically tractable and 'observable' by substitution and analogy (Hartman and Frigg 2006). In this way, a model works as a mediator between theory-building and empirical research (see Chapter 4 by de Haan in this volume). Furthermore, unlike descriptions, unformalised theories and subjective narratives, a model artefact imposes us clarity and fine-grained distinctions, while being at least potentially unequivocally understandable, constructively commented, incrementally developed and collectively tested and reused by other experts (Squazzoni 2012).

While social scientists have traditionally given priority either to abstract theorisation or inductive empiricism, modelling is increasingly seen as a means to extend our imagination and recognise essential elements, thereby making theory and observation dialog more productively. Indeed, yet still controversial, the use of formalised models has recently gained momentum also in the most sceptical disciplines, such as sociology (Gilbert and Abbott 2005; Hedström and Bearman 2009; Hedström and Manzo 2015). Thanks to significant advancement in mathematical and computational research methods and instruments, since the 1990s computational models have been used to examine a variety of relevant topics, such as norms of cooperation, cultural diffusion, opinion dynamics, residential segregation, social networks and socio-ecological systems, with cumulative findings and incremental developments (Macy and Willer 2002; Bianchi and Squazzoni 2015; Bruch and Atwell 2015; see also Flache et al. 2017).

However, one of the main obstacles to the establishment of a modelling tradition in the social sciences has been the dominant inductive, observation-based tradition in quantitative empirical research (Goldthorpe 2000). Note that this attitude has been a widespread response to the crisis of the dominant paradigm of *structural functionalism* and its tendency to develop 'grand theories' (Parsons 1937). This contributed to determine a divide between specialists using 'clean models' for theory development and 'dirty hands' experts working on data-driven empirical research (Hirsch, Michaels, and Friedman 1987), with the former concentrating on the search for generative mechanisms of aggregates, and the latter zeroing in on macro causal relationships between aggregates.

This explains why modelling in the social sciences is generally byword for 'statistical' rather than 'theoretical' endeavours (Skvoretz 1991; Hedström and Manzo 2015). Consider the same meaning that computation has for many social scientists. While social statisticians and quantitative sociologists look at empirical observations by means of systems of equations whose parameters are computationally estimated, theoretical-driven analysts use simulation models to reconstruct entities and dynamic processes that constitute social patterns to be explained (Sørensen 1998). Regardless of the fact that these specialists could use equations and even similar modelling techniques, they have different assumptions regarding the relationships between individuals of a population. The former typically assume independence of observations in order to ensure effective estimations of cause–effect relationships between aggregates, whereas theory-inspired models consider that target populations are composed by interdependent individuals, with certain parameters either justified or estimated according to theoretical assumptions (Macy and Flache 2009).

This is why many consider formalised models as the most effective means to integrate theory and empiricism (Frank 2002; Squazzoni 2012). First, by simplifying complex social phenomena under computational or mathematical constraints (Gilbert and Troitzsch 2005), formalised models help social scientists to establish a link between the intrinsic abstracted nature of theories and fine-grained empirical testing methods (Edling 2002). This is relevant when the complexity of social phenomena, mostly due to the heterogeneous, composite nature of populations and the adaptive tendency of social behaviour, must be considered. Rather than abstracting away these complexity effects or embracing rich, fine-grained empirical descriptions, formalised models help empirical research in deriving rigorous hypotheses from existing theory, exploring possible generalisations of observations or even challenging prevailing theories by testing their predictions (Epstein 2008).

Note that formal modelling embodies the same basic analytic operations that are at the very core of any possible scientific endeavour: dissection and abstraction (Hedström 2005). On the one hand, social complexity is dissected into its basic components so that building blocks of social patterns can be identified and their mechanisms and processes disentangled. Obviously, these components have to be sufficiently abstracted to rule over the context-dependent idiosyncrasy of any particular empirical instance. In this way, modelling allows researchers to avoid the risk of being trapped in idiosyncratic case-based analyses which can hardly relate to higher-level generalisations (see Chapter 4 by de Haan in this volume). On the other hand, these components must be recombined and aggregated step-by-step such that their organisation helps to understand the social pattern of interest. Here, models are 'cognitive' maps helping social scientists to select relevant factors for the understanding and explanation of a phenomenon, while making their selection scrutinised and improvable collectively (Miller and Page 2009).

2 Agent-based models

Agent-based models (ABMs) are a particular class of formal models that is gaining momentum in the social sciences to examine the emergence of complex social patterns from agent interaction. An ABM can be defined as a 'computational method that enables a researcher to create, analyse, and experiment with models composed of agents that interact within an environment' (Gilbert 2008, 2). It represents a social system through a set of *agents* – that is, individual or collective actors (e.g. individuals, organisations, firms, etc.) – interacting by means of certain *micro*-level behavioural rules in a structure of *macro*-level institutional (e.g. a market), relational (e.g. a social network), or spatial (e.g. an urban space) constraints. While modelling in the social science has traditionally considered interaction between variables and aggregates, ABMs allow us to directly address agent interaction to examine its consequences at a macroscopic level.

This is instrumental to explain macro-level processes as generated by micro-level interactions (Elster 1989). As stressed in a seminal contribution by Coleman (1962, 69), by building computational models, we aim to 'put together certain processes at the individual and interpersonal level and then to see what consequences they have at the level of the larger system'. In each iteration according to certain specified initial conditions, a population of interacting agents following certain assumed rules of behaviour generate emergent consequences at the system level, whose spatial-temporal dynamics are observable into a computer. The computer is used as an artificial laboratory in which generative experiments on aggregates are performed by manipulating initial conditions and rules (Squazzoni 2012).

Note that a key characteristic that distinguishes ABMs from other formal modelling techniques used in social sciences, such as statistics, network analysis or analytic mathematical models, is the generative nature of agents and their properties. While traditional formal models usually assume 'representative' agents incorporating different behavioural instantiations in the stochastic variation of synthetic parameters of a population, ABMs explicitly considers a population of heterogeneous, autonomous agents with different typological features (Gilbert 2008; Macy and Flache 2009).

The power and flexibility of ABMs in analysing complex social systems are grounded in the following characteristics (Macy and Willer 2002):

1 Autonomy: aggregate (simulation) outcomes are not fully contained in assumptions about either separate entities or the aggregate whole, because they emerge from autonomous agent decisions, which are not dictated by top-down, centralised decision-making or systemic functions.
2 Interaction and interdependence: agents influence each other in complex ways by directly interacting with each other or by modifying environmental conditions, which in turn can affect agent behaviour.
3 Simplicity: macro-level complexity does not mirror agents' complex cognitive structures, while it can emerge as aggregate effect of simple individual actions.
4 Heterogeneity: a complex diversity of individual properties can be modelled within the same population of agents.

The ABM method is linked to the concept of 'generative explanations' of social phenomena, which could be summarised by the motto 'if you didn't grow it, you didn't explain it' (Epstein 2006). Let us assume a macro-level phenomenon *F*, which is our *explananda*. We will hypothesise certain processes involving a population of agents, their properties, their decision-making rules, the environment providing constraints and opportunities to their decisions, and all interactions between these elements. We will formalise these processes in a (computational) model *M*. We will then run a series of computer simulations of *M* explicitly designed to manipulate certain parameters, either incrementally (e.g. increasing the value of a relevant parameter), additionally (e.g. adding a new conditions) or *par différence* (e.g. comparing identical scenarios that differ only for a single condition). This will permit us to test different configurations of our hypothesised processes systematically. If the simulated dynamics determined by certain configurations of *M* generate an outcome similar to *F*, then we will consider their correlated assumptions as 'sufficient generative conditions' of *F*.

Note that this generative approach is particularly suitable to study complex social systems, where analytically dissecting a system into its components is typically difficult, as the interaction among these components yields non-linear effects at the macro level (Squazzoni 2012). Here, social interaction of heterogeneous, adaptive agents embedded in complex environments typically makes any system's behaviour mathematically non-tractable (Miller and Page 2007). Instead of deriving top-down predictions of the system behaviour by specifying abstract formal models that take for granted the generative conditions and over-impose structure and functions onto actions, ABM allows to explore various micro-level specifications and observe their macro-level consequences from the bottom up. This permits to explore the causes of complex system's behaviour without supposing pre-existing exhaustive knowledge on system's structures, functions and aggregates.

These peculiarities have made ABMs popular in all 'analytical' research programmes in the social sciences, which aim to identify micro-level 'causal mechanisms' of social patterns (e.g. Hedström 2005; Hedström and Bearman

2009; Hedström and Ylikoski 2010). Although there is no consensus yet, a causal mechanism can be defined as a set of conditions and entities which are linked to one another in such a way that they regularly bring about a particular type of social outcome (Machamer, Darden, and Craver 2000). When transposed to the case of social sciences, the term 'social mechanisms' indicates that explanation should refer to agent behaviour and interaction in given structures (Hedström 2005), which are all components of an ABM.

In this respect, the multi-level structure of ABMs is key. As described above, assumptions are made at the level of agents, their properties and the structure of constraints and opportunities within which they act. Then, simulated outcomes are analysed at the emergent macro level as the diachronic execution of agent interaction. This means that ABMs address the so-called 'micro–macro link' (Squazzoni 2012), that is, the way in which the consequences of agent interaction generate new emergent macro-level system's states. However, because of the diachronic and synthetic nature of simulations, ABMs help to touch upon the micro–macro link without conflating ontological and epistemological standpoints. While many follow Coleman (1990) in his idea of studying 'system-level' social patterns as emergent consequences of individual actions, others use ABMs to understand complex macro feedback on agent behaviour without compromising the role of agency (Conte et al. 2001; Squazzoni 2008).

To sum up, ABM has four major strengths compared to standard mathematical or statistical models (Squazzoni 2012; see also Macy and Flache 2009):

1 They mirror human agency by mimicking cognitive properties of decision-making.
2 They easily model heterogeneous populations, with complex agent types endowed with autonomous characteristics (e.g. age, gender, resources, information and behaviour).
3 They consider non-linear agent interaction and its consequences at the macro-level as bottom-up emergent properties of local (non-system-mediated) interactions.
4 They explicitly represent environmental constraints (e.g. space, neighbourhoods or networks).

3 A taxonomy

Depending on the modelling purpose, ABMs can be used in different ways and at different levels of abstraction. Squazzoni (2012) has identified five different types: artificial societies, abstract models, middle-range models, case-based models and applied simulations. While artificial societies and applied simulations have more to do with a synthetic, explorative and policy-oriented approach, the prevalence in the social sciences is of analytical model types, such as abstract and middle-range models (Macy and Willer 2002; Bianchi and Squazzoni 2015).

Artificial societies are models of surrogates of social systems which are reconstructed by the modeller in order to investigate social dynamics which are impossible or considerably difficult to observe empirically. These models are not usually built to test a hypothetical explanation of a well-defined empirically observable social phenomenon. They are rather built to explore the potential evolution of a social system. Hypothesizing trajectories, demonstrating computational possibilities or re-running historical processes even in idealised scenarios with no close correlate with social reality could help to reveal potential, analogical mechanisms that could be helpful to understand social processes. While this type of modelling has played a crucial role during the first stages of development of computational social sciences, it is currently under-developed (e.g. Gilbert et al. 2006). This is also due to the difficult translation of complex simulation outcomes into empirically testable findings (Squazzoni 2012). Nonetheless, the development of artificial societies has demonstrated the potential synthetic nature of ABMs in creating 'artificial worlds', sometimes capable of fruitfully cross-fertilizing disciplines (Epstein and Axtell 1996; Casti 1997).

Abstract models are theoretical ABMs aimed to investigate abstract social phenomena, such as cooperation or opinion dynamics. Unlike empirical models, they do not aim to represent well-circumscribed empirical facts, instances or Weberian 'individual cases' (Boero and Squazzoni 2005). Exactly because of their distance from any empirical content, abstract models are a key for theory-building in the social sciences. ABMs are used here to derive consequences of theories in order to prove their logical consistency or generate testable hypotheses for theory-driven empirical research. While abstract models are surely of crucial importance for the evolution of science, they entail the risk of losing sight of empirically relevant aspects (Troitzsch 2017).

Unlike the above types, *middle-range models* are empirically grounded theoretical models intended to investigate specific instantiations of the same class of empirical phenomena (e.g. social stratification, industrial clusters and job markets). They can be either grounded on empirical evidence by representing 'stylised facts' (Kaldor 1961) or informed by empirical data (Boero and Squazzoni 2005). Yet, these models are not designed to represent all possible idiosyncratic features of an empirical case, but only certain common elements defining a specific class of empirical phenomena. While this type of ABM has been referred to the Weberian concept of 'ideal type' (Weber 1949 [1904]; see also Gilbert and Ahrweiler 2009), the Mertonian concept of 'middle-range' theories indicates that these models can help link theories and empiricism on a common ground (Merton 1968). More precisely, middle-range models target well-specified explanatory objectives, are neither abstract models nor thick descriptive accounts, elicit a fruitful collaboration between theory and empirical research and are key for empirically grounded theory development by helping generalisation of empirical results by comparison, systematisation and aggregation (Hedström 2005). Unfortunately, the theory-empiricism integration of middle-range models is far from being an easy task for social scientists.

Case-based models aim to look at an empirically circumscribed target domain, usually in terms of spatio-temporal features rather than analytically. In this case, all relevant idiosyncratic characteristics of the phenomenon should be considered, including initial conditions and fine-grained contextual factors. However, findings from case-based simulation models are often hardly generalizable across scales and domains (Squazzoni 2012; see also Chapter 4 by de Haan in this volume). Therefore, their relevance depends on the possibility of linking case-based evidence to class features, ideally by scaling up case-based to middle-range models.

Finally, *applied simulations* are detailed, fine-grained representations of certain empirical contexts to guide practical implementation, for problem-solving or assisting policy-making. This type of model has largely been explored in applied socio-environmental sciences and in management and organisation sciences. In some cases, ABMs have been developed collaboratively via stakeholder involvement as part of a joint process of mutual learning (Gilbert et al. 2018). Nevertheless, the fact of requiring specifically collected empirical data for calibration and validation or collaborative endeavour for model construction poses serious limitations to the feasibility of both case-based models and applied simulations.

4 Challenges and problems towards integration of theory and empirical data via modelling

Regardless of the typologies and categorisations just discussed, in a recent review (Bianchi and Squazzoni 2015) we showed that the field of ABM in the social sciences has been growing in the last two decades both in terms of quantity of contributions and methodological maturity (see also Hauke, Lorscheid, and Meyer 2017). Following Macy and Willer (2002), we have clustered all reviewed publications around two topics.

On the one hand, one of the most prolific fields has been the study of the emergence and stability of cooperative regimes and social norms, in which different mechanisms, such as direct reciprocity (e.g. Axelrod 1997), indirect reciprocity (e.g. Bowles and Gintis 2004), trust (e.g. Macy and Skvoretz 1998; Bravo, Squazzoni, and Boero 2012) and reputation (e.g. Conte and Paolucci 2002; Boero et al. 2010) have been examined. These studies showed that a decentralised system of agents can reach evolutionarily stable institutional or normative patterns through self-organised, peer-to-peer endogenous processes which mimic consistent group behaviour in a variety of realistic settings, for example, from the web to the scientific community. Often inspired by behavioural game theory or even incorporating experimental data and findings, these studies have helped to extend our definition of rationality (beyond the straightforwardly, narrow, conventional ideas of material self-interest) and link it to the persistence and resilience of certain (functional) social norms and institutions that make up societal order and our social reality.

On the other hand, other ABMs have investigated the effect of social selection and influence on the emergence of unintended, unpredictable collective patterns, such as residential, social or economic segregation (e.g. Schelling 1971; Van de Rijt, Siegel, and Macy 2009), and the persistence of cultural diversity and opinion polarisation within complex societies (see Flache et al. 2017 for a review). While empirical studies have examined peculiar outcomes in-depth, such as opinion bipolarisation in online settings or ethnic segregation in certain urban neighbourhoods, these studies have examined different social mechanisms of interdependence that apply to a variety of empirical circumstances, at the same time helping us to realise the importance to disentangle micro-motivations to macro-outcomes without conflating levels of analysis.

However, these examples call for measures to integrate hypothesis-driven, top-down theories and bottom-up, inductive empiricism via input parameter calibration, internal consistency verification and output findings validation (e.g. Hedström and Åberg 2005; Manzo 2013; see also Graebner 2018; Muelder and Filatova 2018). While any model entails strategic decisions on continuous trade-offs between abstraction and empirical richness, which could also depend on modelling purposes (e.g. explanation, prediction, illustration, etc.; see Edmonds et al. 2019), calibration, verification and validation of model parameters and outcomes are key to develop robust, transparent and cumulative research collectively (e.g. Boero and Squazzoni 2005; Windrum, Fagiolo, and Moneta 2007; Moss 2008; Thiele, Kurt, and Grimm 2014; Graebner 2018).

For verification, we mean a systematic process aimed to establish the internal consistency between the conceptual model and its computational implementation. This requires model exploration and internal test, which could also have a collective, incremental dimension when combined with replication (Graebner 2018). Calibration is concerned with paramaterising initial conditions and input variables with empirical data (when available), which include either quantitative (e.g. agent population size, resource distribution) or qualitative factors (e.g. agent rules of behaviour). Validation mostly refers to empirical tests on simulation output via empirical data aimed to corroborate findings explanation (Frank, Squazzoni, and Troitzsch 2009).

Methodological and epistemological debates have stimulated the definition of standards for model documentation, access and replication, which are now a collective milestone of the community, though often difficult either to stimulate or enforce (Miodownik, Cartrite, and Bhavnani 2010; Janssen 2017). However, the question of how simulation results can inform scientific arguments about social reality requires to consider the construction of the model itself by an observer (Boero and Squazzoni 2005; Frank, Squazzoni, and Troitzsch 2009).

Depending on the relationship with empirical data, ABMs can have different degrees of explanatory power and fulfil different epistemological functions. Theoretical models can be used to check the internal consistency of one or more theories or derive hypotheses that can be further tested through empirical research. While in principle this can be done via thought experiments, especially when empirical examples or typified cases are simple (Elster 1989), models representing the assumptions and premises of certain hypotheses covering a nexus of expected real mechanisms are often needed to dynamically simulate logical consequences and check whether these matched theoretical predictions.

Here, think about the Schelling's famous model of residential segregation (Schelling 1971) or Sakoda's checkerboard model of social interaction (see Hegselmann 2017). Schelling wanted to help everyone understand that certain unintended consequences of individual behavior depend on aggregation mechanisms of interdependent choices. The example of residential segregation by ethnicity in the United States was used to disentangle the observed spatial segregation from supposed xenophobic preferences among U.S. citizens. In order to spell out this complex interplay, he built a paper-and-pencil prototype of a simple abstract ABM with a two-class population of agents situated in a grid. At each iteration, each agent decided whether to relocate randomly based on a simple threshold preference concerning the proportion of other agents of the same class in the neighbourhood. Schelling mathematically showed that a segregated equilibrium of the population could be reached also by mild integrative preferences, due to the structural interdependence of agents' decisions. Therefore, findings were used to emphasise that residential segregation could be driven more by structural factors without being necessarily the social footprint of xenophobic attitudes. Furthermore, rather than being judged by (lack of) empirical richness and detail, the beauty of this model lies in the fact that it conceptually exemplifies a variety of theoretical cases and empirical instances in which 'analysis can warm against jumping to conclusions about individual intentions from observations of aggregates, or jumping to conclusions about the behaviour of aggregates from what one knows or can guess about individual intentions' (Schelling 1978, 13–14). As Granovetter suggested in the same year (1978, 1442), these models help us to understand the explanatory advantage of taking the 'strangeness' of collective patterns 'out of the heads of actors and put it into the dynamics of situations' and considering that social interaction can make the size of causes and effects disproportional and non-linear in that large scale effects (segregation patterns) do not require causes of the same size (diffusion of xenophobic preferences in the population).

While Schelling's model testifies to a purely theoretical method of ABM building, similar models could also be used as auxiliary tools for empirical research, even without directly calibrating or validating parameters and output with any empirical data (Troitzsch 2017). More specifically, ABMs can be used as generators of empirically testable hypotheses, thereby prearranging theoretical constructs to empirical tests (Squazzoni 2012). In this case, an abstract model of a theory can be simulated to generate certain specific consequences by constraining certain parameters to a defined range of values which are intended to represent a targeted empirical phenomenon. Furthermore, these derived hypotheses can be empirically tested, for instance, via experimental protocols.

For instance, Corten and Buskens (2010) developed a model of a repeated coordination game with network embeddedness which served to generate hypotheses of estimated effects in a well-controlled, artificial scenario. They then tested these computational effects in a laboratory experiment, which aimed to corroborate established causal relationships (see also Corten 2014). These examples suggest that the boundaries between abstract and middle-range models are not fixed and predetermined, as does the over-emphasised difference between ABM and standard mathematical modelling (Edling 2002). At the same time, these flexible epistemological and methodological boundaries are key to integrate theory and empirical research in the social sciences, where general theories and systematic predictions are intrinsically difficult (Bäckman and Edling 1999; Goldthorpe 2000; Squazzoni 2012).

When considering more empirically oriented models, it must be said that also here ABMs can serve different purposes. As mentioned above, testing the sufficient generativity of hypothetical social mechanisms is one of the most important goals of ABM (Epstein 2006; Hedström and Bearman 2009). For instance, a model formulating a specific set of micro-level hypothesized mechanisms can be built to simulate outcomes that can be validated against certain measures of macro-level properties of the empirically observed target. Here, simply validating simulation outcomes without empirically calibrating the model could imply problems of 'multiple realisability'. This means that alternative micro-level specifications of a model or even different assumptions could possibly bring about (and so explain) the same emergent outcome (Sawyer 2005). This is because certain hypothesised social mechanisms could cancel each other out, different initial conditions could trigger in principle the same outcome or the same micro-specification could ideally explain different outcomes and not only that one specifically at work, ideally depending on the co-presence of supplementary conditions (Elster 1989).

In these cases, in order to achieve more explanatory power, empirically calibrating the model at the micro level and/or perform counter-factual manipulations could be beneficial to handle problems of multiple realisability (Boero and Squazzoni 2005). For instance, Gabbriellini (2014) studied the emergent status hierarchies in an online forum by modelling interaction between participants and comparing simulation outcomes with the empirically observed forum's communication network. Moreover, agent interaction was calibrated on empirically collected data on real communication between forum participants. This was instrumental to add constraints to the typical explorative redundancy of any ABM (e.g. heterogeneous agents, non-linear interaction, local information, etc.).

Unfortunately, this epistemological and methodological practice suffers from certain epistemological and practical limitations. First, no process of calibration and validation can logically rule out the possibility of the existence of alternative explanations. Therefore, empirical ABMs cannot in principle replace the theoretical machinery characterising any scientific investigation. On the one hand, modelling is key to constrain the high fly of theory and its speculative nature by over-imposing boundaries of logically plausible and empirically validated claims. In this respect, ABM is instrumental to stimulate well-controlled counter-factual thinking and procedures, which can complement empirical adherence (see below). Secondly, the effort of collecting appropriate data for model calibration and validation is often so demanding in terms of time and resources that a trade-off must be considered between highly empirically detailed studies and more abstract models.

Finally, ABMs can also be used to extend the narrow nature of empirical data and explore possible scenarios, decreasing the (economic and cognitive) costs of qualitative and quantitative parameter manipulations. In this way, simulations of an empirically calibrated ABM can be run to explore hypothetical scenarios and time–space developments that would be impossible or considerably difficult either to explore or observe empirically. This includes using ABMs as an *in silico* laboratory where researchers could run experiments on an empirical phenomenon that could not be easily done in a laboratory. Suppose, for instance, to manipulate model parameters in a controlled way, according to an experimental procedure in which the baseline condition is calibrated on empirical observations and so reproduce exactly the empirical target. Certain counter-factual or 'as if' conditions could be explored to corroborate the validity of findings via alternative 'artificial' scenarios, including varying the initial conditions and so minimizing problems linked to multiple realisability.

For instance, this is what Bravo, Squazzoni, and Boero (2012) have done, first by performing a laboratory experiment with human subjects playing an iterated trust game in order to study the emergence of trust and cooperation in an exchange network. In order to estimate the effect of different initial conditions (i.e. network topologies) on the experiment findings, they calibrated agent with exchange preferences replicating exactly those observed in the experiment. Then, they (re)simulated the game by manipulating various network configurations and comparing outcomes with experimental conditions. This was to explore the dynamics effect of certain structural features that were impossible or difficult to manipulate in the lab, such as network initialisations and complex dynamics. This integrative method (combining experimental models and ABM) represents an innovative use of empirical ABMs, although still relatively under-developed (see also Boero et al. 2010; Andrighetto et al. 2013; Amin, Abouelela, and Soliman 2018). Besides replicating, extending and testing experimental findings towards larger 'artificial' scales, this integration supports counter-factual analysis that could corroborate pre-established (often weak and case-sensitive) causal relationships. Moreover, this approach is particularly promising for policy-making, as it helps study possible consequences of political or organisational interventions on existing social systems, while avoiding ethical implications and economic constraints of traditional experimentations (Gilbert et al. 2018).

References

Amin, Egin, Mohamed Abouelela, and Amal Soliman. 2018. "The role of heterogeneity and the dynamics of voluntary contributions to public goods: An experimental and agent-based simulation analysis." *Journal of Artificial Societies and Social Simulation* 21 (1): 3. http://doi.org/10.18564/jasss.3585.

Andrighetto, Giulia, Jordi Brandts, Rosaria Conte, Jordi Sabater-Mir, Hector Solaz, and Daniel Villatoro. 2013. "Punish and voice: Punishment enhances cooperation when combined with norm signalling." *PLoS ONE* 8 (6): e64941. http://doi.org/10.1371/journal.pone.0064941.

Axelrod, Robert. 1997. *The Complexity of Cooperation: Agent-Based Models of Competition and Collaboration*. Princeton, NJ: Princeton University Press.

Bäckman, Olof, and Christofer Edling. 1999. "Mathematics matters: On the absence of mathematical models in quantitative sociology." *Acta Sociologica* 42 (1): 69–78. http://doi.org/10.1177/000169939904200105.

Bianchi, Federico, and Flaminio Squazzoni. 2015. "Agent-based models in sociology." *Wiley Interdisciplinary Research: Computational Statistics* 7 (4): 284–306. http://doi.org/10.1002/wics.1356.

Boero, Riccardo, Giangiacomo Bravo, Marco Castellani, and Flaminio Squazzoni. 2010. "Why bother with what others tell you? An experimental data-driven agent-based model." *Journal of Artificial Societies and Social Simulation* 13 (3): 6. http://doi.org/10.18564/jasss.1620.

Boero, Riccardo, and Flaminio Squazzoni. 2005. "Does empirical embeddedness matter? Methodological issues on agent-based models for analytical social science." *Journal of Artificial Societies and Social Simulation* 8 (4): 6.

Bowles, Samuel, and Herberg Gintis. 2004. "The evolution of strong reciprocity: Cooperation in heterogeneous populations." *Theoretical Population Biology* 65 (1): 17–28. http://doi.org/10.1016/j.tpb.2003.07.001.

Bravo, Giangiacomo, Flaminio Squazzoni, and Riccardo Boero. 2012. "Trust and partner selection in social networks: An experimentally grounded model." *Social Networks* 34: 481–92. http://doi.org/10.1016/j.socnet.2012.03.001.

Bruch, Elizabeth, and Jon Atwell. 2015. "Agent-based models in empirical social research." *Sociological Methods & Research* 44 (2): 186–221. http://doi.org/10.1177/0049124113506405.

Casti, John L. 1997. *Would-Be Worlds: How Simulation Is Changing the Frontiers of Science*. New York: Wiley.

Coleman, James S. 1962. "Analysis of social structures and simulation of social processes with electronic computers." In *Simulation in Social Science*, edited by H. Guetzkow, 63–9. Englewood Cliffs, NJ: Prentice Hall.

Coleman, James S. 1990. *Foundations of Social Theory*. Cambridge, MA: The Belknap Press of Harvard University Press.

Conte, Rosaria, Bruce Edmonds, Scott Moss, and R. Keith Sawyer. 2001. "Sociology and social theory in agent-based social simulation: A symposium." *Computational and Mathematical Organization Theory* 7 (3): 183–205. http://doi.org/10.1023/A:1012919018402.

Conte, Rosaria, and Mario Paolucci. 2002. *Reputation in Artificial Societies: Social Beliefs for Social Order*. Dordrecht: Kluwer.

Corten, Rense. 2014. *Computational Approaches to Studying the Co-Evolution of Networks and Behavior in Social Dilemmas*. Chichester: Wiley.

Corten, Rense, and Vincent Buskens. 2010. "Co-evolution of conventions and networks: An experimental study." *Social Networks* 32: 4–15. http://doi.org/10.1016/j.socnet.2009.04.002.

Edling, Christofer R. 2002. "Mathematics in sociology." *Annual Review of Sociology* 28: 197–220. http://doi.org/10.1146/annurev.soc.28.110601.140942.

Edmonds, Bruce, Christophe Le Page, Mike Bithell, Edmund Chattoe-Brown, Volker Grimm, Ruth Meyer, Cristina Montañola-Sales, Paul Ormerod, Hilton Root, and Flaminio Squazzoni. 2019. "Different modelling purposes." *Journal of Artificial Societies and Social Simulation* 22(3): 6. http://doi.org/10.18564/jasss.3993.

Elster, Jon. 1989. *Nuts and Bolts for the Social Sciences*. Cambridge and New York: Cambridge University Press.

Epstein, Joshua M. 2006. *Generative Social Science: Studies in Agent-Based Computational Modeling*. Princeton, NJ: Princeton University Press.

Epstein, Joshua M. 2008. "Why model?" *Journal of Artificial Societies and Social Simulation* 11 (4): 12.

Epstein, Joshua M., and Robert L. Axtell. 1996. *Growing Artificial Societies: Social Science From the Bottom Up*. Cambridge, MA: The MIT Press.

Flache, Andreas, Michael Mäs, Thomas Feliciani, Edmund Chattoe-Brown, Guillaume Deffuant, Sylvie Huet, and Jan Lorenz. 2017. "Models of social influence: Towards the next frontiers." *Journal of Artificial Societies and Social Simulation* 20 (4): 2. http://doi.org/10.18564/jasss.3521.

Frank, Ulrich. 2002. *The Explanatory Power of Models: Bridging the Gap Between Empirical and Theoretical Research in the Social Sciences*. Dordrecht: Kluwer.

Frank, Ulrich, Flaminio Squazzoni, and Klaus G. Troitzsch. 2009. "EPOS-epistemological perspectives on simulation: An introduction." In *Epistemological Aspects of Computer Simulation in the Social Sciences: Second International Workshop, EPOS 2006, Brescia, Italy, October 2006. Revised Selected and Invited Papers*, edited by Flaminio Squazzoni, 1–11. Berlin and Heidelberg: Springer.

Gabbriellini, Simone. 2014. "Status and participation in online task groups: An agent-based model." In *Analytical Sociology: Actions and Networks*, edited by Gianluca Manzo, 317–38. Chichester: Wiley.

Gilbert, Nigel. 2008. *Agent-Based Models*. London: SAGE Publications.

Gilbert, Nigel, and Andrew Abbott. 2005. "Introduction." *American Journal of Sociology* 110: 859–63.

Gilbert, Nigel, and Petra Ahrweiler. 2009. "The epistemologies of social simulation research." In *Epistemological Aspects of Computer Simulation in the Social Sciences: Second International Workshop, EPOS 2006, Brescia, Italy, October 2006. Revised Selected and Invited Papers*, edited by Flaminio Squazzoni, 12–28. Berlin and Heidelberg: Springer.

Gilbert, Nigel, Petra Ahrweiler, Pete Barbrook-Johnston, Kavin Preethi Narasimhan, and Helen Wilkinson. 2018. "Computational modelling of public policy: Reflections on practice." *Journal of Artificial Societies and Social Simulation* 21 (1): 4. http://doi.org/10.18564/jasss.3669.

Gilbert, Nigel, Matthijs den Besten, Akos Bontovics, Bart G.W. Craenen, Federico Divina, A.E. Eiben, Robert Griffioen, Gyorgy Hévízi, Andras Lõrincz, Ben Paechter, Stephan Schuster, Martijn C. Schut, Christian Tzolov, Paul Vogt, and Lu Yang. 2006. "Emerging artificial societies through learning." *Journal of Artificial Societies and Social Simulation* 9 (2): 9.

Gilbert, Nigel, and Klaus G. Troitzsch. 2005. *Simulation for the Social Scientist*. 2nd ed. New York: Open University Press.

Goldthorpe, John H. 2000. *On Sociology. Numbers, Narratives, and the Integration of Research and Theory*. Oxford: Oxford University Press.

Graebner, Claudius. 2018. "How to relate models to reality? An epistemological framework for the validation and verification of computational models." *Journal of Artificial Societies and Social Simulation* 21 (3): 8. http://doi.org/10.18564/jasss.3772.

Granovetter, Mark. 1978. "Threshold models of collective behavior." *American Journal of Sociology* 83 (6): 1420–43.

Hartmann, Stephan, and Roman Frigg. 2006. "Models in science." In *The Stanford Encyclopedia of Philosophy*, edited by E. N. Zalta. Stanford: Stanford University. https://plato.stanford.edu/archives/sum2018/entries/models-science/.

Hauke, Jonas, Iris Lorscheid, and Matthias Meyer. 2017. "Recent development of social simulation as reflected in JASSS between 2008 and 2014. A citation and co-citation analysis." *Journal of Artificial Societies and Social Simulation* 20 (1): 5. http://doi.org/10.18564/jasss.3238.

Hedström, Peter. 2005. *Dissecting the Social: On the Principles of Analytical Sociology*. Cambridge: Cambridge University Press.

Hedström, Peter, and Yvonne Åberg. 2005. "Quantitative research, agent-based modelling and theories of the social." In *Dissecting the Social: On the Principles of Analytical Sociology*, edited by Peter Hedström, 114–44. Cambridge: Cambridge University Press.

Hedström, Peter, and Peter Bearman, eds. 2009. *The Oxford Handbook of Analytical Sociology*. Oxford: Oxford University Press.

Hedström, Peter, and Gianluca Manzo. 2015. "Recent trends in agent-based computational research: A brief introduction." *Sociological Methods & Research* 44 (2): 179–85. http://doi.org/10.1177/0049124115581211.

Hedström, Peter, and Petri Ylikoski. 2010. "Causal mechanisms in the social sciences." *Annual Review of Sociology* 36: 49–67. http://doi.org/10.1146/annurev.soc.012809.102632.

Hegselmann, Rainer. 2017. "Thomas C. Schelling and James M. Sakoda: The intellectual, technical, and social history of a model." *Journal of Artificial Societies and Social Simulation* 20 (3): 15. http://doi.org/10.18564/jasss.3511.

Hirsch, Paul, Stuart Michaels, and Ray Friedman. 1987. "'Dirty Hands' versus 'Clean Models'. Is sociology in danger of being seduced by economists?" *Theory and Society* 16 (3): 317–36. http://doi.org/10.1007/BF00139485.

Janssen, Marco. 2017. "The practice of archiving model code of agent-based models." *Journal of Artificial Societies and Social Simulation* 20 (1): 2, 8. http://doi.org/10.18564/jasss.3317.

Kaldor, Nicholas. 1961. "Capital accumulation and economic growth." In *The Theory of Capital*, edited by Friedrich A. Lutz and Douglas C. Hague, 177–222. London: St. Martin's Press.

Machamer, Peter, Lindley Darden, and Carl F. Craver. 2000. "Thinking about mechanisms." *Philosophy of Science* 67 (1): 1–25. http://doi.org/10.1086/392759.

Macy, Michael W., and Andreas Flache. 2009. "Social dynamics from the bottom up: Agent-based models of social interaction." In *The Oxford Handbook of Analytical Sociology*, edited by Peter Hedström and Peter Bearman, 245–68. Oxford: Oxford University Press.

Macy, Michael W., and John Skvoretz. 1998. "The evolution of trust and cooperation between strangers: A computational model." *American Sociological Review* 63: 638–60. http://doi.org/10.2307/2657332.

Macy, Michael W., and Robert Willer. 2002. "From factors to actors. computational sociology and agent-based modeling." *Annual Review of Sociology* 28: 143–66. http://doi.org/10.1146/annurev.soc.28.110601.141117.

Manzo, Gianluca. 2013. "Educational choices and social interactions: A formal model and a computational test." In *Class and Stratification Analysis: Comparative Social Research, vol. 30*, edited by Gunn E. Birkelund, 47–100. Bingley: Emerald.

Merton, Robert K. 1968. *Social Theory and Social Structure*. New York: The Free Press.

Miller, John H., and Scott Page. 2007. *Complex Adaptive Systems: An Introduction to Computational Models of Social Life*. Princeton, NJ: Princeton University Press.

Miodownik, Dan, Britt Cartrite, and Ravi Bhavnani. 2010. "Between replication and docking: 'Adaptive agents, political institutions, and civic traditions' revisited." *Journal of Artificial Societies and Social Simulation* 11 (1): 5. http://doi.org/10.18564/jasss.1627.

Moss, Scott. 2008. "Alternative approaches to the empirical validation of agent-based models." *Journal of Artificial Societies and Social Simulation* 11 (1): 5.

Muelder, Hannah, and Tatiana Filatova. 2018. "One theory-many formalizations: Testing different code implementations of the theory of planned behaviour in energy agent-based models." *Journal of Artificial Societies and Social Simulation* 21 (4): 5. http://doi.org/10.18564/jasss.3855.

Parsons, Talcott. 1937. *The Structure of Social Action*. New York: The Free Press.

Sawyer, R. Keith. 2005. *Social Emergence: Societies as Complex Systems*. Cambridge: Cambridge University Press.

Schelling, Thomas C. 1971. "Dynamic models of segregation." *Journal of Mathematical Sociology* 1: 143–86. http://doi.org/10.1080/0022250X.1971.9989794.

Schelling, Thomas C. 1978. *Micromotives and Macrobehaviour*. New York: W. W. Norton.

Skvoretz, John. 1991. "Theoretical and methodological models of networks and relations." *Social Networks* 13 (3): 273–300. http://doi.org/10.1016/0378-8733(91)90009-I.

Sørensen, Aage B. 1998. "Theoretical mechanisms and the empirical study of social processes." In *Social Mechanisms: An Analytical Approach to Social Theory*, edited by Peter Hedström and Richard Swedberg, 238–66. Cambridge: Cambridge University Press.

Squazzoni, Flaminio. 2008. "The micro-macro link in social simulation." *Sociologica* 2 (1): 5–13. http://doi.org/10.2383/26578.

Squazzoni, Flaminio. 2012. *Agent-Based Computational Sociology*. Chichester: Wiley.

Thiele, Jan C., Winfried Kurth, and Volker Grimm. 2014. "Facilitating parameter estimation and sensitivity analysis of agent-based models: A cookbook using NetLogo and R." *Journal of Artificial Societies and Social Simulation* 17 (3): 11. http://doi.org/10.18564/jasss.2503.

Troitzsch, Klaus. 2017. "Axiomatic theory and simulation: A philosophy of science perspective on Schelling's segregation model." *Journal of Artificial Societies and Social Simulation* 20 (1): 10. http://doi.org/10.18564/jasss.3372.

Van de Rijt, Arnout, David Siegel, and Michael W. Macy. 2009. "neighborhood chance and neighborhood change: A comment on bruch and mare." *American Journal of Sociology* 114: 1166–80. http://doi.org/10.1086/592200.

Weber, Max. 1949 [1904]. "Objectivity in social science and social policy." In *The Methodology of the Social Sciences*, edited by Edward A. Shils and Henry Finch. New York: The Free Press.

Windrum, Paul, Giorgio Fagiolo, and Alessio Moneta. 2007. "Empirical validation of agent-based models: Alternatives and prospects." *Journal of Artificial Societies and Social Simulation* 10 (2): 8.

Part 2

State of the art

6 Modelling the multi-level perspective

The MATISSE agent-based model

Jonathan Köhler

1 Introduction

The field that is now known as sustainability transitions (STRN 2017; Köhler et al. 2019) is based on the assessment that in order to achieve a sustainable society in the future, far-reaching changes in socio-technical-environmental systems are essential. Its origins are in the 'neo-Schumpeterian' theory of Kondratiev or 'long' waves of industrial and economic change. Freeman and Louçã (2002) develop a theoretical framework using these ideas, in which industrial revolutions are understood as co-evolutionary processes in socio-technical systems, including political, scientific, economic, technological and cultural subsystems. A fundamental insight of this theory is that historical examples show that radical change originates in niches (Freeman and Louçã 2002). The multi-level perspective on transitions (MLP: Geels and Schot 2010) has been developed as a theoretical framework for analysing such processes and has adopted this structure (Köhler 2012). Loorbach and Rotmans (2010) show that these features mean that the socio-technical-environmental systems addressed by transitions theory are complex adaptive systems.

Transitions research has the ambition to understand and assess potential transition pathways (e.g. Köhler, Turnheim, and Hodson 2018), for which modelling is a useful approach (Halbe et al. 2015; Holtz et al. 2015; Li, Trutnevyte, and Strachan 2015; Chapter 4 by de Haan in this volume). Köhler (2019) argues that models for transition pathways representing the structure of niches, regime(s) and landscape interacting in a co-evolutionary process can address sustainability transitions more effectively than the techno-economic models used in, for example, IPCC (2018). Köhler et al. (2018) review approaches to modelling transitions. There are several approaches that can incorporate the complex adaptive systems nature of transitions, including system dynamics, models from complexity science and socio-ecological systems. There are also some models using computational social science agent-based modelling (CSS ABM), building on the use of ABM in the social sciences (Chapter 5 by Bianchi and Squazzoni in this volume). Li and Strachan (2017) is an example of this approach for transitions in energy systems.

The MATISSE model of sustainability transitions was developed to implement the theoretical framework of the MLP. The conceptual framework of the model is developed in Haxeltine et al. (2008). The model is described in detail in Bergman et al. (2008). It is a general prototype model and was initially used to represent four historical case studies described in (Geels and Schot 2007). These were (1) the transformation pathway in Dutch public hygiene from cesspools to sewer systems, (2) the de-alignment and realignment pathway from horse-drawn carriages to automobiles in the U.S., (3) the technological substitution pathway from sailing ships to steam ships in the UK and (4) the reconfiguration pathway from traditional factories to mass production in the U.S. The MATISSE model has also been applied for forward-looking analysis of transition pathways to mobility (Köhler et al. 2009; Köhler, Turnheim, and Hodson 2018; Moallemi and Köhler in press 2019).

This general structure of the model, enabling different cases to be analysed, addresses the criticism that transitions studies are too heavily dependent on analysis of single case studies and that a structure for comparative analysis of case studies is lacking (STRN 2017).

The objectives of the present chapter are to show how the concepts of sustainability transitions have been interpreted in the MATISSE model and how the general structure of the model enables it to be used for different applications. These different applications require case-specific interpretations of the MATISSE model structure. Two applications to households' sustainable mobility and low-emissions shipping are used to illustrate interpretations of the categories in the model. The shipping case is an example of choices over technologies for ship propulsion (i.e. a clearly defined engineering function). Sustainable mobility is a broader concept, involving both technologies and lifestyle decisions. The discussion shows how this approach to modelling can incorporate barriers to change that cannot be directly addressed by techno-economic models.

2 Development of the MATISSE model: interpretation of the MLP

The first version of the model was developed for the MATISSE project. The model was developed as a completely new structure, and therefore the project team worked out the concept in a series of workshops. The idea was to implement the framework of the MLP to analyse possible pathways of niches and regimes in transitions processes.

The MLP (Geels and Schot 2010) has three categories or levels: landscape, regime and niches. It is a 'mid-range' theory; that is, it is positioned as an aggregated framework at a level in between the micro or individual actions and decisions and the macro or whole society/economy. Niches may or may not 'grow', and the regime (usually only one, although the interactions between multiple regimes are now being discussed, e.g. STRN [2019]) may or may not decline or change. The macro-level landscape represents changes in society generally, which may exert pressure on the regime and/or support the development of niches.

The interactions between these three levels are dynamic and determined amongst other factors by the different rates of change in the three levels (Rotmans and Loorbach 2009). The landscape (e.g. culture) changes slowly, while niches are small social structures and can change quickly. The regime is in between. It has a large, established technology and institutions that cannot change as quickly as niches. Furthermore, regimes are characterised by lock-in through the established, mature technologies and institutional structures. The dynamics in the MLP are determined by the interacting behaviours of the regime and niches. A critical aspect of this is that the niches are initially very small in comparison to the regime. However, because they can grow, they can have a decisive impact on the overall system – a property of complex adaptive systems (Rotmans and Loorbach 2009).

While this framework is widely used in transitions research, this is where the agreement ends. As a deliberately mid-level framework, the detail of the landscape, regime and niches is left open. Most early transitions research used detailed descriptions of particular cases to explore the structure of regimes, niches and interactions between levels. However, a simulation model requires a precise definition of its structure, which is one of the advantages of quantitative modelling. A model of the MLP therefore has to include a representation of the niches and the regime, with explicit hypotheses about their (sub-)system properties and their decision-making.

Because the MLP is explicitly non-linear, a dynamic simulation approach was adopted. The MLP describes the object of research as a 'socio-technical system', so a goal was to structure the model as a system model. At the same time, the idea of a regime as consisting of structure, culture and practices (Haxeltine et al. 2008) was considered. However, there was no literature that examined the detailed structure of a regime in systems terms, even though the idea of a socio-technical regime as a system comprising science, culture, technology, policy and markets/users sub-systems (Geels and Schot 2010; Freeman and Louçã 2002) is a central part of the MLP structure. There was also no literature on how to interpret cultural changes in transitions in terms of numerical simulations. However, what is clear in the MLP is that transitions dynamics are determined by the interactions between niches and a regime, given landscape pressures. Taking an average over all decision makers, as is common in large-scale techno-economic and macroeconomic simulation models, prevents a model from directly representing these system properties. Therefore, an agent-based approach is necessary to represent the dynamics in the MLP framework (Köhler 2019; Li 2017, Chapter 7 by Holtz and Chappin in this volume). Because the niches and regimes can be considered as entities in the socio-technical system, but with different properties, the socio-technical field under consideration was represented by agents for the niches and regime within a system, as shown in Figure 6.1. In this sense, the model is a dynamic system model, but it is not a 'system dynamics' model with stock and flow variables.

Since landscape pressures include cultural changes (Geels and Schot 2010), the model includes input (i.e. exogenous assumptions) about general changes in demand and behaviour, as a part of the representation of the 'landscape'. The idea of a socio-technical regime having practices was applied to the structure of the niches and regime as described below.

A further consideration was how to interpret the idea of niches 'challenging' the regime and the regime 'opposing' niches (Geels and Schot 2010). This implies that there is strategic behaviour by the regime and niches. This aspect of the MLP was described for historical cases, but was also not elaborated as a theoretical structure. Therefore, the empirical case literature was reviewed and two different states of the niches were identified – one proto-phase where the niche is very small and just tries to establish a stable structure, and then a further stage where the niche has grown and starts to actively challenge the regime. Haxeltine et al. (2008) therefore included a further entity – the empowered niche – that tries to directly use its resources to take support from the regime as opposed to a niche, which just tries to grow by gaining more supporters. The empowered niche was removed in Köhler et al. (2018) because it was considered that a (small) niche should be able to interact directly with the regime. Also, this did not represent an explicit feature of the MLP framework. The implementation of strategic behaviour of niches and regimes is described below.

For the MATISSE model, Haxeltine et al. (2008) consider regimes as a single entity in the model with an internal structure and behavioural rules. This means that a regime is treated as a single agent, in the sense of agent-based modelling. Niches are considered to be 'proto'-regimes. This means that they have the same structure as a regime, but are 'smaller' initially and have different behavioural rules. The size of a regime and niches is interpreted as an abstract 'strength'. The regime is by definition the strongest agent and dominates the system, while niches have much less strength. An agent's strength determines its behaviour (strategy in practices space as described below), its interactions with other agents, and (partly) its attractiveness to supporters. This strength is determined by the number of supporters who adopt the regime or a niche. Thus, the MATISSE structure has an additional type of agent to the regime and niches. These supporters represent the structure of demand. In the MATISSE structure of Köhler et al. (2009), these are consumer households choosing a mobility lifestyle. In the new version of the model (see Section 6.3.2), these are shipowners ordering new ships. The landscape is represented as factors that influence the regime, niches and supporters, but are exogenous variables. These factors may change as part of a scenario but are not influenced by the regime or niches. This is an approximation, as the cultural landscape can change in response to the development of a new regime, but it is assumed that this process is slow enough in terms of the flow of time in the model to be approximated through exogenous changes. This framework is shown in Figure 6.1.

The internal structure of a regime (and therefore in the MATISSE system, the niches as well) is based on the theory of Rotmans (2005), which introduces the idea of regimes having structure, culture and practices. Rotmans

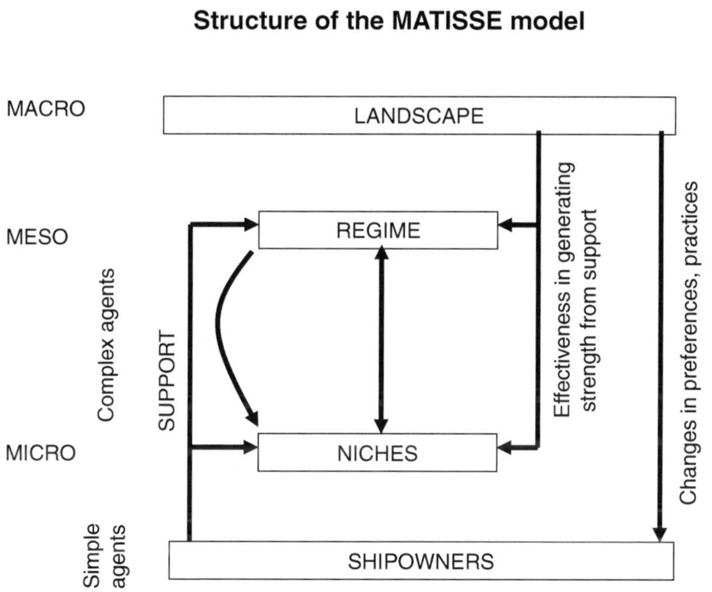

Figure 6.1 Agents in the MATISSE model.

Source: Based on Köhler et al. (2009).

and Loorbach (2009 p. 185) state that a regime is 'a conglomerate of structure (institutional and physical setting), culture (prevailing perspective), and practices (rules, routines, and habits)'. Following this, Haxeltine et al. (2008) use the idea of practices to implement the behavioural rules in the model. Bergman et al. (2008) describe in detail the implementation of practices in the model.

The model uses the concept of practices as dimensions through which agents position themselves in society and over which behaviour is defined for the supporters. This is analogous to the concept of preferences over different goods in microeconomics. Practices are each represented as values along axes, constituting a multi-dimensional practices space. Agents – regime, niches and supporters – are differentiated by their positions in the multi-dimensional practices space. This is described in the Appendix. Preferences of the supporters are defined in terms of their positions. They are also assigned a direction in practice space to represent changes in preferences over time to allow for the influence of, for example, changes in the landscape.

Supporters assess the distances between their desired position in the practice space to the positions of the regime and niches and choose to support the niche or regime with the smallest distance. The niches and regime can also move in this practice space. In this way, the characteristics of the 'solutions' which include technologies, but also may be a complete system of provision, can change. The regime follows a strategy of changing its practices with the objective of maximising its support. There are no limitations to how far along a dimension the regime can move, although it may take a long time. In this way, the characteristics – practices – of a regime can change radically such that the whole socio-technical system of the regime may change, also considered a transition in Geels and Schot (2007). The niches are restricted in the direction in which they can move in some dimensions (e.g. a niche pursuing sustainable mobility will not increase its emissions). The niches and regime choose directions of movement in the practice space according to their strategy and restrictions on the direction and speed of change.[1]

Each agent type (regime, niche) has a different heuristic for its movement in the practice space. These are derived from the policy-driven party dynamics of Laver (2005). The regime is an 'aggregator', changing its practices towards the average of the regime supporters' positions in the practice space, with the goal of maximising support. The rate of change of the regime is slow compared to the niches. Additionally, when the regime's support falls below a threshold it changes to aggregate all supporters. This represents the regime's tendency to be locked in to its practices and to optimise rather than innovate. Niches are 'hunters'. They move in the same direction as long as their strength increases; otherwise, they move in another random direction. Niches are restricted in their movement to a certain sub-space of the practices space. The niches and regime co-evolve through their movements in the practice space in response to the distribution of the supporters in the practice space and the changes over time in this distribution.

This structure generates as the main result the distribution of support for the regime and niches through time; that is, the pathways of adoption of the niches and the regime.

The other main aspect was that the different transitions pathways involve differing strategic behaviours of the regime and niches. One typical behaviour of large incumbent firms faced by a new technology is that they buy out the niche firms. Another is that firms copy technological developments. These behaviours are included: niches can combine with each other or with the regime, allowing for a range of niche and regime strategies for growth or survival. All agents are represented in all the multiple practices dimensions and may change their practices through time such that the regime and niche can (partly) adopt the characteristics of another niche or regime. In reality, a regime or niche can use more than one technology, such that the model represents two niches coming together as a gradual process, as well as an amalgamation of two niche/regime agents.

The MLP has been successfully used to analyse many different examples of transitions, and therefore the MATISSE model can also be used to analyse transitions as changes in socio-technical systems in terms of the landscape factors, the regime and niches that are identified through case study analysis. The first version of the model was tested by calibrating the model to reproduce the transitions pathways of historical transitions (Bergman et al. 2008). These were: cesspools to sewer systems in the Netherlands, horse-drawn carriages to automobiles in the U.S., sailing ships to steam ships and craft factories to mass production.

The interpretation of this generalised structure for particular cases will now be discussed, using two contrasting examples of transitions.

2.1 Use of data and assumptions to calibrate and parameterise the model

The first decision to be made when designing a MATISSE application is to determine what the regime and niches are. In the shipping example, these are different technologies for powertrains for a ship. The published literature can easily be used to show that the regime is diesel engines, and there are plenty of reviews of the different technical possibilities (see Section 3.2). In the mobility application, the niches in particular but also the regime are more complicated to define. Köhler, Turnheim, and Hodson (2018) undertook case studies to define the niches to be examined and also to define the practice dimensions to be included.

The MATISSE model is designed to be relatively small and simple, rather than a highly detailed model based on statistical estimates of parameter values from data. The practice dimensions are normalised to a range of 0–100. For practice variables that are measured quantities, such as emissions or costs, the niche and regime values are assigned so that the relationships between the average values in the data are maintained. The regime values are taken as a basis, assigned to 80 if the regime is expected to reduce its value or if niches all have similar or lower values or to 50 if the changes in the future and positions of the niches could be higher or lower. The values of quantitative variables such as use of information and communication technology (ICT) to reduce travel demand or perceived convenience or operational change are assessed based on qualitative survey data and interview data. The distribution of different supporters and their practice parameters are estimated from recent data about the practices and the distribution of technologies in the regime and niches. Changes in the practice values from the landscape can also be either measurable or qualitative. The changes are interpreted from scenario narratives; for example, if a scenario includes increasing fossil fuel prices, the petrol/diesel regime and other fossil fuels increase their operational costs. If in the mobility case, a scenario includes the development of recharging infrastructure for battery electric vehicle (BEVs), the perceived convenience of BEVs moves towards that of internal combustion engine (ICE) vehicles for which there is already a widely available infrastructure.

3 Two contrasting examples: households' sustainable mobility and low-emissions shipping

3.1 Sustainable mobility lifestyles

This example illustrates possible transitions to sustainable mobility in passenger transport at the level of households. Mobility behaviour depends on more than a technology choice over alternative types of car. Rather, it is dependent on culture and the availability of alternatives given the built environment in addition to transport system technologies (Köhler 2006). Conventional transport modelling considers the cost of the transport mode and the cost of the time used in a trip (Hensher and Button 2007) but does not consider the other aspects of urban structure and transport system infrastructures or the prevailing perspective on the social meaning of the car. The MATISSE model on mobility was developed to consider these broader aspects in households' decisions about mobility. In this application, the supporters of Figure 6.1 are households deciding on transport modes, including considerations of the structure of the built environment and their attitudes towards alternative transport modes and reducing travel demand.

The regime is the ICE privately owned car, but households that own a car may use other modes; for example, trains to commute into large cities, buses for some local trips. There are niches based on alternative powertrains for cars: ICE–battery hybrid, biofuels, hydrogen and BEV niches, in which mobility is still centred on ownership of a private vehicle. Other niches involve different structures of supply and use of transport services: the public transport niche is mainly mechanised public transport; active modes use mainly cycling and walking with mechanised transport for long-distance trips. This is assumed to be public transport or some form of car sharing and not the ownership of a motor car. Mobility based on car sharing is also included as a separate mobility lifestyle.

In this way, the mobility niches used fall into two groups. The first has the ownership of a car as the main mode of transport, with the possibility of different power trains or fuels. A transition from the ICE privately owned car to one of these alternatives only involves buying a car with a different powertrain or buying different fuels.

The other group has a behaviour that is different in a more complex way. They do not own a car. This requires access to alternative infrastructures and systems for day-to-day and longer-distance trips. Use of conventional public transport is most convenient in densely populated urban conglomerations, where the public transport is comprehensive and intensive. For almost all trips, households can get to any destination – local or longer distance – without long waiting times or very high costs. If a household lives in an extensive, suburban area or in the countryside, the public transport provision may be minimal or even lacking completely. This makes it difficult to change from a car-based mobility lifestyle without moving into a town. Decisions about mobility are therefore part of decisions about where to live. There is also an important cultural aspect; the ownership of a car provides social status and personal security. Owning your own car is felt to be a symbol of freedom, reaching back to the history of the touring and racing niches of the early days of the car (Geels 2005; Köhler 2012). Environmental performance as a factor in the decision is also included.

Practices represent the factors that come into the decision on which transport systems to use. The perception of the transport system, the built environment and the possibility of reducing travel through remote ICT are included in addition to the price and time of conventional models. The perception of the transport regime or niche is intended to represent considerations such as waiting times or the 'range uncertainty' effect of battery vehicles with a limited range (Egbue and Long 2012). The built environment factor represents whether a household wishes to live an urban, city-centre lifestyle in a dense urban environment and often in a flat, a suburban lifestyle with one's own house, garden and garage, or in the quiet, but remote country. The use of an ICT factor represents behaviours including the use of ICT for home working, networked communication instead of business travel or the purchase of goods over the internet instead of in a shop. The practice dimensions are then (Köhler et al. 2009; Köhler, Turnheim, and Hodson 2018):

- CO_2 emissions
- Cost (i.e. price)
- Use of mechanised individual transport (mostly cars)
- Use of ICT to reduce travel demand
- Use of public mechanised transport (train, bus, taxi)
- Built environment (extensive urban structure with suburbs, compact with high-rise buildings and public transport)
- Perceived user convenience (requirements for and availability of refuelling infrastructure, whether there is the main transport mode immediately available for long as well as short distance, or whether there is a waiting time and/or restrictions on range without changing modes)

The cost and emissions practices have a higher weight than the other dimensions in the households' choice decision. This is implemented by multiplying the differences in the cost and emissions practice dimensions between the household and the niche or regime under consideration by a factor of two.

The model is stochastic in that the supporter household agents are initially assigned as a distribution across different behaviour groups over the practice space, with a random normal distribution of the supporters in a group around a central value for the group. The central value is initially that of the regime or relevant niche. In Köhler et al. (2009) and Köhler, Turnheim, and Hodson 2018) these behaviour groups are: car drivers, green car drivers, public transport users and cyclists and pedestrians. The model is calibrated to empirical data for the quantitative indicators: emissions, cost, use of mechanised individual transport and use of public mechanised transport.

The other practice variables (ICT, built environment and perceived convenience) were not estimated using a statistical procedure, but assessed for the different household groups through reviews of survey data and case study analysis (Köhler et al. 2009; Köhler, Turnheim, and Hodson 2018). This makes the distribution of values for these parameters highly uncertain; this uncertainty can be addressed through the method of exploratory modelling (Moallemi and Köhler in press 2019).

The model is calibrated for all of the practice dimensions for the niches and regime as well as the supporter-households to annual data and runs in time steps of one year. The supporters are initially distributed in the practice space such that the regime and niches support shares model the data on trip distributions. The decisions modelled are not day-to-day choices of trip modes, but rather the occasional choice of what transport modes will be available. In each time step, a proportion of households (random normal distribution centred on 10% of households) make a choice among the different possible mobility lifestyles. These are whether to buy a car, whether the household will live remotely from or close to the locations of work, shopping, entertainment, etc., and the consequent need to commute and over what distance, whether to live somewhere where there is readily accessible and frequent public transport or good pedestrian access to local services, etc.

The model is run to generate scenarios of transition pathways by assuming gradual changes each year in the desired practices of the household groups. These changes are exogenous and thus form part of the 'landscape' in the model. The landscape can also include exogenous changes in the technology characteristics; for example, energy policy to tax CO_2 increasing the operating cost of ICE and ICE–hybrid vehicles. Figure 6.2 shows the example for a model calibrated on 2015 data for the UK of a transition to BEVs. The results are shown for a representative sample of 1000 households, initially distributed across the different behavioural groups of households, representing the distribution of the modal split data in terms of numbers of trips per year in the UK. The *y*-axis shows the shares of the regime and niche technologies/lifestyles. In the run of Figure 6.2, the households initially supporting the ICE regime represent over 70% of the population of households. The desired practices of the ICE households are assumed to gradually change, moving away from the initial values of the ICE regime. This can be interpreted as a scenario reflecting a change in the culture of society through a higher importance placed on climate

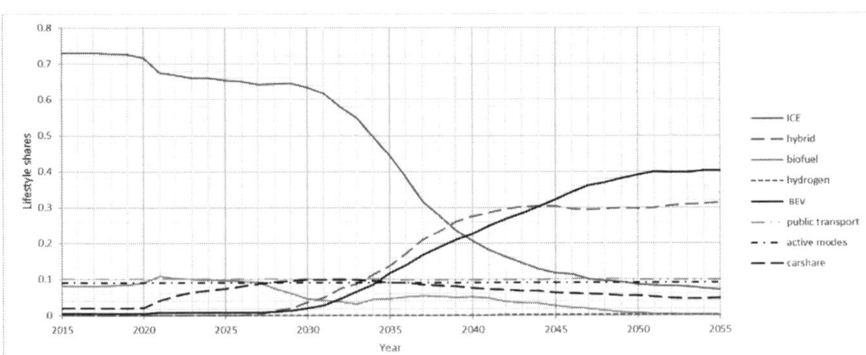

Figure 6.2 A pathway for a transition to BEVs in the UK, using the MATISSE-KK model from the EU PATHWAYS project.

change and its impacts. The desired cost remains the same, but the household desire reduced CO_2 emissions. At the same time, their perception of the convenience of alternative modes changes towards a moderately increased acceptance of reduced range of their vehicle, a reduced use of individual vehicles and an increased use of public transport. These changes are reinforced in this simulation by an assumed increasing desire to live in a compact urban environment and an increasing use of ICE to substitute actual trips. The combined effect of all these assumed landscape changes is to move the distribution of desired household practices away from the ICE regime.

There is a slight growth in car sharing and a very small increase in public transport use to 2025. This is because the regime fails to react quickly enough to the changing attitudes of households. After that, it is assumed that there is an increased shift in households' attitudes towards increased use of ICT (e.g. home working, video conferencing, internet shopping) to reduce travel demand and improved environmental performance. The resulting scenario is that after 2030 the BEV and hybrid niches take off, as their reduced emissions compared to ICE make them more attractive to more households. Biofuels lose their moderate share in the model run, as their price stays high. In the longer run, BEVs compete with hybrids, and by 2050 BEVs have a higher share because their environmental performance is better and therefore closer to the demand for very low emissions from mobility from most households. Narrative scenarios interpreting such changes are discussed in Köhler, Turnheim, and Hodson (2018).

These results can be explained in terms of the model behaviour: behavioural changes in the households and the reactions of the niches and regime to these changes. The improvements in battery technology (movements of the hybrid and BEV niches along the cost and user convenience dimensions) shift the hybrid and BEV cost practices closer to the ICE value and shift the perceived user convenience practice values closer to the convenience of the private ICE car. The regime reacts by slightly improving environmental performance of ICE vehicles, but is increasingly distant from the demands of households for major improvements. The changes in households' desired practices are not drastic enough to move close enough to the practices of active modes for these to take off. The practices offered by public transport are assumed to remain similar to the initial values, in particular in terms of perceived convenience and the built environment.

There is no radical change in the practices of public transport (e.g. organisation of intermodal connections) or the attitudes of ICE-supporting households towards public transport or the perceived convenience of ICE cars.

Using the same version of the model, but calibrated to Netherlands data, (Köhler, Turnheim, and Hodson 2018) develop three scenarios for the Netherlands of transitions to BEVs, to a new public transport regime and to a mobility system where lifestyles with active modes are dominant. Moallemi and Köhler (in press 2019) present scenarios for the UK of a transition to a public transport regime.

This very wide range of possible outcomes shows how the structure of the MATISSE model generates dynamic pathways that show properties typical of complex adaptive systems and the theoretical framework of the MLP. The model incorporates not only the multiplicity of technologies available and their costs through time, but also allows the interpretation of results in terms of qualitative factors: user perceptions of the different niches and regime and urban built environment. Köhler, Turnheim, and Hodson (2018) explain how the idea of a 'bridging process' (Turnheim et al. 2015) is applied to combine transitions case studies with the modelling of transitions pathways using the MATISSE model. In Köhler, Turnheim, and Hodson (2018) the results are quantitative pathways (as in Figure 6.2) combined with scenario narratives that are developed in an iterative process with the modelling to depict consistent qualitative and quantitative transition pathway scenarios.

3.2 Low-emissions shipping

This example introduces the application of this model structure to transitions pathways for low-emissions shipping in the MATISSE-SHIP model. Emissions from shipping are a significant part of global GHG emissions (around 2–3%: (Nelissen et al. 2016; Köhler, Nelissen, and Traut 2017) and are also a major contributor to SO_2 and NO_x emissions in coastal areas and harbours (Köhler, Kirsch, and Timmerberg 2018). Furthermore, ships are large and infrequent investments, such that change in the industry through the introduction of new technology is slow. Most emissions happen outside national waters and regulations for international waters, where most GHGs from ships are emitted, are implemented through international agreements in the International Maritime Organisation (IMO). SO_2/NO_x emissions that affect air quality are important in coastal regions and harbours, such that national regulations are effective.

This socio-technical-environmental system has different characteristics and structure compared to households' mobility. In terms of decisions in the system, it is more limited in scope. The main relevant decision is the choice of powertrain technologies in new ships, which can also have implications for ship operations and speeds. Ship operating speeds are a vital factor in fuel use, and hence emissions for fossil fuels. At higher speeds for conventional ships, the wave-making resistance increases according to a power law, such that for small container ships, an increase in operating speed from 15 knots (nautical miles per hour) to 17.6 knots doubles the power requirement (MAN 2011).

The market for ships is global with relatively few ships in comparison to land transport vehicles (102,154 registered vessels: IMO 2014). Diesel engines dominate all markets. The only established niche in commercial shipping is liquid natural gas (LNG), which is used in LNG carriers and is being adopted for cruise vessels operating in areas with SO_2/NO_x regulation. Diesel engines can be combined with other prime movers in electric drive trains in cruise vessels, platform service vessels and other specialist service vessels and tugs (Köhler, Kirsch, and Timmerberg 2018). Because ships are large, there is a very wide range of propulsion technologies that are feasible in engineering terms, ranging from atomic power (in aircraft carriers, icebreakers and military submarines) to sails. LNG, compressed natural gas (CNG), other low-flashpoint liquid and gaseous fuels (LfL such as ethanol, methanol, ammonia, dimethyl ether) are all in operation in commercial ships. LNG and LfL fuels are commercially viable for specialist LNG and gas carriers. Hydrogen in combustion engines is feasible and hydrogen fuel cells are in operation in military submarines. Batteries (and shore-supplied electricity in ports or 'cold-ironing') are in commercial operation (DNV-GL 2018; Köhler, Kirsch, and Timmerberg 2018).

The other fundamental difference to the mobility case is that shipping is an intermediate rather than final demand. The supporters, agents making decisions about which types of power system to adopt, are ship owners or shipping lines and not households. Therefore, there is a higher priority placed on the economics of operation and a lower impact of the culture of the technology on investment decisions. However, there are also non-cost barriers to the adoption of alternative technologies. Infrastructure for alternative fuels is perceived to be a major barrier to alternative fuels (Köhler, Nelissen, and Traut 2017); the perceived risks of adopting new technologies are also a major barrier (Nelissen et al. 2016; Köhler, Nelissen, and Traut 2017).

Therefore, the use of the MATISSE structure is appropriate, as it allows for the representation of differing decisions by a distribution of supporter agents and for more decision variables than just cost and emissions. The MATISSE structure of landscape, regime, niches and supporters is easily interpreted in terms of entities in the shipping socio-technical system. The practices of the regime and niches are defined in terms of propulsion technologies, analogous to ICE cars and the alternative powertrains in the mobility example. The supporters are shipping lines who order ships for different markets or 'trades'. As with the household supporters in the mobility case, the shipping lines are assigned in groups representing market segments with different behaviours (cruise and ferry, LNG passenger, bulker/tanker, wind-assisted bulker, service: PSV/tugs/fishing vessels, LNG services) over the practice space, with a normal distribution of these supporters in a group around a central value for the group.

For the case of shipping, the regime is considered to be propulsion by diesel motors. The alternatives considered for niches are:

- Liquid natural gas, including dual- and triple-fuel engines (LNGDF)
- LfL fuels
- Hydrogen
- Wind
- Biofuels
- Synthetic fuels (so-called power-to-liquid, PtL)

Atomic power is not included as it is not currently under active consideration in the industry (Köhler, Kirsch, and Timmerberg 2018).

From the above discussion, practices included in the MATISSE-SHIP model are:

- GHG intensity (CO_2 emissions/tonne km or passenger km)
- Fuel cost Euro/tonne
- Operational speed (adoption of slow steaming to reduce power requirements and emissions)
- Local air emissions/tonne km or passenger km (NO_x, SO_2, particles)
- Capital cost/Megawatt
- Technological novelty, requiring new engineering and management skills and the need for operational change (reduced operational speed and weather-optimised routing, even if this involves extended transit times compared to diesel propulsion at contemporary operational speeds)
- Requirement for new bunker infrastructure.

The model is initialised with data from IMO (2014) for the numbers of ships in 2012 and (Nelissen et al. 2016; Köhler, Kirsch, and Timmerberg 2018) for costs and the other practice variables. The two largest groups of ship owners operate or charter bulker/tanker and service/fishing vessels.

The model starts from 2015 with only a few LNGDF demonstration vessels apart from the LNG carrier fleet. Shipping lines face a change in regulation from 2020, when a global limit on SO_2 emissions agreed at the IMO enters into force. This is included in the model as a rapid change of practices of ship owners towards lower local air emissions. This means that large numbers of ships have to change, such that in the results in Figure 6.3, the capital cost

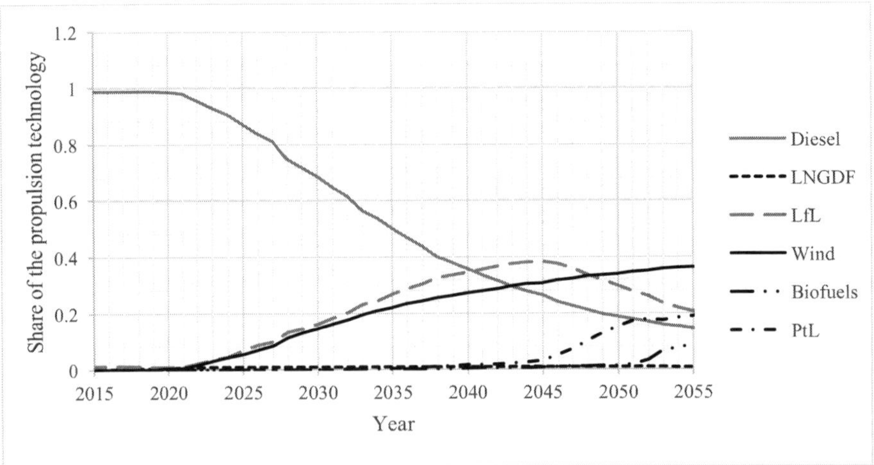

Figure 6.3 A scenario of a transition pathway to wind propulsion in shipping using the MATISSE-SHIP model.

advantage of LfL proves to be decisive. This can be explained in terms of the investment decisions of ship owners as follows.

The current assessments in the industry are that LNGDF, which is mainly methane and has almost no SO_2 emissions, will be the main alternative to HFO and diesel fuels. However, LNG has to be cooled for storage to −162°C. This requires specialised tanks and fuel-handling systems, which are as expensive in total as the engines themselves. LfL fuels do not have to be cooled to these extreme temperatures such that although they do require special handling, they have a certain cost advantage over LNG. LNG is currently popular in the industry as an alternative because it is an established operational technology in LNG carriers (Köhler, Kirsch, and Timmerberg 2018). Therefore, LfL in the model is calibrated to have a slightly lower initial value for capital cost.

LfL fuels, like LNG, already have an industrial supply chain, which can be extended to bunkering fuel logistics chains for ships. An LNG supply chain is already under development, but is at an early stage given the small number of ships that require it so far (Köhler, Kirsch, and Timmerberg 2018). The technologies involved can easily be adapted to other LfL fuels, such that the lack of LfL ships is not a major disadvantage compared to LNGDF ships (LNG carriers burn fuel that evaporates from the main cargo, so do not require a bunker infrastructure). This means that in the model the LNG and LfL niches are given a similar desired practice value for new bunker structure.

The scenario in Figure 6.3 assumes that fossil fuel prices increase such that the operating cost of diesels increases and that more effective regulation of GHG emissions is introduced through the period from 2020 to 2030. The alternative fuels to diesel therefore become more competitive in terms of operating costs and in terms of the desired levels of GHG intensity. The main disadvantages of wind are that sailing ships cannot sail as fast as ships with engines for propulsion and cannot be as large as conventional ships. Therefore, more, smaller ships are required compared to diesel-powered ships, which will increase crew costs and also investment and administrative costs compared to fewer, larger ships. The variability of the winds means also that their operation is more complex and the voyage times have a wider range than conventionally powered ships. Overall, these effects are expected to be more than offset by the reduction in fuel costs, such that in the model wind is assigned a lower operating cost than all other propulsion systems. These factors prove to be enough to overcome the disadvantage of wind – the reluctance of ship owners to adopt a very different technology involving operational change in this scenario. The resulting scenario sees a slow decrease in diesel new builds. There is development of LfL instead of LNGDF, because it is cheaper than diesel and reduces NO_x and SO_x emissions. Wind is also adopted, starting with the bulker and tanker markets operating with smaller ships on routes with favourable winds (e.g. coastal routes).

In the long run, GHG emissions reduction is assumed to become more important compared to SO_2 and NO_x reductions. This is implemented as a further reduction in the desired levels of CO_2, such that emissions-free wind propulsion expands. Other carbon neutral fuels (carbon-neutral electricity to fuel PtL and biofuels are also adopted for markets where higher speed is very important, such as ferries).

4 Discussion: the contribution of the MATISSE models

4.1 Understanding transitions

The objective of the MATISSE model is to investigate the processes of radical change – transitions – in socio-technical systems. It uses an agent-based approach to allow for differences in decision-making between different agents. This allows the differences in adaptation strategy to changing market conditions and regulation between the niche(s) and the regime to be represented, which is an important point in the MLP theoretical framework. Also, in contrast to most eco-innovation models, it allows for a range of choice behaviours on the demand side. This is important because niches are established and grow through addressing the needs or preferences of user groups (supporters in the MATISSE structure) which are significantly different to the majority of users. Cian et al. (2018) discuss the limitations of representative consumer based choice models for representing actors and institutions in quantitative systems models for climate change policy.

The model does not represent the detailed internal structure of what constitutes a regime or a niche. However, it does represent their interactions and strategies of change in terms of the technology and/or service system provided. It also represents the structure of demand in a detailed way. This enables the dynamics of transition pathways to be analysed in terms of the choice of technologies or of more general choices (e.g. mobility lifestyles) and the dynamics of their overall development. These co-evolutionary processes are very complex, even with the simplified representation of the niches and regimes used. The models generate complex patterns of competition and growth of niches and regime in a way not captured by most other models of technological diffusion or change.

The two examples presented demonstrate that the MATISSE structure can be applied to different socio-technical systems where niches and a regime can be identified. The inclusion of decision variables covering more than the cost and technical performance enable the non-cost barriers to change (e.g. perception of the need to adopt new practices in using the system) and broader factors such as the characteristics of the built environment in the mobility example to be addressed. This is not possible with techno-economic models, even if they do have a complex, non-linear structure, such as the recent UK transport model UKTCM (Brand, Tran, and Anable 2012).

The examples also demonstrate that such models can be calibrated to empirical data, even though the methodology of interpreting data for qualitative variables requires a careful and structured analysis, as described in (Köhler, Turnheim, and Hodson 2018). Bianchi and Squazzoni (Chapter 5 in this volume) emphasise the difficulty of calibrating models of social processes to empirical data.

The results generated by the MATISSE structure require careful interpretation, because the niche and regime categories do not have an established structure in the transitions literature. Commonly accepted indicators for the state or performance of niches and regimes have not yet been established (STRN 2017). Therefore, the meaning of the practice variables, and indeed the structure of the niches and regime, is not completely common between the different examples and therefore detailed comparison of the two cases can only be limited. Geels and Schot (2007) propose a typology of patterns of transition pathways and the results of the MATISSE model can be mapped into these overall patterns.

The model does not predict costs and the output quantities of shares of lifestyle in the mobility model, and technologies in the shipping model require careful interpretation as well. Even in the relatively limited technological case of shipping the aggregation of technologies to an average performance for a niche or a regime covers a wide range of actual units (e.g. from very large tankers of up to 500,000 tonnes deadweight down to 1000 tonnes or less). An implication of this is that the overall size of the system in terms of, for example, total transport volumes or total emissions is not directly calculated.

The MATISSE structure of multiple practice dimensions to represent different aspects of demand-side choices and the features of a regime or niche allows for a broader approach to decision-making than just the costs and emissions performance variables of eco-innovation models. While costs and emissions are fundamental to choices and policy-making, they are not the only factors that have an impact on choices. Lifestyle and cultural factors may also directly affect decisions. Such factors are not included in the microeconomic choice theory used in (almost all) eco-innovation models (Cian et al. 2018). An exception is Van Sluisveld et al. (2016), which analyses the impacts of lifestyle changes using the IMAGE integrated assessment model for climate change policy analysis.

The MATISSE models have been shown to be capable of capturing complex transition dynamics in a way that is not possible with most quantitative system models. This is because the MATISSE model structure is co-evolutionary and because both the supply side and demand side change and interact in the co-evolutionary system. This means that more complex patterns of growth and decline of socio-technical systems can be modelled than with most eco-innovation models (Köhler 2019). In particular, processes of competition between niches are simulated as well as the niche–regime interaction. While the overall results may be a typical diffusion 'S-curve' (Köhler et al. 2018), this competition between niches as well as with the regime may determine the overall outcome of a transition.

4.2 How the MATISSE model is used to generate scenarios for transition pathways

Elsewhere in this book (see Chapter 13 by Moallemi, de Haan and Köhler in this volume), the distinction is made between simulation as prediction (model with future scenario of the exogenous rest of world calculates the system development) versus exploratory simulation (model results found over parameter ranges and many scenarios to generate a portfolio of system developments). The MATISSE model is applied in a somewhat different way in the present examples. The model is intended to search for conditions under which a transition to sustainability can take place. The MATISSE model has a co-evolutionary structure with the niches and regime changing their practices to affect the supporters' decisions as well as the supporters' decisions being affected by the changes in supporters' desired preferences. Therefore, there are a very wide range of possible dynamics and outcomes in the model. This very wide range is restricted as follows.

The model is initialised to recent data (currently 2015). This includes assumptions of how the supporters' desired practices will change in the future. These variables cannot be calibrated on empirical data, because behaviours are not only strongly path dependent but are also strongly contingent on future events (perception of climate change in society, changes in cultures, patterns of economic growth, etc.) The rate of change of desired practices is varied until a 'plausible' rate of decline of the regime is generated. This plausibility as assessed from the historical rate of turnover of the stock of technology. This is assumed to remain similar (there is a random element in the decision-making of supporters), which determines the approximate feasible rate of adoption of new technologies or in the mobility example lifestyles.

However, there are still a wide range of outcomes of the model that do include a transition, so the single scenario runs for these two example models can only be regarded as an illustration of a possible outcome, rather than a statement of what is probable in the future. Moallemi and Köhler (in press 2019) address this through an exploratory approach to find and describe ranges of parameterisations and the resulting ranges of transitions pathway scenarios.

This can be considered as a partial backcasting approach, in which a fundamental aspect of the state of the model at the end of the number of time steps considered is assumed – a transition has taken place. However, the state of the different niches is not assumed, and so the model runs are an open exploration of the transition dynamics.

The model could also be run in a conventional forward-looking mode for a simulation model, with the empirically calibrated initial state and the scenario developed as a set of assumptions about the exogenous changes in culture, policies, economics, etc.

An extension of this approach could be the use of the model to explore the properties of the co-evolutionary system that is being modelled. Holland (1995) proposed that simulation models of complex adaptive social systems could be used for 'policy flight simulators' in a gaming mode to develop understanding of possible impacts of modellers' choices about, for example, policy assumptions. An exploration of the outcomes from changes in practices of the supporters or the niches/regime could be developed from narrative scenarios of policy changes in climate policy for a carbon price. These could lead to changes in the relative costs of the regime fuel(s), and the niche alternatives that could be used to develop an understanding of the patterns of policy response of the model. The problem that Holland's suggestion does not address is that there are very many possible outcomes from such open models. A 'gaming' approach, with manual limited changes in decision parameters, will not provide a clear understanding of the behaviour of the model except over a very limited parameter space. This could be addressed by using the exploratory approaches discussed above and in Chapter 13 by Moallemi, de Haan and Köhler in this volume.

4.3 Limitations of the MATISSE models

The MATISSE model structure, like the MLP framework on which it is based, has been criticised for not being detailed enough (STRN 2017). It does not explain the detailed structures of niches or the regime, and therefore does not provide a detailed explanation of the social and technical processes of change. A corollary of this is that the models can only be approximately calibrated to historical data. The results of the model runs must therefore be interpreted as illustrations of possible futures, rather than predictions which are expected to be close to what will actually happen.

A further criticism is common to all agent-based models (ABMs) of social systems. The large number of decision makers all have decision-making parameters that cannot be objectively measured, but can only be inferred from observed behaviour (see also Chapter 7 by Holtz and Chappin in this volume). While this is true of preferences in economic models, the large number of choice agents with different parameters make the problem more important in ABMs. While there are a few ABMs that have taken extensive surveys to estimate the distributions of psychological decision parameters, the MATISSE approach is to initialise the decision parameters based on historical choices and then explore possible changes to these parameters as scenarios.

5 Summary and future directions

The way in which the MATISSE model was developed to implement the MLP framework for analysing transitions has been described. Two applications: a transition to households' sustainable mobility and a transition to low-emissions shipping have been illustrated. These two examples involve different interpretations of what a regime and niche are. The sustainable mobility case looks at households' choice of mobility lifestyles, involving not just transport modes, but also the geography of town planning, ways to reduce transport demand and attitudes towards public and private ownership of transport. The low-emissions shipping case involves a technology transition, although operational aspects of shipping are also considered. Therefore, the MATISSE structure has been shown to be adaptable to different cases.

Two possibilities for further extension of the MATISSE framework have been mentioned in the discussion. Firstly, the development of a facility for exploratory modelling, to allow structured exploration of the many structural uncertainties in both the model parameterisations and in the alternative potential scenarios generated by the models.

Learning curves for technology development are also important, as this is a fundamental feature of the dynamics of technology innovations and their uptake. It is a dynamic through which initially uncompetitive niche technologies reduce their costs and improve their performance to compete with established technologies. Finally, since the models are often designed to analyse the possibility of transitions to low-emissions systems, calculations of the emissions and their reductions would be useful for the development of policy lessons.

Another idea is to contribute to the development of indicators that enable comparisons between different transitions case studies, possibly through identifying those practices that are common between cases as a starting point. New insights might be developed by analysing the conditions under which the perceived novelty of radically different technologies is not a barrier that prevents the growth of a niche. Finally, the MATISSE model framework can be extended by including interactions between different regimes and also by a more complex system representation of niches and regimes. The processes for amalgamation between niches or niche and regime could be examined in future research on niche and regime strategies and interactions.

The MATISSE model has therefore considerable potential for development and for contributing to the development of the field of transitions modelling.

Acknowledgements

The contribution of the EU FP6 MATISSE project and PATHWAYS FP7 project (grant no. 603942) to the funding and development of the MATISSE-KK model is acknowledged.

Note

1 The two versions of the model described here do not include a learning curve. Technological development is endogenous through the decisions of the regime and niches to improve, for example, their emissions or cost characteristics.

References

Bergman, Noam, Alex Haxeltine, Lorraine Whitmarsh, J. Köhler, Michel Schilperoord, and J. Rotmans. 2008. "Modelling socio-technical transition patterns and pathways." *Journal of Artificial Societies and Social Simulation* 11 (3): 7. http://jasss.soc.surrey.ac.uk/11/3/7.html.

Bianchi, Frederico, and Flaminio Squazzoni. 2019. "Modelling and social science: Problems and promises." In *Modelling Transitions – Virtues, Vices, Visions of the Future*, edited by Enayat Moallemi and Fjalar de Haan. This volume: Routledge.

Brand, Christian, Martino Tran, and Jillian Anable. 2012. "The UK transport carbon model: An integrated life cycle approach to explore low carbon futures." *Energy Policy* 41: 107–24.

Cian, Enrica de, Shouro Dasgupta, Andries F. Hof, Mariësse A.E. van Sluisveld, Jonathan Köhler, Benjamin Pfluger, and Detlef P. van Vuuren. 2018. "Actors, decision-making, and institutions in quantitative system modelling." *Technological Forecasting and Social Change*. http://doi.org/10.1016/j.techfore.2018.10.004.

DNV-GL. 2018. "Assessment of Selected Alternative Fuels and Technologies." Hamburg. www.dnvgl.com, checked on 11/12/2018.

Egbue, Ona, and Suzanna Long. 2012. "Barriers to widespread adoption of electric vehicles: An analysis of consumer attitudes and perceptions." *Energy Policy* 48: 717–29.

Freeman, Chris, and Francisco Louçã. 2002. *As Time Goes by: From the Industrial Revolutions to the Information Revolution*. Oxford: Oxford University Press.

Geels, Frank W. 2005. "The dynamics of transitions in socio-technical systems: A multi-level analysis of the transition pathway from horse-drawn carriages to automobiles (1860–1930)." *Technology Analysis & Strategic Management* 17 (4): 445–76. http://doi.org/10.1080/09537320500357319.

Geels, Frank W., and Johan Schot. 2007. "Typology of sociotechnical transition pathways." *Research Policy* 36 (3): 399–417. http://doi.org/10.1016/j.respol.2007.01.003.

Geels, Frank W., and Johan Schot. 2010. "The dynamics of socio-technical transitions: A socio-technical perspective." In *Transitions to Sustainable Development: New Directions in the Study of Long Term Transformative Change*, edited by John Grin, Jan Rotmans, and J.W. Schot. Routledge studies in sustainability transitions. New York: Routledge.

Halbe, Johannes. 2019. "Participatory modelling in sustainability transitions research." In *Modelling Transitions – Virtues, Vices, Visions of the Future*, edited by Enayat Moallemi and Fjalar de Haan. This volume: Routledge.

Halbe, Johannes, Dominic Reusser, Georg Holtz, Marjolijn Haasnoot, Annette Stosius, Wibke Avenhaus, and Jan Kwakkel. 2015. "Lessons for model use in transition research: A survey and comparison with other research areas." *Environmental Innovation and Societal Transitions* 15: 194–210.

Haxeltine, Alex., Lorraine Whitmarsh, Noam Bergman, Jan Rotmans, Michel. Schilperoord, and Jonathan Köhler. 2008. "A Conceptual Framework for transition modelling." *International Journal of Innovation and Sustainable Development* 3 (1–2): 156–61.

Hensher, David A., and Kenneth J. Button. 2007. *Handbook of Transport Modelling*. Vol. 1. Bingley: Emerald Group Publishing Limited.

Holland, John Henry. 1995. *Hidden Order: How Adaptation Builds Complexity: Ulam Lectures Series*. Cambridge, MA: Helix Books/Perseus Books.

Holtz, Georg, Floortje Alkemade, Fjalar de Haan, Jonathan Köhler, Evelina Trutnevyte, Tobias Luthe, Johannes Halbe, et al. 2015. "Prospects of modelling societal transitions: Position paper of an emerging community." *Environmental Innovation and Societal Transitions* 17: 41–58.

Holtz, Georg, and Emile J.L. Chappin. 2019. "Considering actor behaviour: Agent-based modelling of transitions." In *Modelling Transitions – Virtues, Vices, Visions of the Future*, edited by Enayat Moallemi and Fjalar de Haan. This volume: Routledge.

IMO. 2014. *Third IMO GHG Study 2014: MEPC 67/INF.3*. www.imo.org/en/OurWork/Environment/PollutionPrevention/AirPollution/Pages/Greenhouse-Gas-Studies-2014.aspx.

IPCC special Report on global warming of 1.5 °C: Special Report. 2018. Genève: IPCC. www.ipcc.ch/sr15/.

Köhler, Jonathan. 2006. "Transport and the environment: The need for policy for long term radical change: A literature review for the DTI FORESIGHT project on Intelligent Infrastructure Systems." *IEE Proceedings Intelligent Transport Systems* 153 (4): 292–301.

Köhler, Jonathan. 2012. "A comparison of the neo-Schumpeterian theory of Kondratiev waves and the multi-level perspective on transitions." *Environmental Innovation and Societal Transitions* 3: 1–15.

Köhler, Jonathan. 2019. "Advances in modelling sustainable innovation: From technology bias to system theories and behavioural dynamics." In *Handbook of Sustainable Innovation*, edited by F. Boons and A. McMeekin. Abingdon: Routledge.

Köhler, Jonathan, Fjalar de Haan, Georg Holtz, Klaus Kubeczko, Enayat Moallemi, George Papachristos, and Emile Chappin. 2018. "Modelling sustainability transitions: An assessment of approaches and challenges." *Journal of Artificial Societies and Social Simulation* 21 (1). http://doi.org/10.18564/jasss.3629.

Köhler, Jonathan, Frank W. Geels, Florian Kern, Jochen Markard, Elsie Onsongo, Anna Wieczorek, Floortje Alkemade, et al. 2019. "An agenda for sustainability transitions research: State of the art and future directions." *Environmental Innovation and Societal Transitions*. http://doi.org/10.1016/j.eist.2019.01.004.

Köhler, Jonathan, Daniela Kirsch, and Sebastian Timmerberg. 2018. *Teilstudie "Studie über die Marktreife von Erdgasmotoren in der Binnen- und Seeschifffahrt"*. Wissenschaftliche Beratung des BMVI. www.bmvi.de/SharedDocs/DE/Artikel/G/MKS/studie-marktreife-erdgasmotore-schifffahrt.html?nn=214206.

Köhler, Jonathan, Dagmar Nelissen, and Michael Traut. 2017, May. "Fighting the windbreak." *The Naval Architect*: 26–32.

Köhler, Jonathan, Bruno Turnheim, and Mike Hodson. 2018. "Low carbon transitions pathways in mobility: Applying the MLP in a combined case study and simulation bridging analysis of passenger transport in the Netherlands." *Technological Forecasting and Social Change*. http://doi.org/10.1016/j.techfore.2018.06.003.

Köhler, Jonathan, Lorraine Whitmarsh, Björn Nykvist, Michel Schilperoord, Noam Bergman, and Alex Haxeltine. 2009. "A transitions model for sustainable mobility." *Ecological Economics* 68 (12): 2985–95. http://doi.org/10.1016/j.ecolecon.2009.06.027.

Laver, Michael. 2005. "Policy and the dynamics of political competition." *American Political Science Review* 99 (2): 263–81. http://doi.org/10.1017/S0003055405051646.

Li, Francis G.N. 2017. "Actors behaving badly: Exploring the modelling of non-optimal behaviour in energy transitions." *Energy Strategy Reviews* 15: 57–71.

Li, Francis G.N., and Neil Strachan. 2017. "Modelling energy transitions for climate targets under landscape and actor inertia." *Environmental Innovation and Societal Transitions* 24: 106–29.

Li, Francis G.N., Evelina Trutnevyte, and Neil Strachan. 2015. "A review of Socio-Technical Energy Transition (STET) models." *Technological Forecasting and Social Change* 100: 290–305.

Loorbach, Derk, and Jan Rotmans. 2010. "Towards a better understanding of transitions and their governance: A systemic and reflexive approach." In In *Transitions to Sustainable Development: New Directions in the Study of Long Term Transformative Change*, edited by John Grin, Jan Rotmans, and J.W. Schot. Routledge studies in sustainability transitions. New York: Routledge.

MAN. 2011. *Basic Principles of Ship Propulsion*. https://spain.mandieselturbo.com/docs/librariesprovider10/sistemas-propulsivos-marinos/basic-principles-of-ship-propulsion.pdf?sfvrsn=2.

Moallemi, Enayat, Fjalar de Haan, and Jonathan Köhler. 2019. "Exploratory analysis of transitions: An emerging approach for coping with uncertainties in transitions research." In *Modelling Transitions – Vices and Virtues, Lessons and a Look Ahead*, edited by Enayat Moallemi and F. de Haan. This volume: Routledge.

Moallemi, Enayat A., and Jonathan Köhler. 2019. "Coping with uncertainties of sustainability transitions using exploratory modelling: The case of the MATISSE model and the UK's mobility sector." *Environmental Innovation and Societal Transitions*. http://doi.org/10.1016/j.eist.2019.03.005.

Nelissen, D., M. Traut, J. Köhler, W. Mao, J. Faber, and S. Ahdour. 2016. *Study on the Analysis of Market Potentials and Market Barriers for Wind Propulsion Technologies for Ships*. www.cedelft.eu/publicatie/study_on_the_analysis_of_market_potentials_and_market_barriers_for_wind_propulsion_technologies_for_ships/1891.

Rotmans, Jan. 2005. "Societal innovation: Between dream and reality lies complexity." *DRIFT Working Paper*. http://doi.org/10.2139/ssrn.878564.

Rotmans, Jan, and Derk Loorbach. 2009. "Complexity and transition management." *Journal of Industrial Ecology* 13 (2): 184–96. http://doi.org/10.1111/j.1530-9290.2009.00116.x.

STRN. 2017. *Research Agenda for the Sustainability Transitions Research Network*. https://transitionsnetwork.org/about-strn/research_agenda/.

Turnheim, Bruno, Frans Berkhout, Frank Geels, Andries Hof, Andy McMeekin, Björn Nykvist, and Detlef van Vuuren. 2015. "Evaluating sustainability transitions pathways: Bridging analytical approaches to address governance challenges." *Global Environmental Change* 35: 239–53.

van Sluisveld, Mariësse A.E., Sara H. Martínez, Vassilis Daioglou, and Detlef P. van Vuuren. 2016. "Exploring the implications of lifestyle change in 2 °C mitigation scenarios using the IMAGE integrated assessment model." *Technological Forecasting and Social Change* 102: 309–19.

Publication bibliography

DNV-GL (2018): Assessment of selected alternative fuels and technologies. Hamburg. Available online at www.dnvgl.com, checked on 11/12/2018.

Appendix

Explanation of the practice space in the MATISSE model

Adapted from Bergman et al. (2008).

Figure 6.A1 schematically shows a two-dimensional practices space, which might be, for example, Px CO2 emissions and Py cost of transport. The complex agents (regime and niches) and the supporter agents are shown separately for clarity, but actually occupy positions along the same Px and Py practices axes. Supporter agents are points in the space, while in the figure the size of the regime and niche ovals is proportional to their relative support. The model is stochastic in that the simple agents are initially assigned over the practice space, with grouped distributions (e.g. car drivers, a smaller set of 'green mobility' households). There is also a random element in the initial distribution to allow for variation in agents' behaviours. In the model the supporter agents choose to adopt the practices (transport characteristics in the current example) of the

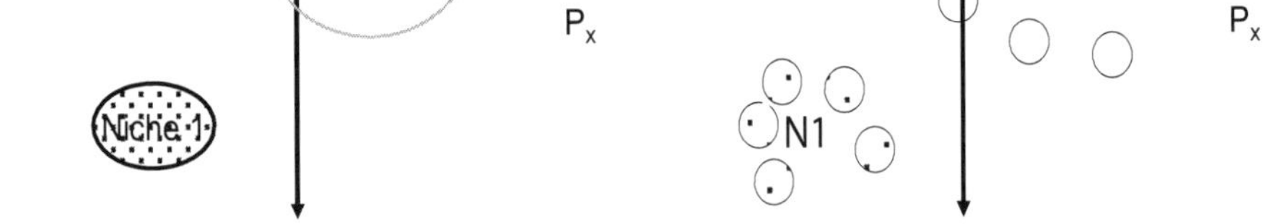

Figure 6.A1 Two illustrations of a two-dimensional practices space, with practice axes PX and PY.

Left: regime and niches, which can move in the space and interact with each other. Right: the consumer agents showing supporters scattered in the practices space, patterns show the agent they support, clear = regime (R), diagonal lines = niche 2 (N2), points = niche 1 (N1).

niche or regime agent that is closest to them. The distance from the supporter's position to that of the niche or regime under consideration is calculated for all of the practice dimensions and these distances are summed as a vector to calculate an overall vector distance between the supporter and the niche or regime in the (multi-dimensional) practice space. The positions of the supporters in the practices space change depending on landscape signals, so the regime and niches have to move, not only to grow, but often just to maintain their support. The movement strategies, that is, the development strategies of the regime and niches, are explained in Bergman et al. (2008)

7 Considering actor behaviour

Agent-based modelling of transitions

Georg Holtz and Émile J. L. Chappin

1 Introduction

A transition is a fundamental shift in a socio-technical system such as the energy, mobility or food system that emerges from co-evolutionary developments in the social, technical, political, regulatory and market domains. Transitions are large-scale processes that span up to several decades and take place across nested spatial levels ranging from the local to the global levels. It is a widely shared conviction of transition researchers that transitions are open, path-dependent processes with uncertain outcomes, and not pre-determined (e.g. by technological determinants). That is, transitions essentially are driven by and emerge from actors' decisions, behaviours and interactions, whereas the pace and direction of transitions are beyond the control of single actors or small actor groups. Furthermore, a broad range of institutions[1] permeate socio-technical systems and form the (changing) context in which those actors act.

Boxes 7.1 and 7.2 illustrate the broad range of actors and institutions involved in a transition, using the example of the transition of the German electricity system towards a mostly renewables-based electricity system.[2] The involved actors are not only power-producing companies, but also technology developers, grid operators, large and small consumers (and prosumers), policy makers, regulators, traders, etc. All of them take decisions from their respective perspectives, and their behaviours are driven by their particular interests and motivations. They operate in many markets (various electricity markets, fuel markets, emission markets) which are governed by a complex institutional context that provides frames for actors' perspectives, and guides and constrains their behaviours. Furthermore, systems are strongly interconnected (electricity being a backbone for other infrastructures and services) and span national borders, which leads to cross-policy and cross-border effects (Chappin et al. 2017).

Box 7.1 Actors in the German electricity transition

To give a non-exhaustive list, the German electricity system is governed and operated by the following actors: big utility companies operating mostly large-scale electricity production plants; small-scale plant owners, including private households, farmers and cooperatives; regional supply companies; municipality utilities; a variety of consumer types ranging from private households to large industrial consumers; transmission grid operators; distribution grid operators; specific actor groups who operate at electricity markets and the interface between markets and the physical system, such as electricity brokers, balancing group managers and aggregators; power plant producers, craftsmen and technicians; the grid regulator; legislators and executive authorities on local, regional, national and European levels; capital suppliers; insurance firms; a variety of associations representing different interest groups; NGOs; community-level initiatives; research institutes and universities. The apparent need for deeper integration of different energy sectors – in particular integration of the electricity sector with the heat, mobility and industry sectors – further increases the range of actors and (diverging) interests that shape the electricity system.

Box 7.2 Institutions in the German electricity transition

Using the classification of Scott (2001) and following up on the example of the German electricity transition, institutions include *regulative institutions*, examples of which are: technical standards for power generation units, grid operation and appliances; market access and market clearance rules for the different electricity markets; regulations and procedures that safeguard grid stability, such as re-dispatch of power plants and provision and utilisation of balancing energy; and taxes, fees and subsidies (e.g. feed-in tariff) that shape the economic calculations of providers and consumers. *Normative institutions* are less 'hard coded', but influence actors' behaviour as their behaviour is subject to scrutiny by others who hold particular expectations about those behaviours. Examples include investors and companies' expectations about capital return rates for investments; engineers' rules of 'good practice'; and behavioural norms such as to turn off the light when leaving a room. Last, but not least, *cultural-cognitive institutions*, covering the 'taken-for-granted' common beliefs and shared understanding that create frames through which meaning is made, shape actors' behaviour in more subtle yet fundamental ways. Examples include: provision of electricity always follows demand[1]; nuclear power will be phased out and renewables will form the core of a future sustainable electricity system; electricity is a marketable good; and other guiding principles such as a liberalised market.

[1] This frame was deeply ingrained in the fossils-based electricity system, but is meanwhile revised, and the role of 'demand side management' in a future electricity system is widely accepted.

From this perspective, a 'transition' may emerge as an outcome of the decisions, behaviours and interactions of the multitude of actors involved. A better understanding of the emergent dynamics that result from the interactions of those actors with their respective powers and interests in the specific institutional context would not only be of scientific interest, but support the steering of transitions towards vital societal goals for maintaining and increasing societal welfare in the coming decades.

Agent-based modelling (ABM) is a modelling method whose particular strength is representing heterogeneous actors and their interactions, and letting systemwide structures and dynamics emerge from those actors' interactions. Behaviours of actors are represented through 'decision rules', whose designs are flexible. A rich set of modelling practices has emerged, and this provides for the possibility to represent a diversity of decision-making modes and rich institutional contexts. The strength of ABM – simulating heterogeneous agents making decisions and interacting with each other – fits well with the understanding of transitions as emerging from the interactions of heterogeneous actors in an institutional context. This makes ABM a key approach to modelling transitions.

The abundance of different actors and institutions shaping the transition process and the multitude of theoretical perspectives for describing and understanding them, however, poses considerable challenges to the design of ABMs.

In this chapter, we provide a brief introduction to ABM, investigate the conceptualisation of actors and institutions involved in transitions, examine the challenges that arise from their multitude for agent-based modelling of transitions and discuss four different approaches to tackle these challenges. We point out examples for each approach.

2 Agent-based modelling

ABM allows for the study of complex systems from a bottom-up perspective and has been applied in a broad range of disciplines, ranging from molecular self-assembly, biology and ecology to economics, sociology, anthropology and cognitive science (Macal and North 2010). In the following we briefly outline some core characteristics of ABM. We refer the interested reader to Bianchi and Squazzoni (Chapter 5 in this volume), who provide a more extensive overview and a taxonomy of using agent-based models in the social sciences.

Agent-based models define elements – the agents – that interact with each other and their environment. Agents thereby may be heterogeneous in their attributes and behavioural rules. When applied to social systems, at their core, agent-based models consist of agents that represent actors that interact with each other and their environment. Simulation of agents' interactions over time then allows for generating emergent phenomena on the level of a group, organisation or other collection of actors – be they spatial or temporal patterns or characteristic statistical distributions of variables of interest – based on the boundedly rational behaviour and social interactions of these actors (Gilbert and Troitzsch 2005; Heckbert, Baynes, and Reeson 2010).

As such, ABM provides a means to identify possible explanations for emergent system-level behaviour by identifying underlying mechanisms, that is, by providing a causal reconstruction of the processes that account for the emergent phenomenon (Epstein and Axtell 1996), and hence to enhance understanding of system behaviour. Used in this way, agent-based models can also contribute to theory building (Chapter 4 by de Haan in this volume). Furthermore, agent-based models are also a promising approach for policy support (Chappin 2011; Farmer and Foley 2009). As Bankes (2002) justifies, ABM can be used for policy support because of (1) the unsuitability of competing modelling formalisms to address the problems of social science, (2) agents as a natural ontology for many social problems and (3) its power to demonstrate emergent phenomena (i.e. to relate actors' behaviours and system-level developments). Besides understanding and policy support, ABM can be used for other purposes, such as developing an explicit and systematic description of a complex socio-technical system. Different purposes come with different challenges and data needs. It is, however, beyond the scope of this chapter to discuss those in detail.

The ABM method does not impose any restrictions concerning the level of abstraction of the analysis. It is suitable for modelling a particular case in great detail, as well as for developing a model that abstracts from a wide range of empirical phenomena that share some common features (Boero and Squazzoni 2005; Sun et al. 2016). Furthermore, agent-based models can be developed for a variety of empirical scales, ranging from small groups up to the supra-national level. ABM is therefore applicable to a broad range of research questions that relate to various empirical scales. Concerning the modelling guideline, that model design should follow the model's purpose, it can be observed that the flexibility of the ABM method facilitates a vast range of model designs; that is, the method does not per se impose strong limitations to tailoring the model to fulfil a specific purpose. There are, however, delicate interdependencies between a model's purpose, its design and resulting complexity, data availability and validity.

We conclude that, in principle, ABM is well suited for developing a bottom-up representation of a transition, and in particular for the representation of social aspects (heterogeneous actors, various types of decisions) that are more difficult to grasp with other modelling methods (Hoekstra, Steinbuch, and Verbong 2017). Furthermore, ABM allows for the study of the co-evolution of technical and social elements of a system and facilitates tracing back transition dynamics to actors and their interactions.

3 Conceptualisation of actors in transitions

An essential step for agent-based modelling of transitions is to define agents, based on a selection of the real-world social entities they shall represent. Often, and most intuitively, agents represent actors. This section therefore provides a brief overview of the knowledge base modellers can draw on for selecting and describing actors involved in transitions.

In the social sciences, actors have been studied and described on different levels of aggregation, ranging from individuals, corporate actors (e.g. organisations, firms) to collective actors (e.g. consumer types). Individual actors have been discussed as playing a role in transitions, for example, as 'front runner', 'change agent', 'champion' or 'policy entrepreneur'. Corporate actors include a set of individuals who are linked in an organisational structure and through organisational routines, and jointly act according to the goals and interests of the organisation. For understanding their role in transitions, organisations can often be treated as one single (corporate) actor. Organisations involved in transition processes include public authorities, firms, social movements or research organisations. Collective actors cover larger numbers of individuals in similar roles that act and interact in an unorganised way (e.g. consumers in a market). Their behaviours sometimes can simply be aggregated. However, caution needs to be applied because their interactions can also lead to non-linear system-level dynamics – for example if network effects lead to a self-reinforcing diffusion of an innovation.

Actors can also be distinguished through the differentiation of categories or backgrounds to which those actors belong – such as policy makers, firms, users, social movements or civil society. Actors belonging to different categories most likely differ in their perspectives, powers, interests, etc. They have different behavioural options and furthermore are influenced by their respective institutional contexts. Furthermore, actors can assume multiple roles in the transition, for example Schot, Kanger, and Verbong (2016) distinguish five roles of users (user-producers, user-legitimators, user-intermediaries, user-citizens and user-consumers). Each role reflects a different type of action on the system. Given this diversity, describing and understanding actors and their behaviours requires sensitivity to their particular characteristics and context. The social sciences, including the different strands of economics, (social-)psychology and sociology provide a wide range of concepts and theories, including, for example, utility, preferences, values, interests, cognitive frames, expectations, routines, social practices, norm-following behaviour, heuristics, strategies, power and learning.[3] For modelling actors belonging to different categories who assume particular roles, one should carefully reflect on the selection of appropriate theories of decision-making and behaviour.

To support this reflection and selection, a clear conceptualisation of actors and their behaviours in the context of transitions would be a valuable starting point for agent-based modelling of transitions. Such a conceptualisation – that maps the 'jungle' of possibilities and provides insights into which actor theories are most suitable for the representation of which type of actor in which role, and for understanding their behaviours in the context of transitions – could support the modeller and give orientation when taking decisions about a selection of actors, an appropriate level of aggregation and the most appropriate variables and processes for describing their behaviours.

However, developing a systematic account of the actors involved in transitions is challenging due to the diversity of sectors and countries/regions analysed by transition researchers, which exhibit partly idiosyncratic sets of actors and institutions. Conceptualising actors and their interactions has only recently begun in transitions research, and different perspectives have been taken to systematise actors.

Fischer and Newig (2016) conduct a comprehensive review on the treatment of actors and agency in transitions research. They find that the often-stated hypothesis that actors have been neglected in favour of more abstract system concepts cannot be confirmed on a general level. Rather, they find a diversity of approaches to classify actors and identify different typologies to group actors involved in transitions: actors related to the levels of the multi-level perspective,[4] actors related to institutional domains (e.g. government, market, civil society) and actors related to levels of governance (local, regional, national, global). They furthermore discuss intermediaries as a specific class of actors, because some scholars see them as important actors for information distribution and mediation during a transition.

Avelino and Wittmayer (2016) observe that most contributions in transition studies which refer to actors are troubled by conceptual ambiguity. They develop a conceptualisation of actors who exercise power in transitions and their shifting power relations. They distinguish between four categories of actors (state, market, community and third sector) along three axes (informal–formal, for profit–non-profit, public–private) that imply different 'logics' of the respective sectors. They furthermore distinguish between actors at different levels of aggregation (individual actors, organisational actors and sector-level actors) to facilitate different framings ranging from perceiving a sector (e.g. 'the state') as one actor to seeing those sectors as sites in which more specific individual and organisational actors interact.

Wittmayer et al. (2017) argue that the transitions literature does not attend to the fact that social roles can themselves be changing during a transition, and that transitions research, to date, even 'lacks a suitable vocabulary to analyse the (changing) interactions and relations of actors as part of a sustainability transition' (p. 46). They propose the concept of 'roles' to study changes in the social fabric and of shared values, norms and beliefs.

De Haan and Rotmans (2018) develop a theoretical transitions framework that includes a conceptualisation of actors and agency as one of its three main pillars. Actors are perceived as being value driven, and a typology of four different transformative actor roles ('front runners', 'connectors', 'topplers' and 'supporters') is proposed. Those actors are suggested to be in the centre of three different types of 'alliances' ('initiatives', 'networks', 'movements') between actors.

We conclude that a huge wealth of concepts from the wider social sciences is available that developers of agent-based models of transitions can draw on, but that the narrower field of transitions research (so far) provides no widely agreed conceptualisation of actors that would provide guidance and orientation for the selection of most appropriate concepts and theories for modelling actors in the context of transitions. Different in-roads to conceptualise actors in transitions have, however, been made by some scholars, which could serve as starting points for developing agent-based models.

4 Conceptualisation of institutions in transitions

A core observation and theoretical pillar of transition research is that the established dominant system shows inertia to radical change and rather follows a path of incremental changes. Besides factors such as vested interests, sunk investments and longevity of some technical artefacts and infrastructures, a main source for this inertia is seen in the institutions that govern the perceptions and behaviours of actors.

The theoretical development of the role of institutions in transition research as condensed in the multi-level perspective draws on insights from evolutionary economics and enriches those based on institutional theory as well as Giddens' structuration theory (Fünfschilling 2014). The core concept of the 'regime' originated in evolutionary economics, where the 'technological regime' or 'technological paradigm' referred to shared cognitive routines – such as common beliefs and assumptions amongst engineers about which technological advances to pursue and certain sets of problem-solving activities of firms – that result in an innovation process that follows an internal logic and streamlines innovation efforts along certain pathways (Nelson and Winter 1985; Dosi 1982). This breaks with neo-classical assumptions about rational actors, as well as simple supply and demand dynamics. Instead, institutional structures – in this case cognitive routines – become highly responsible for the evolution of technological innovation.

This conceptualisation has been widened in subsequent work by transition scholars, in particular the field of relevant actors has been broadened to encompass not only engineers and firms, but also policy makers, users, associations and other actors. All of these actor groups are seen as being guided by their own set of institutions, or 'regimes'. Building on Scott (2001), Geels (2004) goes on to widen the focus on cognitive routines of early evolutionary economic work and to distinguish between regulative, normative and cognitive institutions, which provide different kinds of coordination and structuration of the activities of the different actor groups (see Box 7.2 for examples). The *regulative* dimension refers to explicit formal rules, which constrain or provide incentives to particular behaviours, and which are backed legally (e.g. taxes, trade laws, patent laws). The *normative* dimension covers rules whose basis of compliance is social obligation and which follow the logic of 'how we do things' (e.g. role expectations). *Cognitive* institutions provide the frames through which meaning is made, which are often taken for granted and culturally supported (e.g. concepts, beliefs, jargon/language). Relating those different types of institutions to the different actor domains results in a matrix that classifies a multitude of specific institutional domains (e.g. formal rules in markets; normative rules in the policy regime) that can be ascribed a certain relevance for understanding transitions, and to whose elaboration different literatures (e.g. sociology of technology, innovation studies, cultural studies) contribute.

The overarching 'socio-technical regime', that lies at the heart of transition studies, provides some meta-coordination to the development of this diversity of institutions (Geels 2004). The reason for this is that institutions are linked to each other, forming rule systems, not only within each actor-group related regime, but also between those regimes.

Overall, empirical and conceptual work on transitions emphasises the importance of institutions for understanding transitions. Geels (2004) provides a classification scheme that crosses societal domains with three different types of institutions which is useful as a starting point to provide orientation and overview of the many areas in which institutions become relevant in transitions. For specific conceptualisations of the institutions in those areas, the modeller must refer to scholars from different literatures. More recent work (e.g. Fünfschilling 2014) elaborates on the understanding of the dynamics of institutional structures over time.

5 Challenges for modelling transitions with ABM

As outlined, ABM opens a range of unique possibilities for representing heterogeneous, bounded-rational actors,[5] and thus for the relaxation of certain fundamental assumptions regarding the optimality of actor behaviour that underlie other types of models, in particular optimisation models. The strength of ABM to allow for relaxation of fundamental assumptions often is a source for criticism at the same time. While other modelling methods have fewer degrees of freedom, their implicit assumptions are widely shared and typically not challenged time and again for each newly developed model. In contrast, agent-based models vary greatly in their selection and representation of actors and institutions. Each newly developed model may be criticised for the specific assumptions on which it relies, because due to the fragmented and qualitative insight base from the social sciences, those assumptions could typically also be made in a different way. A core challenge for agent-based modelling in all types of fields therefore is to underpin the 'correctness' or usefulness of a model.

Agent-based modelling research started with abstract models to showcase and explore fundamental principles (e.g. Epstein and Axtell 1996). Since then it has increasingly been recognised that empirical embeddedness of agent-based models – that is, calibration and validation of models based on empirical data – matters for drawing policy-relevant conclusions. Traditionally, validation is understood as testing whether the model captures reality sufficiently well through comparison of the model and its results with empirical data (Windrum, Fagiolo, and Moneta 2007). This includes demonstrating the correspondence between model entities and processes that constitute the micro-foundation of the model with real-world entities and processes. It furthermore includes demonstration of the ability of the model to replicate emergent empirical patterns on the system level. Conventional validation is, however, often challenging for ABM.[6] Degrees of freedom for designing agent-based models are large, and theoretical and empirical knowledge to guide model design is often limited when it comes to specific details of the model design. A certain level of subjectivity with regards to the model structure chosen by the model builder therefore is unavoidable for ABM. A lack of empirical data to test and parameterise the model often results in a number of alternative plausible model structures and free parameters.

Those challenges for the design and validation of agent-based models are very pronounced when modelling transitions at full scope. In particular – as outlined in the previous sections – there are many different actors and institutions involved in transitions, resulting in a vast number of possibilities for the selection of actors and institutions and their formal representation. Transition research does not provide much guidance for this task. Furthermore, data for rigorous testing of alternative model designs is typically very limited or not available at all. In particular, transitions typically feature such a level of idiosyncrasy that system level data is required to parameterise the model for this particular transition, and no data is left for rigorous validation (e.g. Bergman et al. 2008).

Based on our discussion so far, we argue that transitions – understood in the meaning and wide scope adopted in this book and including the multiplicity of involved actors and institutions outlined above – include too many different actors and institutions for explicitly representing *all* of them at a great level of detail into a model with a reasonable level of complexity. Some strategy for the selection and abstraction of actors and institutions to be represented in agent-based models is needed. Based on examples from the literature and our own experience in the field, we have identified four different strategies for the selection and formalisation of actors, their behaviours, and of the institutions guiding actor behaviour. We discuss those in the next section.

6 Modelling strategies

Against the background of the multitude of actors and institutions shaping transitions, we discuss in the following four strategies for dealing with the resulting challenges, based on examples from the literature. The first three strategies focus on the identification of a useful sub-system that facilitates reduction of the model's scope. This simplifies a selection of actors and institutions. The fourth strategy aims at abstraction from individual actors and institutions.

6.1 Focus on sub-problems

This first strategy includes definition of a sub-system of smaller scope that is suitable for capturing a relevant sub-process of the overall transition. This means the resulting model refrains from trying to capture the transition in its full scope, but reduces the scope in terms of temporal or spatial scale, topic or otherwise; often using a suitable combination of dimensions in which the focus is narrowed. The narrower focus then allows making a selection of actors and institutions to be included in the model, leading to a model of manageable complexity.

For example, Friege et al. (Friege, Holtz, and Chappin 2016; Friege 2016) focus on a particular sub-topic of the energy transition, namely the insulation of buildings. The rationale of this focus is that improving the energy efficiency of buildings through increasing the insulation rate of the existing building stock is vital for a success of the German energy transition, and that this sub-problem shows persisting inertia to existing policy interventions (such as providing financial incentives). Besides a reduced topical focus, the model furthermore limits its scope to a city, based on the reasoning that relevant interactions among actors take place on this scale, while national policies can still be included as external influences. On the temporal scale, the model runs are limited to a time span of 10 years, which was found to be a sufficiently long time span to capture relevant aspects of the target system, such as the age distribution of buildings. At the same time, the time frame is short enough to reasonably abstract from co-evolutionary developments with other parts of the energy system. In sum, the model reduces its scope in terms of topic, spatial scale and temporal scale to achieve a well-defined sub-problem.

In a similar fashion, many other model studies also focus on sub-problems of transitions. Indeed, the border between 'transitions modelling' and other fields of modelling becomes blurred when the focus of the model is narrowed down. In their systematic literature review of ABM and socio-technical energy transitions, Hansen, Liu, and Morrison (2019) observe that 'the thematic contribution of the selected literature to the study of energy transitions lies more in addressing varied sub-components of the energy system than in modelling whole transitions' (p. 46). This indicates that most modellers who are interested in (certain aspects of) transitions are following the strategy to choose a more narrow focus.

The advantage of this strategy obviously is that a model of manageable complexity can be designed while the level of abstraction of the representation of actors and institutions may remain low, facilitating a comparably straight-forward translation of topical knowledge and (appropriate) social science theories into model components. It should be noted, however, that such models are (typically) not designed from a perspective of radical system change. Therefore, their results require some interpretation concerning their contribution to transitions research. Furthermore, insights from transition research about the co-evolutionary nature of transitions let us conclude that a narrow topical focus implies a limited temporal scale for which the chosen sub-system may be assumed to be sufficiently stable to allow abstraction from its interactions with the wider societal system.

6.2 Focus on building blocks

Another strategy, that is similar to the previously discussed one, also reduces the scope of the modelled system. The difference, however, is that the delineation of the target system does not follow a 'topical logic', but a 'transition theory logic'; that is, the delineation is rooted in transitions theory. Holtz (2011) suggests that breaking down the 'big story' of a transition into sub-processes would be a useful starting point for the definition of more specific modelling studies of a reduced scope. Existing transition frameworks used by empirical researchers could be used to define such 'building blocks'. For example, Holtz (2012) combines the multi-level perspective with the multi-phase model (Rotmans, Kemp, and van Asselt 2001) to dissect a transition into developments that occur in particular phases and on/across particular levels – such as niche-creation in the pre-development phase, or niche-regime interactions in the acceleration phase. The multi-pattern approach (de Haan and Rogers 2019) would also be a useful framework to define sub-processes of the overarching transition.

A model that follows this strategy is presented by Lopolito, Morone, and Taylor (2013). With their agent-based model they focus on the process of niche creation, which is a core concept of transition thinking as reflected in the multi-level perspective and strategic niche management (Schot and Geels 2008). In their model they include three mechanisms that are viewed as underlying niche creation (converging expectations, network building and empowerment, knowledge development and diffusion).

The advantage of this strategy over the previous one is that it facilitates an easier incorporation of and exchange with empirical and conceptual transition research and provides a straightforward way to feed model results into conceptual and theoretical discussions in the wider transitions community. A challenge is that the pre-existing knowledge concerning essential elements and processes of the suggested building blocks generating transitions is limited in many cases. While there is quite some knowledge available on niche creation, there is, for example, much less known about niche–regime interactions. A main challenge thus is the selection of the essential actors, institutions and technologies for the modelling of building blocks. This challenge, however, at the same time is an opportunity for modelling to contribute to conceptual and theoretical advancements when the model purpose is rather understanding and theory building than policy advice.

6.3 Focus on key actors and institutions

A third approach to define an appropriate sub-system is selecting key actors and institutions, and developing a model that represents (only) those at high level of detail. The rationale for selecting the actors is focussing on actors who have command over large resources and whose decisions are thus key for the future development of the whole societal subsystem of interest. Institutions are selected according to their influence on these key actors.

EMLab-Generation is an example which simulates investments of electricity producers in power generation capacity given a complicated set of EU and national policies on CO_2, on renewables and on security of supply (Chappin et al. 2017). The model is rooted in economic theory of investments in energy markets, it models policies to a very high degree of detail and studies effects across borders with policy differences. The model captures key processes in the energy transition: investments in energy supply, penetrating renewables in a system in which a strong regime exists, protected by vast assets and large companies. EMLab-Generation is the result of a collaborative modelling effort spanning four PhD theses, with foci on different policies (climate policies, capacity mechanisms) and against different assumptions with respect to investment behaviour. The work led to insights into how the core climate policy instrument in the EU, the emissions trading scheme, could be improved and contributed to policy changes. It proved very useful in complementing prominent modelling work in the area of energy transition scenario development, which has serious blind spots in the conditions under which such scenarios would or would not materialise; that is, it complements scenario development work (*what* would be a lowest-cost pathway and end state for the energy transition) with what-if type of analysis (*how* transition paths for the energy transition may unfold).

The advantage of this strategy of focusing on key actors and institutions – similar to the strategy 'focus on sub-problem' – is that a model of manageable complexity can be designed while the representation of actors and institutions may be very detailed and close to empirical information. The main difference to the 'focus on sub-problem' strategy and an additional advantage is that a long time horizon and a large spatial scale can be covered. The EmLab-Generation model, for example, simulates the path dependence in scenarios up to 2050 and covers multiple national electricity systems, simulating the effects of differences in policies.

A challenge for this strategy is that the explanatory power of a model that focuses on a few actors only hinges on the persisting power of those actors to shape the system's developments. In the typology of transition pathways developed by Geels and Schot (2007) that synthesises insights from analysis of historical transitions into a typology of four different pathways, powerful incumbent actors survive in the 'transformation' and the 'reconfiguration' pathways. That means that assuming persisting influence of a few key actors may be a valid in some cases, but not in others – while it is typically not known ex-ante which pathway a specific transition will follow.

Another challenge for this strategy is defining (dynamic and co-evolutionary) influences on the selected key actors – in particular institutional and societal change – as exogenous input to the model. In the example of the EmLab-Generation model this includes *inter alia* the phase out of particular electricity generation technologies such as coal due to societal pressure, and the penetration of the system with small-scale decentralised renewables power production plants. This challenge can be addressed if the variables that represent those influences are changed between multiple runs (see Chapter 13 by Moallemi et al. in this volume).

6.4 *Defining large-scale social entities*

Acknowledging the large amount and diversity of actors and institutions involved in transitions, this strategy nevertheless aims at representing transitions in their full scope. To achieve this, the main approach is abstracting from single actors and rules. Hence, such a model represents emergent social entities on higher levels of aggregation, and uses stylised representations of their interactions.

A prime example of this type of model is the MATISSE model, which is presented and discussed at length by Köhler (Chapter 6 in this volume). Building on an extended multi-level perspective, this model focuses on a regime and niches, and develops a range of attributes and processes to represent these social entities and their interactions. The niches and regimes aggregate and abstract from a number of different actors including producers (along the respective value chains), intermediaries and governmental actors who support particular solutions. Niches and regimes also subsume the institutions that provide the 'glue' between the different actors in a regime/niche.

The benefits of this strategy arise from its ability to combine a 'full scope' perspective with the inclusion of social aspects[7] and to unleash the power of formalisation and computational experiments to study transitions on this level of aggregation and from this angle.

A main challenge for this strategy resides in the foundation of the specific design of model components and processes. To our knowledge, theories that provide a mechanistic understanding of the behaviours and interactions of large-scale social entities such as regimes and niches are scarce and not formalised. The resulting need to 'make up' model components and processes opens room for critique. For example, Holtz (2011) observes that '(d)ue to a lack of useful theories, most of the mechanisms generating the dynamics of the MATISSE model were conceived based on the intuition of the transition researchers involved' (p. 175). Nevertheless, we argue that due to its particular benefits, this model design strategy has its merits, if its limitations are well balanced with the purpose of the modelling exercise.

7 Conclusions

Our discussion has shown that ABM is well suited for representing processes that emerge from the behaviours and interactions of bounded-rational actors and (changing) institutions that guide and constrain those actors' behaviours. This ability of ABM provides a unique opportunity to relax rigid assumptions with regards to homogeneity and rationality of actors that are often implicit in widely used economic approaches to the modelling of innovation for sustainability. ABM allows development of a bottom-up representation of (some selected sub-process of) a transition, and in particular representing social aspects and the co-evolution of technical and social elements of a system.

The multitude of actors and institutions involved in transitions requires that the modeller makes a selection of the involved actors and institutions for representation in the model, or boldly abstracts from single actors and institutions. Empirical and conceptual work in transition research does not provide a clear and widely shared conceptualisation of actors and institutions, which leaves considerable freedom to the modeller, but also creates the necessity to explain and substantiate the choices made. Some transition scholars have made some in-roads towards the conceptualisation of actors and institutions that may serve the modeller as guidelines when designing a model.

We have identified four strategies for selection and abstraction of actors and institutions that follow different rationales and have different benefits and challenges. They are: (1) focus on sub-problem, (2) focus on building blocks derived from transition concepts, (3) focus on key actors and institutions and (4) define large-scale social entities. The selection of a strategy for a particular modelling task needs to be in line with the purpose that this particular model shall serve. When developing an agent-based model, the interdependencies between a model's purpose, its design and resulting complexity, data availability, and validity are always very delicate, and providing general insights and guidance is difficult. It is therefore beyond the scope of this chapter to provide a more extensive guideline which strategy is best used for what purpose and context.

The existing literature of ABM of transitions shows a strong bias towards the strategy of focussing on a sub-problem of the overall transition. This includes many models that have not been developed explicitly as 'transition models', but can be classified as belonging to this category within the frame of reference we have chosen in this chapter. While there are indeed plenty of opportunities to define and model highly relevant sub-problems, we also see high potential in following the other strategies we have identified. In particular we see a high potential for the strategies 'focus on building blocks' and 'define large-scale social entities' to establish closer collaboration between transition modellers and researchers working with qualitative methods, and to advance the conceptual and theoretical base of transition research. The strategy of 'focussing on key actors and institutions' is promising for policy advice about (possible) long-term developments, in particular in combination with exploratory modelling. We therefore encourage ABM modellers to develop modelling exercises following those other strategies.

Notes

1 The term 'institutions' is used here in its sociological meaning. In the institutional school of thought, there is, however, no universally agreed definition of the term. We use it to broadly refer to formal and informal rules and regulations that structure and give meaning to social life.

2 We use this example of a transition for illustration of our arguments because we are particularly knowledgeable about it. However, we propose that the points made in this chapter about the multitude of actors and institutions and the diversity of theoretical perspectives for describing them are not limited to this particular transition.

3 Jackson (2005) provides an excellent overview that provides an introduction to a wide range of theories.

4 The multi-level perspective (Geels et al. 2017; Rip and Kemp 1998) is a widely adopted framework to analyse transitions.

5 The rational actor is assumed to identify all possible behaviours, to determine all consequences for each possibility, to evaluate all those consequences, and then to choose the

solution that optimises utility. Real actors face uncertainty about the future, have limited access to information and may have limited resources and cognitive ability for determining and evaluating consequences. Due to these reasons, fully rational decision-making is not possible in practice. Simon (1957) proposed the term 'bounded-rationality' to highlight the limitations of practical decision-making.

6 It is beyond the scope of this chapter to discuss challenges and approaches for traditional model validation and alternative approaches to validation at greater length, in particular as those are strongly linked not only to the model design but also the model purpose. The interested reader is referred to Moss (2008).

7 Capturing fundamental social aspects of the dynamics of transition processes is a main limitation of the process understanding implicit in other modelling approaches that are able to capture a 'full scope' perspective (see Chapter 3 by Köhler and Holtz in this volume).

References

Avelino, Flor, and Julia M. Wittmayer. 2016. "Shifting power relations in sustainability transitions: A multi-actor perspective." *Journal of Environmental Policy & Planning* 18 (5): 628–49.

Bankes, Steven C. 2002. "Agent-based modeling: A revolution?" *Proceedings of the National Academy of Sciences* 99 (suppl 3): 7199–200.

Bergman, Noam, Alex Haxeltine, Lorraine Whitmarsh, Jonathan Köhler, Michel Schilperoord, and Jan Rotmans. 2008. "Modelling socio-technical transition patterns and pathways." *Journal of Artificial Societies and Social Simulation* 11 (3). http://jasss.soc.surrey.ac.uk/11/3/7.html.

Bianchi, Federico, and Flaminio Squazzoni. 2019. "Modelling and social science: Problems and promises." In *Modelling Transitions – Virtues, Vices, Visions of the Future*, edited by Enayat Moallemi and Fjalar de Haan. This volume: Routledge.

Boero, Riccardo, and Flaminio Squazzoni. 2005. "Does empirical embeddedness matter? Methodological issues on agent-based models for analytical social science." *Journal of Artificial Societies and Social Simulation* 8 (4): 6. http://jasss.soc.surrey.ac.uk/8/4/6.html.

Chappin, Émile Jean Louis. 2011. *Simulating Energy Transitions* (PhD diss.), Delft University of Technology.

Chappin, Emile Jean Louis, Laurens J. de Vries, Joern C. Richstein, Pradyumna Bhagwat, Kaveri Iychettira, and Salman Khan. 2017. "Simulating climate and energy policy with agent-based modelling: The Energy Modelling Laboratory (EMLab)." *Environmental Modelling & Software* 96: 421–31.

Dosi, Giovanni. 1982. "Technological paradigms and technological trajectories: A suggested interpretation of the determinants and directions of technical change." *Research Policy* 11 (3): 147–62.

Epstein, Joshua M., and Robert L. Axtell. 1996. *Growing Artificial Societies: Social Science From the Bottom Up*. Cambridge, MA: MIT Press.

Farmer, Doyne J., and Duncan Foley. 2009. "The economy needs agent-based modelling." *Nature* 460 (7256): 685–6.

Fischer, Lisa-Britt, and Jens Newig. 2016. "Importance of actors and agency in sustainability transitions: A systematic exploration of the literature." *Sustainability* 8 (5): 476.

Friege, Jonas. 2016. "Increasing homeowners' insulation activity in Germany: An empirically grounded agent-based model analysis." *Energy and Buildings* 128: 756–71.

Friege, Jonas, Georg Holtz, and Émile J.L. Chappin. 2016. "Exploring homeowners' insulation activity." *Journal of Artificial Societies and Social Simulation* 19 (1). http://jasss.soc.surrey.ac.uk/19/1/4.html.

Fünfschilling, Lea. 2014. *A Dynamic Model of Socio-Technical Change: Institutions, Actors and Technologies in Interaction* (PhD diss.), University of Basel.

Geels, Frank W. 2004. "From sectoral systems of innovation to socio-technical systems: Insights about dynamics and change from sociology and institutional theory." *Research Policy* 33 (6–7): 897–920.

Geels, Frank W., and Johan Schot. 2007. "Typology of sociotechnical transition pathways." *Research Policy* 36 (3): 399–417.

Geels, Frank W., Benjamin K. Sovacool, Tim Schwanen, and Steve Sorrell. 2017. "Socio-technical transitions for deep decarbonization." *Science* 357 (6357): 1242–4.

Gilbert, Nigel, and Klaus G. Troitzsch. 2005. *Simulation for the Social Scientist.* 2nd ed. McGraw-Hill Education, Berkshire: Open University Press.

Haan, Fjalar J. de. 2019. "Making it a science – Aspirations and hesitations of transitions research." In *Modelling Transitions – Virtues, Vices, Visions of the Future*, edited by Enayat Moallemi and Fjalar de Haan. This volume: Routledge.

Haan, Fjalar J. de, and Briony C. Rogers. 2019. "The multi-pattern approach for systematic analysis of transition pathways." *Sustainability* 11 (2): 318.

Haan, Fjalar J. de, and Jan Rotmans. 2018. "A proposed theoretical framework for actors in transformative change." *Technological Forecasting and Social Change* 128: 275–86.

Hansen, Paula, Xin Liu, and Gregory M. Morrison. 2019. "Agent-based modelling and socio-technical energy transitions: A systematic literature review." *Energy Research & Social Science* 49: 41–52.

Heckbert, Scott, Tim Baynes, and Andrew Reeson. 2010. "Agent-based modeling in ecological economics." *Annals of the New York Academy of Sciences* 1185: 39–53.

Hoekstra, Auke, Maarten Steinbuch, and Geert Verbong. 2017. "Creating agent-based energy transition management models that can uncover profitable pathways to climate change mitigation." *Complexity* 2017.

Holtz, Georg. 2011. "Modelling transitions: An appraisal of experiences and suggestions for research." *Environmental Innovation and Societal Transitions* 1 (2): 167–86. https://doi.org/10.1016/j.eist.2011.08.003.

Holtz, Georg. 2012. "The PSM approach to transitions: Bridging the gap between abstract frameworks and tangible entities." *Technological Forecasting and Social Change* 79 (4): 734–43.

Jackson, Tim. 2005. "Motivating Sustainable Consumption. A review of evidence on consumer behavior and behavioural change." *A Report to the Sustainable Development Research Network*. www.sustainablelifestyles.ac.uk/sites/default/files/motivating_sc_final.pdf.

Köhler, Jonathan. 2019. "Modelling the multi-level perspective: The MATISSE agent-based model." In *Modelling Transitions – Virtues, Vices, Visions of the Future*, edited by Enayat Moallemi and Fjalar de Haan. This volume: Routledge.

Köhler, Jonathan, and Georg Holtz. 2019. "Transitions modelling: Status, challenges and strategies." In *Modelling Transitions – Virtues, Vices, Visions of the Future*, edited by Enayat Moallemi and Fjalar de Haan. This volume: Routledge.

Lopolito, Antonio, Piergiuseppe Morone, and Richard Taylor. 2013. "Emerging innovation niches: An agent based model." *Research Policy* 42 (6–7): 1225–38.

Macal, Charles M., and Michael J. North. 2010. "Tutorial on agent-based modelling and simulation." *Journal of Simulation* 4 (3): 151–62. https://doi.org/10.1057/jos.2010.3.

Moallemi, Enayat A., Fjalar J. de Haan, and Jonathan Köhler. 2019. "Exploratory modelling of transitions: An emerging approach for coping with uncertainties in transitions research." In *Modelling Transitions – Virtues, Vices, Visions of the Future*, edited by Enayat Moallemi and Fjalar de Haan. This volume: Routledge.

Moss, Scott. 2008. "Alternative approaches to the empirical validation of agent-based models." *Journal of Artificial Societies and Social Simulation* 11 (1): 5. http://jasss.soc.surrey.ac.uk/11/1/5.html.

Nelson, Richard, and Sidney Winter. 1985. *An Evolutionary Theory of Economic Change*. Cambridge, MA: Harvard University Press.

Rip, Arie, and René Kemp. 1998. "Technological change." In *Human Choice and Climate Change – Resources and Technology*, edited by S. Rayner and E.L. Malone. Columbus: Battelle Press.

Rotmans, Jan, René Kemp, and Marjolein Van Asselt. 2001. "More evolution than revolution: Transition management in public policy." *Foresight* 3 (1): 15–31.

Schot, Johan, and Frank W. Geels. 2008. "Strategic niche management and sustainable innovation journeys: Theory, findings, research agenda, and policy." *Technology Analysis & Strategic Management* 20 (5): 537–54.

Schot, Johan, Laur Kanger, and Geert Verbong. 2016. "The roles of users in shaping transitions to new energy systems." *Nature Energy* 1 (5): 16054.

Scott, W. Richard. 2001. *Institutions and Organisations: Foundations for Organisational Science*. 2nd ed. London: SAGE Publications.

Simon, Herbert. 1957. *Models of Man*. New York: Wiley.

Sun, Zhanli, Iris Lorscheid, James D. Millington, Steffen Lauf, Nicholas R. Magliocca, Jürgen Groeneveld, et al. 2016. "Simple or complicated agent-based models? A complicated issue." *Environmental Modelling & Software* 86: 56–67.

Windrum, Paul, Giorgio Fagiolo, and Alessio Moneta. 2007. "Empirical validation of agent-based models: Alternatives and prospects." *Journal of Artificial Societies and Social Simulation* 10 (2): 8.

Wittmayer, Julia M., Flor Avelino, Frank van Steenbergen, and Derk Loorbach. 2017. "Actor roles in transition: Insights from sociological perspectives." *Environmental Innovation and Societal Transitions* 24: 45–56.

8 System dynamics methodology and research

Opportunities for transitions research

George Papachristos and Jeroen Struben

1 Introduction

Our society is on an unsustainable pathway exemplified by anthropogenic climate change, deforestation, human displacements and species extinction (IPCC 2018). Altering this pathway requires a combination of technical, organisational, economic, institutional, social-cultural and political efforts that are increasingly referred to as socio-technical transitions to an environmentally sustainable economy (Van den Bergh, Truffer, and Kallis 2011). Transitions research has developed frameworks and methods (e.g. Geels, Berkhout, and van Vuuren 2016; Papachristos 2018a)[1] for understanding historical and contemporary transitions and for exploring system interventions related to governance and sustainability transitions (Smith, Voß, and Grin 2010). While qualitative, narrative-based case studies dominated early research, modelling and simulation approaches have recently gained ground. These latter approaches are particularly suited to explore interdependencies across different system elements (Bergman et al. 2008; Papachristos 2011, 2014a; Holtz et al. 2015; Walrave and Raven 2016; Köhler et al. 2018) and to perform counter-factual analysis (Sterman 2002) that allows posing 'what-if' questions and avoiding potential selection biases that limit the power of pure empirical analysis.

In this chapter we discuss how system dynamics (SD) modelling and simulation (Forrester 1958; Sterman 2000) can contribute to and complement – rather than substitute for – other methods in transitions research. SD is an established modelling and simulation research methodology that spans several subject areas (Sterman 2018). The phenomena discussed in the transitions literature are aligned with the foundational ideas of SD – a problem orientation, broad model boundaries across multiple realms and an 'endogenous point of view' towards processes of system change that are driven by actor decisions (Richardson 2011). Thus, SD is well suited to addressing transitions research questions that relate to high-leverage policies (Van den Bergh, Truffer, and Kallis 2011). The chapter is structured as follows. Section 2 briefly overviews transitions research and some critical contemporary research issues. Section 3 briefly summarises SD history and foundations. Section 4 discusses the core SD practices and how they can provide entry points for transitions research. Section 5 discusses exemplary SD work relevant to transitions research across three levels of analysis: macro, meso and micro. Emphasis is on the meso level, as this level is also central in transitions research (Geels 2002; Geels and Schot 2007). In Section 6 we discuss some key opportunities and challenges for applying SD methods in transitions research.

2 A brief overview and critical questions for transitions research using the multi-level perspective

Several frameworks have been developed and applied to transition case studies: the multi-level perspective (MLP; Geels and Schot 2007), the functions of (technological) innovation systems approach (Bergek et al. 2008; Hekkert et al. 2007), the transition management approach (Rotmans, Kemp, and van Asselt 2001) and strategic niche management (Kemp, Schot, and Hoogma 1998). We focus on the widely used MLP framework to illustrate how SD can contribute to transitions research. However, we believe our conclusions are also relevant for other transition frameworks.

A central concept of the MLP is the socio-technical system. It facilitates analysis of the rules that align and coordinate the activities of actor groups and the users who reproduce system elements and, in this way, contribute to system trajectory stability in regimes (Geels and Schot 2007; Geels et al. 2016). This meso-level regime is of primary interest because transitions are defined as shifts from one regime to another. The MLP has two additional analytical concepts that are defined in relation to the regime: niches and landscape. Niches are defined as incubation spaces at the micro level that shield, nurture and empower new innovations (Smith and Raven 2012). Active shielding might come through public or private interventions and incentives for technology support. Passive shielding arises in spaces through no direct actor involvement. Nurturing and empowerment involve the articulation of actor expectations and visions that guide learning processes, attract attention and legitimate the protection, nurturing and development of social networks. Learning processes improve the competitiveness of the technology along multiple dimensions (Kemp, Schot, and Hoogma 1998): technical, market and user, cultural, infrastructure, industry, regulations, societal.

The socio-technical landscape provides the macro-level context that influences niche and SD. The landscape encompasses the technical, material and macro-economic societal backdrop, which includes industrywide processes, climate change and geographic formations but also demographic trends, political ideologies and societal values (Geels and Schot 2007; Geels 2011). The implication is that landscape trends, for example, climate change or peak oil, can destabilise socio-technical systems.

MLP-based transitions research focuses on how the nature, timing and intensity of interactions between landscape pressures, the buildup of niche innovations

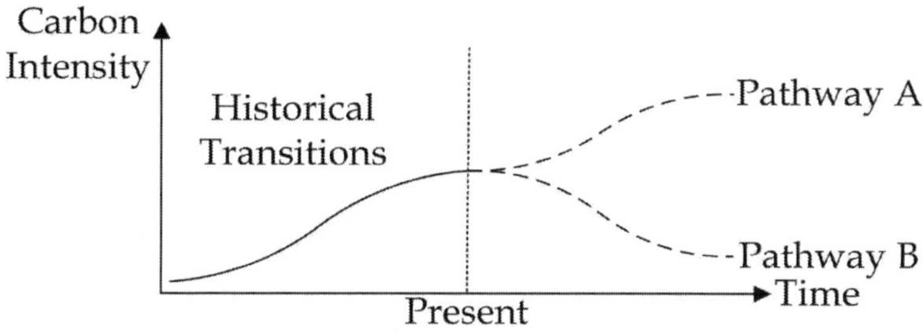

Figure 8.1 Differences in historical and contemporary transition paths.

Source: Adapted from Papachristos (2014a).

by groups of actors, internal or external to the focal regime, as well as internal regime tensions may unfold over time, and enable or constrain a transition process (Geels and Schot 2007; Papachristos, Sofianos, and Adamides 2013). In this respect, future sustainability transitions are unlikely to resemble past ones (Fouquet and Pearson 2012; Papachristos 2014a). Historic pathways led systems towards greater scale, consumption and carbon-intensity states (Pathway A in Figure 8.1). Contemporary pathways must be towards states of a fundamentally different nature (Pathway B in Figure 8.2): a low-carbon state of less growth, less consumption of resources, cyclical flows of goods and choices driven by natural resource constraints.

Current MLP-based transitions research suggests that regime-disruptive processes – either in niches or in regimes – can be nurtured and reinforced to bring about a transition, while, simultaneously, dominant regimes destabilise, unravel and decline or continue to coexist, along with the rise of new niches (Turnheim and Geels 2012). A way to conceptualise this challenge is to think of socio-technical system trajectories as driven by the aggregate balance of reinforcing and disrupting forces (the plus and minus signs in Figure 8.2). Disruptive dynamics endogenous to the regime or arising from a niche (the minus sign in Figure 8.2) must rise and overcome those that support and maintain the regime in balance (the plus sign in Figure 8.2).

3 A brief history and foundations of system dynamics

The field of SD explores how system structure generates system behaviour within diverse contexts: societal, technological, managerial, urban and ecological. The field of SD emerged with founder Jay W. Forrester's publications on industrial dynamics that explained the persistence of business cycles and outlines the principles for formulating dynamic system models to explain social system behaviour, involving distinct actors responding to perceived changes within their environment – and thus, directly and indirectly, to each

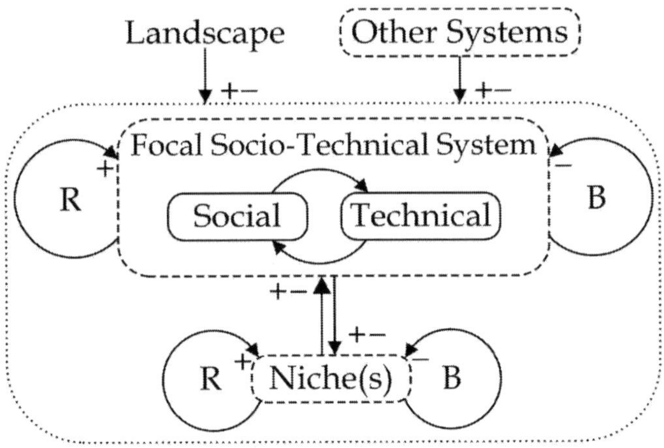

Figure 8.2 Reinforcing and disrupting loops in sociotechnical systems.

Source: Adapted from Papachristos (2014a, 2018b).

other (Forrester 1958, 1961). The basic structural elements of systems are (1) a closed system boundary with feedback loops as the basic structural elements, (2) stock variables to represent system accumulation processes, (3) flow-rate variables to represent change within the feedback loops and (4) a goal for the system, the discrepancy with its state and action based on this (Forrester 1968, 1969).

Regularities that appeared across application contexts turned SD into a theory of system structure and an approach to policy design (Forrester 1968). In line with this, over time, the field has developed a range of systems-thinking approaches and advanced process mapping, modelling, and computer simulation tools to (1) hypothesise, test and refine endogenous explanations of system change and (2) use those explanations to guide policy and decision-making (Richardson 2011). The focus on actors and real-world problems in SD model development necessitates engagement with system actors, problem owners, stakeholders and policy makers and knowledge of cognitive and social psychology, economics and other social sciences.

Since its foundation, SD has grown into a vibrant field around the world (see *systemdynamics.org*). Seminal published works include *Urban Dynamics* (Forrester 1969), *World Dynamics* (Forrester 1971) and *Limits to Growth* (Meadows et al. 1972). The latter two can be seen as a precursor to sustainability studies, where SD scholars explored the complex dynamics of economic growth and their long-term impact on the environment, population and food on a global scale. The key insights were that timely and system-wide interventions aimed at halting material growth were critical to achieving sustainable pathways with high quality of life. Since then, the pressures studied in these early works have become too strong to be ignored, and have been shown to follow the business-as-usual scenario of the original limits-to-growth studies (Turner 2014).

4 The relevance of system dynamics to transition studies

This section discusses the practices that are critical to SD research on systems exhibiting complex dynamic behaviour (Sterman et al. 2015) and why their application can provide insights into transitions research of socio-technical processes. Complex dynamic behaviour arises through the interaction of multiple non-linear positive and negative feedbacks and the delays they involve (Forrester 1961). In socio-technical systems, such feedbacks involve actors' decision-making processes and other information flows, as well as the various physical elements, resources and institutions (Sterman 2000). Because such systems – whether within the micro, meso or macro levels or across those – can produce path-dependent and condition-specific behaviour, observed behaviour patterns are not something to be assumed but to be explained. System disequilibrium is the norm, not the exception (Sterman 2000).

Crucially, systems can produce counter-intuitive behaviour from the point of view of the actors within the system and resist change because their feedback structure is reproduced by the actors embedded in it, and they observe and understand only part(s) of the system (Sterman 2000). For example, the organisational capability trap or competence trap (Levitt and March 1988) results from self-reinforcing pressures to increase firm performance, leading actors to favour short-term solutions – overtime work, less maintenance and training – over long-term ones, such as process improvement and capabilities development (Repenning and Sterman 2002). The capability trap research shows that intentionally rational responses to problems in complex systems often do not help and can make the situation worse. Instead, identification of high-leverage interventions requires actors to understand the system-level interactions. Capability trap research shows that interventions to improve system performance must often be sustained longer and with stronger commitment in contrast to the short-term solutions that actors tend to favour.

4.1 Relevant system dynamics practices for transitions research

The SD field has developed a number of methodological practices to analyse and improve the dynamic complex behaviour of socio-technical systems. In addition to its practical relevance to transitions research, one can note an epistemological connection, as both fields align with critical realism (Sayer 2000). SD uses grounded methods to endow a model with operational thinking, and it provides a formal approach to represent social mechanisms (Lane 2001). In particular, SD aims to identify enduring mechanisms and structures that produce what humans come to cognise at the 'empirical' level (Sayer 2000; Mingers 2014). In line with this, within MLP one can find calls to identify transition mechanisms (Geels 2002), although this has yet to receive much attention (Papachristos 2018a; Sorrell 2018). The following sections highlight five SD practices that aim to identify high-leverage system interventions.

Explore model boundaries

The system boundary is critical because model behaviour and policy recommendations are often more sensitive to model boundaries than to parameter uncertainty or to the level of model aggregation (Sterman 2000). SD modellers are trained to explore the system boundary of the problem under study, and to consider feedbacks that are spatio-temporally distant from the symptoms of the problem. This approach is relevant to transition processes that, by their very nature, have broad spatio-temporal boundaries (Geels and Schot 2007). The research challenge is that people tend to associate effects with proximal causes in space and time and to omit distal and delayed impacts (Sterman 1994). This challenge is also relevant for transitions researchers, as they may form their mental models of systems with narrow boundaries that leave out interactions that contribute to the system behaviour they are interested in (Sterman 2000). The result may be policy resistance and the tendency to implement policies that work in the short term but not over the long term.

Capture interdependency, endogeneity and feedback

Methodologically, SD and transitions research aim to develop endogenous theories about processes (Geels and Schot 2007; Sterman 2000). To do this, they use qualitative case studies as a research method (Geels 2002; Papachristos 2012; Morrison and Oliva 2018). Qualitative data are recognised as important for modelling decision-making processes (Forrester 1961). Transition studies provide a wealth of case studies that can be complemented with SD (Geels, Berkhout, and van Vuuren 2016; Moallemi, Aye et al. 2017, Moallemi, de Haan et al. 2017; Papachristos 2018a, 2018b).

SD researchers often map the factors that drive reinforcing/disrupting and balancing processes in causal loop diagrams (CLDs) and represent explicitly the structural and agent system elements that may produce endogenously the dynamics under study. The transfer of CLDs into a simulation model enables the exploration of complex system behaviour. SD models are differential equation models that capture the physical and institutional system structure that governs information flows and incentives, and the behavioural decision rules of the real-world actors, aggregated into compartments or levels within which they are assumed to act homogeneously. This can contribute to the aim of transitions research to map and explain endogenous change (Geels and Schot 2007; Geels et al. 2016).

Represent system actor behaviour

The representation of real-life actors at a compartment level and how their behaviour generates system-level patterns has been at the core of SD practice since the foundational SD work on industrial dynamics (Forrester 1958, 1961). Forrester (1961) argued that it is possible to identify the structural elements of this process and the policies that guide related decisions despite the fact that a decision-making process is non-linear, noisy and influenced by the perceptual and cognitive limitations of the decision makers. This is important, as research shows that people do not follow the expected or rational course of action, which has large implications for system behaviour (Lane 2017). The representation of actor decision-making in SD is consistent with the strong theoretical and empirical foundations of the 'Carnegie School' on managerial decision-making (March and Simon 1958; Cyert and March 1963; Sterman et al. 2015). This explicit representation of actors, their bounded rationality and their actions is central within the MLP, explaining trajectory stability as well as individual and coordinated efforts for regime shifts (Geels and Schot 2007).

Identify important system delays

Delays influence actor decisions and shape system behaviour. Delays arise between actions and the response they evoke (e.g. competitive moves and market responses), but also simply because system accumulation processes require time. Industrial dynamics gave rise to SD research on the difficulties humans have to understand the behaviour of systems with stock accumulation (Sterman 1989; Cronin, Gonzalez, and Sterman 2009), with direct implications for climate change policy (Sterman 2008). The study of delays is directly relevant for transitions research on system interactions that shape system trajectories, and on niche accumulation processes that are integral to transition processes, such as the emergence of recharging standards for electric vehicles (Bakker, Leguijt, and van Lente 2015).

Focus on problems, systems improvement, and actor involvement

The problem-oriented approach guiding SD research facilitates the generation of policy insights towards system change (Auvinen et al. 2015; Lane 2017). The value of an SD study materialises when such insights bring about the desired effect (Forrester and Senge 1980). However, well-intended change efforts may lead to side effects, because many actors have the ability to resist change efforts in complex socio-technical systems. To avoid such resistance, models are often developed with active involvement of policy makers and/or many other key stakeholders (Forrester 1961). Model development in a group context is used to negotiate inter-subjective meaning, create a shared description of reality, facilitate group problem-solving and catalyse commitment to action (Lane 1992, 1999; Vennix 1996). Through elicitation and mapping techniques, stakeholders develop an increased understanding of the problem and develop a consensus on joint action. A significant outcome of a successful modelling effort is that the mental model(s) of the key stakeholders involved will change and align, to identify and allow the high-leverage interventions and possibly even a redefinition of goals. This is precisely the kind of approach advocated in reflexive transition governance (Voss, Bauknecht, and Kemp 2006).

Together these practices not only characterise the SD approach but also distinguish it from other complexity modelling approaches used in transitions research, in particular agent-based modelling (ABM). (See Chapter 6 by Köhler for an agent-based model and Chapter 3 by Köhler and Holtz, Chapter 5 by Bianchi and Squazzoni, Chapter 7 by Holtz and Chappin and Chapter 13 by Moallemi, de Haan and Köhler in this volume for overviews of different modelling challenges and opportunities). Contrasting the SD feedback approach to explain dynamics arising from compartment-level interactions, ABM emphasises emergence as explanation of phenomena at the group, organisation, or system level, arising as heterogeneous decision makers interact with each other and with their environment (Epstein 2007). In principle the SD and ABM approaches can be seen as extreme points on a continuum – one can disaggregate SD models by increasing the number of compartments – highlighting the tradeoffs in benefits from such disaggregation with its computational and cognitive costs (Rahmandad and Sterman 2008). In particular, disaggregation, while allowing richer behaviour patterns, imposes stronger conceptual and empirical demands, on behavioural assumptions of the actors and on data collection and fitting, and weakens the particular leverage these core SD practices provide to generate dynamic understanding and insights.

5 Modelling transitions at different levels for different aims

For analytical clarity we now discuss illustrative SD research that corresponds respectively to the macro, meso and micro levels of analysis of the MLP framework. These examples may be the basis for future research, including work that cuts across levels.

5.1 Macro level: national/supra policy-making and collaboration

SD has been applied to long-term, large-scale change issues that the transition community also deals with. For example, the *Limits to Growth* study (Meadows et al. 1972; Table 8.1, #1) addressed sustainability and transition issues. The simulation-based approach demonstrated that sustainable pathways are possible but that more technology or efficiency would not suffice, and that timely and unpopular system interventions were critical. The study generated much debate and drew criticism from mainstream researchers (Cole 1973; Nordhaus 1973). Yet, the pioneering vision and scientific strength of the work is evident, as most

Table 8.1 System dynamics research with key questions, method and mappings to the multi-level perspective

#	*SD Work*	*Phenomenon*	*Key Questions*	*SD Method*	*Level of Analysis*
1	*Limits to Growth* (Meadows et al. 1972)	Global sustainability	How can we achieve a transition towards sustainable high-quality lifestyle?	Simulation model (policy)	Macro: socio-technical to landscape (global footprints)
2	C-ROADS (Sterman et al. 2013)	Global policy interventions for GHG emission reductions	What are implications of GHG commitments?	Management flight simulator	Macro: socio-technical to landscape (global GHG emissions)
3	EN-ROADS (Sterman 1982; Fiddaman 2007)	Energy system transformation	How can we achieve a transition towards a sustainable energy system? What mix of policies are high leverage? How long does it take?	Management flight simulator with participative process	Macro: global GHG emissions and energy production/demand
4	Alternative fuel vehicle market transformation (Struben and Sterman 2008)	Market transformation	What are the barriers and high-leverage (collective) strategies for market transformation success?	Suite of simulation models (policy)	Meso: market
5	Nutritious food market transformation (Struben, Chan, and Dubé 2014; Papachristos and Adamides 2016)	Market transformation	What are the barriers and high-leverage (collective) strategies for market transformation success?	Simulation model (policy)	Meso: market
6	Airline business model (Casadesus-Masanell and Ricart 2010; Cosenz and Noto 2018)	Organisational business model transformation	Combination of business model dynamics with system dynamics modelling for strategy design.	Causal loop diagrams	Micro-meso: niche–system interactions
7	Capability traps (Repenning 2001; Repenning and Sterman 2002; Sterman 2015a)	Organisational capability development	Why is it that many firms don't succeed in transforming? How long does it take to overcome worse-before-better dynamics from transforming capabilities?	Causal loop diagrams and conceptual simulation model	Micro-meso: niche–system interactions
8	Policy support for community safety (Rouwette, Bleijenbergh, and Vennix 2016)	Community development	How to implement policies that help safety in city neighbourhoods?	Participative modelling	Micro: community
9	*World Climate* (Sterman 2015b)	Citizen awareness and mobilisation around GHG emissions	What are implications of waiting with efforts GHG commitments?	Participative process, using management flight simulator	Micro: citizen mobilisation (global GHG emissions)

Note: GHG = greenhouse gases; SD = system dynamics.

of the global indices follow its business-as-usual scenario (Turner 2014), and experts and policy makers around the world are beginning to realise that the advice the work offers is fundamental to global sustainability.

Subsequent SD work focused on the climate system (Fiddaman 2002, 2007) and showed that the poor understanding of the relationship between greenhouse gas (GHG) emissions and their likely climate impacts can influence equally the public and policy makers (Sterman 2008). This work has led to an SD-based model application for UN climate negotiations (Sterman et al. 2013). The model can reproduce large climate models results and facilitate the rapid assessment of UN delegate proposals for emissions abatement (Table 8.1, #2). The model has been widely used by delegates during multiple Conference of the Parties (COP), the supreme decision-making body of the United Nations Framework Convention on Climate Change meetings.

A similar model has been developed to explore how we can achieve our energy transition and climate goals through sustainable, win-win changes in our energy use, consumption and policies (Table 8.1, #3), The model focuses on how changes in global GDP, energy-efficiency improvement, generation capacity, energy and carbon prices, fuel mix, consumer adoption and other factors influence carbon emissions, energy access and global temperatures (Fiddaman 2002, 2007).

5.2 Meso level: market and organisational transformations

Market and platforms

Technology standards are part of socio-technical systems, as they enable complementarities between technological components and innovations, and contribute to their inertia (Geels and Schot 2007). Standards enable modular technological architectures or platforms (Gawer and Cusumano 2014). Standards are part of socio-technical regimes, so platforms should be considered as well because there can be no separate demand for individual components without a core platform system. Thus, a platform is a technological artefact that binds a socio-technical system together and, as such, platform governance and competition (Papachristos and Van de Kaa 2018; Van de Kaa, Papachristos, and de Bruijn, 2019) may contribute to transition inertia (Papachristos 2017).

A team of researchers developed a suite of SD simulation models to explore platform category–level dynamics in the case of alternative fuel vehicles (AFVs; Struben and Sterman 2008; Table 8.1, #4). The failure of AFV programmes to transform the transportation sector in recent decades is often attributed to the low attractiveness of AFVs, high costs, immature technology and lack of choice compared with fossil-fuelled internal combustion engine vehicles. AFV adoption is suppressed by the powerful feedback processes that operate between the dominant automakers, the petroleum industry, the transport networks, settlement patterns, technologies and institutions (Figure 8.3). For example, energy producers, automakers and governments will not invest in AFV technology and

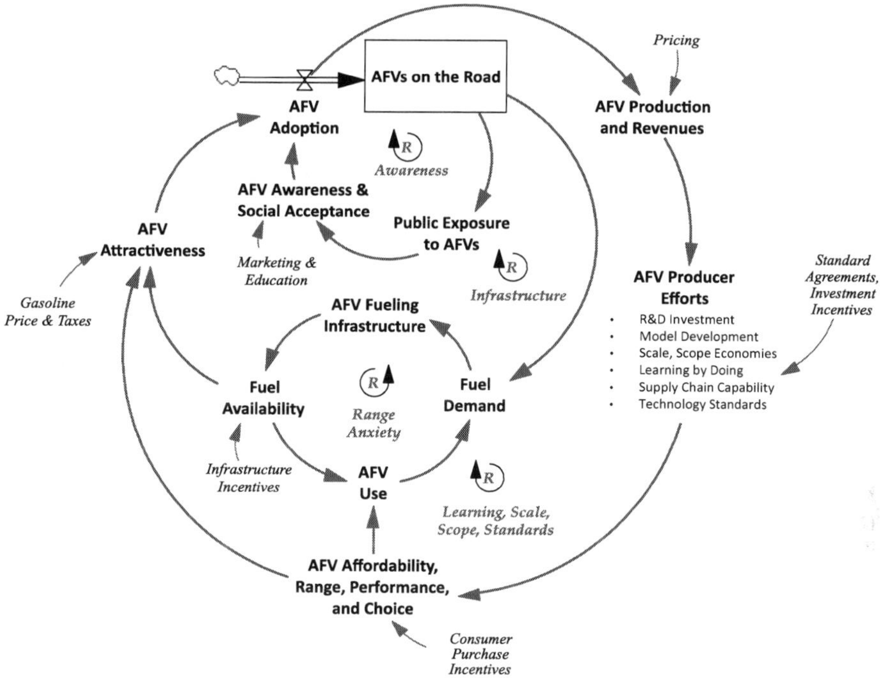

Figure 8.3 Causal loop diagram of alternative fuel vehicle (AFV) adoption.

Source: Adapted from Struben and Sterman (2008); Sterman et al. (2013).

infrastructure that is incompatible with the current fuel supply chain without the prospect of a large AFV market. In turn, consumers will only adopt AFVs with the development of infrastructure and complementary technological, financial and norm standards (Figure 8.3, *Infrastructure* loop). The availability of fuel and support services affects AFV adoption and use (*Range Anxiety* loop). Low AFV adoption in areas with sparse fuelling infrastructure keeps AFV fuel demand low, and this in turn reduces the profitability and deployment of fuelling infrastructure in these areas and further suppresses AFV diffusion (Struben and Sterman 2008).

Multiple simulation-based analyses of AFV dynamics have explored policies to stimulate AFV adoption. Results showed that sustained success requires early adoption of technology standards and much larger and longer marketing campaigns and subsidies for vehicles and infrastructure than is typical in most markets (Struben and Sterman 2008). Further, current systems may be reconfigured through intermediate phases involving multiple niche innovations – for example, hybrid electric vehicles – but, under some conditions, such efforts may inhibit large-scale transitions (Keith, Sterman, and Struben 2017). Finally, based on this model, researchers have developed an application to enable scenario exploration by industry decision makers, governments and others (Keith, Naumov, and Sterman 2017).

Similar simulation-based SD analyses of market transformation problems have highlighted the importance of alignment of policy makers, NGOs and industry stakeholders, for example in the context of transitions to nutritious food (Struben, Chan, and Dubé 2014; Papachristos and Adamides 2016; Table 8.1, #5).

At the organisational level, SD work has looked at why, and how, firms might stay in their current performance trajectory, try to improve and change or decline. Illustrative examples directly relevant to transitions research involve both business model and organisational capabilities as unit of analysis.

Business model change

A business model (BM) refers to how a firm operates and creates value for its stakeholders (Casadesus-Masanell and Ricart 2010, 195). A BM change shifts the value proposition of the firm and the relationship to the customer, providing a source of market disruption (Teece 2010). BM change may be fundamental to reorienting organisational trajectories and the current socio-technical systems out of their current lock-in state. A BM perspective on transitions shifts the focus from the classic view of technology introduction in niches that compete against incumbent technologies (Geels and Schot 2007) towards disruptive BMs introduced by incumbents or new entrants that compete against legacy BMs (Wainstein and Bumpus 2016).

BM change in a single organisation can bring about concomitant business changes to other organisations that will aggregate up to the organisational field level. A BM change involves a change in the value offered in the market and in how it is generated. The value offered may change the preferences of users/customers and may be seen as a threat by competitors, who can potentially try to innovate on or imitate the BM if it is successful. A change in how value is generated with other supply-side actors may involve a change in their BMs, as in the case of a supply chain. The end result is likely to be change in the BMs of immediate collaborators and immediate competitors of an organisation. Hence, the question of disruptive and sustainable BM design and operation in relation to incumbent BMs relates to the question of bringing about and accelerating socio-technical transitions (e.g. Adamides, Papachristos, and Pomonis 2012; Papachristos and Adamides 2014; Papachristos 2014b).

The potential of SD application is large: for example, in the case of Ryanair (Casadesus-Masanell and Ricart 2010; Table 8.1, #6) highlights how SD, with CLDs as a representational tool, can facilitate research on BM strengths, weaknesses and interactions with other BMs. Thus, using SD one can explore BM change in a competitive environment, and the reorientation of socio-technical systems away from locked-in carbon-intensive trajectories towards a low-carbon state.

Organisational capability change

Organisations often forgo opportunities to improve their processes and capabilities even when there is clear return on investment (Lovins 2012). Organisations tend to get stuck in a capability trap (Table 8.1, #7) that is difficult to escape. Most organisations face pressures that lead managers to cut costs rather than improve processes, as they often lack the necessary slack to implement process- and sustainability-related programmes (Sterman 2015a). When an organisation tries to escape the capability trap, it inevitably exhibits worse before better behaviour and has to persist until its performance improves (Sterman 2015a). This behaviour arises because improving organisational capabilities and processes require re-allocating resources from value-generating processes to process-improvement activities, which temporarily reduces organisational performance. This weakens the belief that the organisation can sustain the sustainability investments needed to succeed; therefore, managers focus more on short-term improvements. The capability trap has direct implications for socio-technical systems research, as it arises in transition-related settings: the oil and chemical industries (Repenning and Sterman 2002), energy efficiency (Sterman 2015a), product development (Repenning 2001) and corporate strategy (Rahmandad 2012).

5.3 Micro level: communities in transition

Socio-technical system reorientation requires political pressure or public support to enact necessary policies and changes that span individual, city, state, national and supranational levels. Sustainability transitions are not conducive to "Manhattan project" approaches whereby experts provide advice or technical solutions (Sterman 2015a). Policy makers and the public need to understand the implications of possible decisions for climate and jointly act in a timely manner. Active societal learning and not just technology innovation is required to catalyse science-based environmental activism (Sterman 2015a, 2015b; see also Chapter 9 by Rojas and de Haan in this volume).

Efforts at the community level are required where there is a mismatch between macro-level institutions and local contexts or when public resources are not available (Dóci, Vasileiadou, and Petersen 2015). In such cases, users, technology suppliers and other local stakeholders can engage and co-construct niches. In developing countries, such niches can have a higher impact on cost reductions in renewable energy than overarching global technology curves (Huenteler, Niebuhr, and Schmidt 2014). In this way, community-led niches can influence renewable energy diffusion.

SD group model building (Vennix 1996) can help to shield, nurture and empower renewable energy communities vis-à-vis the established energy system (Smith and Raven 2012). This approach has already been applied at the community level, for example, in interventions to improve public safety within neighbourhoods (Table 8.1, #8). Through this process, SD can facilitate BM development for the community and negotiate the tensions between old and new BMs, incumbent and new market actors, the centralised versus the distributed technological paradigm and the societal shift from a passive to an active user role in the BM value chain (Wainstein and Bumpus 2016).

The strong scientific consensus on climate change has been available for years, but further progress on sustainability transitions depends on public education and mobilisation (Sterman 2008, 2011, 2015a). Humans develop mental models with narrow boundaries; they tend to promote a wait-and-see attitude on the public and policy-making side when the consequences of actions stretch out in space and time (Cronin, Gonzalez, and Sterman 2009). Mental model limitations cannot be remedied simply through information, education and citizen mobilisation about the climate; this change requires the provision of experiential learning environments (Sterman 2011). *World Climate* has been developed around the C-ROADS simulation platform (freely available at *climateinteractive.org*) and allows policy makers and citizens to experience the challenges of reducing global climate emissions in a role-play game setting that emulates the UN negotiations process for abatement commitments (Table 8.1, #9). *World Climate* has been played in over 80 countries, by over 45,000 participants. Such role-playing games, when designed for public engagement, can be pivotal for helping citizens realise the need for lifestyle changes related to sustainability, understand their role in this process and mobilise them to initiate critical action – within communities or through entrepreneurship. In this way, such systems thinking tools serve as action-learning tools for transitions research at the micro level.

6 Conclusions

This chapter focused on the contribution of SD research for transitions research and how SD can complement other transitions research methods. We argued for this potential in two ways. First, we highlighted the conceptual and methodological affinity of SD to transition studies, and we pointed to complementary foundational elements and key practices of SD research. Second, we discussed illustrative SD work that corresponds to, and provides insights relevant to the macro-, meso- and micro-level issues that transitions research aims to address. At the macro level, we highlighted how SD research on the global impact of climate change negotiations can catalyse change at the international level. At the meso level, we provided examples of how organisational and industry change may overcome obstacles towards sustainable pathways through the development and diffusion of alternative technological platforms, and the transformation of organisational capabilities and BMs. Finally, we pointed to research about learning at the micro levels, because transitions cannot be accelerated solely in a top-down way, but require bottom-up social change to complement and catalyse sustainability-related innovations. These illustrations could be the basis for future research, including work that cuts across levels (see Chapter 6).

Our approach in this chapter suggests how SD research could be valuable to address questions of interest to transitions researchers. In addition, one can consider the application of particular SD practices to certain steps of transitions research. For example, at the research questions stage, SD can help structure and scope the project, accompany the development of case-based work with CLDs and conceptualise system mechanisms in terms of reinforcing and balancing feedback loops (Figure 8.2), or demonstrate particular dynamics informed by management literature or group modelling with involved stakeholders.

Applying SD to transitions research involves also limitations and trade-offs. For example, transitions research questions often lead to analytical representations that are geographically explicit because individuals or organisations often participate within regional or international networks. The choice of models for policy analysis should consider the benefits as well as computational and cognitive costs of such disaggregation. Importantly, whereas geographical and social network considerations combine well with SD, such finer levels of analyses as well as generally make it harder to generate dynamic insights – a core strength of SD.

In summary, SD research in its first 60 years of existence (Sterman 2018) proved that it can catalyse systems learning, and thus it can be an indispensable approach for transitions research and change. SD began as a radically different view to change our understanding of systems, and the time is ripe to do this in transitions research, perhaps best initiated through cross-field collaborations.

Note

1 See, for example, the dialogues within the multi-disciplinary community (formalised into the Sustainable Transitions Research Network [STRN]) that has emerged to explore multiple aspects of research on the sustainability challenge (Van den Bergh et al. 2011).

References

Adamides, Emmanuel, George Papachristos, and Nikolaos Pomonis. 2012. "Critical realism in supply chain research: Understanding the dynamics of a seasonal goods supply chain." *International Journal of Physical Distribution & Logistics Management* 42 (10): 906–30.

Auvinen, Heidi, Sampsa Ruutu, Anu Tuominen, Toni Ahlqvist, and Juha Oksanen. 2015. "Process supporting strategic decision making in systemic transitions." *Technological Forecasting and Social Change* 94: 97–114.

Bakker, Sjoerd, Pieter Leguijt, and Harro van Lente. 2015. "Niche accumulation and standardization – The case of electric vehicle recharging plugs." *Journal of Cleaner Production* 94: 155–64.

Bergek, Anna, Staffan Jacobsson, Bo Carlsson, Sven Lindmark, and Annika Rickne. 2008. "Analyzing the functional dynamics of technological innovation systems: A scheme of analysis." *Research Policy* 37: 407–29.

Bergman, Noam, Alex Haxeltine, Lorraine Whitmarsh, Jonathan Köhler, Michel Schilperoord, and Jan Rotmans. 2008. "Modelling socio-technical transition patterns and pathways." *Journal of Artificial Societies and Social Simulation* 11 (3): 7.

Bianchi, Federico, and Flaminio Squazzoni. 2019. "Modelling and social science: Problems and promises." In *Modelling Transitions – Virtues, Vices, Visions of the Future*, edited by E. Moallemi, F. de Haan. This volume: Routledge.

Casadesus-Masanell, Ramon, and Joan E. Ricart. 2010. "From strategy to business models and onto tactics." *Long Range Planning* 43: 195–215.

Cole, H. Sam. 1973. *Models of Doom: A Critique of the Limits to Growth*. 1st hardcover ed. New York: Universe Publishing.

Cosenz, Federico, and Guido Noto. 2018. "A dynamic business modelling approach to design and experiment new business venture strategies." *Long Range Planning* 51: 127–40.

Cronin, Matthew A., Cleotilde Gonzalez, and John D. Sterman. 2009. "Why don't well-educated adults understand accumulation? A challenge to researchers, educators, and citizens." *Organizational Behavior and Human Decision Processes* 108: 116–30.

Cyert, Richard, and James G. March. 1963. *A Behavioral Theory of the Firm*. Englewood Cliffs, NJ: Prentice Hall.

Dóci, Gabriella, Eleftheria Vasileiadou, and Arthur C. Petersen. 2015. "Exploring the transition potential of renewable energy communities." *Futures* 66: 85–95.

Epstein, Joshua M. 2007. *Generative Social Science: Studies in Agent-Based Computational Modeling*. Princeton, NJ: Princeton University Press.

Fiddaman, Thomas. 2002. "Exploring policy options with a behavioral climate-economy model." *System Dynamics Review* 18 (2): 243–67.

Fiddaman, Thomas. 2007. "Dynamics of climate policy." *System Dynamics Review* 23 (1): 21–34.

Forrester, Jay W. 1958. "Industrial dynamics: A major breakthrough for decision makers." *Harvard Business Review* 26: 37–66.

Forrester, Jay W. 1961. *Industrial Dynamics*. Cambridge, MA: MIT Press.

Forrester, Jay W. 1968. *Principles of Systems*. Waltham, MA: Pegasus Communications.

Forrester, Jay W. 1969. *Urban Dynamics*. Cambridge, MA: MIT Press.

Forrester, Jay W. 1971. *World Dynamics*. Cambridge, MA: MIT Press.

Forrester, Jay W., and P.M. Senge. 1980. "Tests for building confidence in system dynamics models." In *System Dynamics: TIMS Studies in the Management Sciences*, edited by A.A. Lagasto, J.W. Forrester, and J.M. Lyneis. Oxford: North-Holland, 209–28.

Fouquet, Roger, and Peter J.G. Pearson. 2012. "Past and prospective energy transitions: Insights from history." *Energy Policy* 50: 1–7.

Gawer, Annabelle, and Michael A. Cusumano. 2014. "Industry platforms and ecosystem innovation." *Journal of Product Innovation Management* 31: 417–33.

Geels, Frank W. 2002. "Technological transitions as evolutionary reconfiguration processes: A multi-level perspective and a case-study." *Research Policy* 31: 1257–74.

Geels, Frank W. 2011. "The multi-level perspective on sustainability transitions: Responses to seven criticisms." *Environmental Innovation and Societal Transitions* 1 (1): 24–40.

Geels, Frank W., Frans Berkhout, and Detlef P. van Vuuren. 2016. "Bridging analytical approaches for low-carbon transitions." *Nature Climate Change* 6: 576–83.

Geels, Frank W., Florian Kern, Gerhard Fuchs, Nele Hinderer, Gregor Kungl, Josephine Mylan, Mario Neukirch, and Sandra Wassermann. 2016. "The enactment of socio-technical transition pathways: A reformulated typology and a comparative multi-level analysis of the German and UK low-carbon electricity transitions (1990–2014)." *Research Policy* 45: 896–913.

Geels, Frank W., and Johan Schot. 2007. "Typology of sociotechnical transition pathways." *Research Policy* 36: 399–417.

Hekkert, Marko P., Roald A.A. Suurs, Simona O. Negro, Stefan Kuhlmann, and Ruud E.H.M. Smits. 2007. "Functions of innovation systems: A new approach for analysing technological change." *Technological Forecasting and Social Change* 74 (4): 413–32.

Holtz, Georg, Floortje Alkemade, Fjalar de Haan, Jonathan Kohler, Evelina Trutnevyte, Tobias Luthe, Johannes Halbe, George Papachristos, Emille Chappin, Jan Kwakkel, and Sampsa Ruutu. 2015. "Prospects of modelling societal transitions: Position paper of an emerging community." *Environmental Innovation and Societal Transitions* 17: 41–58.

Holtz, Georg, and Emille L. Chappin. 2019. "Considering actor behaviour: Agent-based modelling of transitions." In *Modelling Transitions – Virtues, Vices, Visions of the Future*, edited by E. Moallemi and F. de Haan. This volume: Routledge.

Huenteler, Joern, Christian Niebuhr, and Tobias S. Schmidt. 2014. "The effect of local and global learning on the cost of renewable energy in developing countries." *Journal of Cleaner Production* 128: 1–16.

IPCC. 2018. *Global Warming of 1.5°C. An IPCC Special Report on the Impacts of Global Warming of 1.5°C Above Pre-industrial Levels and Related Global Greenhouse Gas Emission Pathways, in the Context of Strengthening the Global Response to the Threat of Climate Change, Sustainable Development, and Efforts to Eradicate Poverty*. World Meteorological Organization, Geneva, Switzerland.

Keith, David R., Sergey Naumov, and John D. Sterman. 2017. "Driving the future: A management flight simulator of the US automobile market." *Simulation and Gaming* 48 (6): 735–69.

Keith, David R., John D. Sterman, and Jeroen Struben. 2017. "Supply constraints and waitlists in new product diffusion." *System Dynamics Review* 33 (3–4): 254–79.

Kemp, Rene, Johan Schot, and Remco Hoogma. 1998. "Regime shifts to sustainability through processes of niche formation: The approach of strategic niche management." *Technology Analysis and Strategic Management* 10: 175–95.

Köhler, Jonathan 2019. "Modelling the multi-level perspective: The MATISSE agent-based model." In *Modelling Transitions – Virtues, Vices, Visions of the Future*, edited by E. Moallemi and F. de Haan. This volume: Routledge.

Köhler, Jonathan, Fjalar de Haan, Georg Holtz, Klaus Kubeczko, Enayat A. Moallemi, George Papachristos, and Emille Chappin. 2018. "Modelling sustainability transitions: An assessment of approaches and challenges." *Journal of Artificial Societies and Social Simulation* 21 (1): 8.

Köhler, Jonatha, and Georg Holtz. 2019. "Transitions modelling: Status, challenges and strategies." In *Modelling Transitions – Virtues, Vices, Visions of the Future*, edited by E. Moallemi and F. de Haan. This volume: Routledge.

Lane, David C. 1992. "Modelling as learning: A consultancy methodology for enhancing learning in management teams." *European Journal of Operational Research* 59: 64–84.

Lane, David C. 1999. "Social theory and system dynamics practice." *European Journal of Operational Research* 113: 501–27.

Lane, David C. 2001. "Opportunities generated by the agency/structure debate and suggestions for clarifying the social theoretic position of system dynamics." *System Dynamics Review* 17: 293–309.

Lane, David C. 2017. "Behavioural system dynamics: A very tentative and slightly sceptical map of the territory." *Systems Research and Behavioral Science* 34 (4): 414–23.

Levitt, Barbara, and James G. March. 1988. "Organizational learning." *Annual Review of Sociology* 14: 319–40.

Lovins, Amory 2012. *Reinventing Fire*. White River Junction, VT: Chelsea Green.

March, James G., and Herbert Simon. 1958. *Organizations*. Cambridge, MA: John Wiley & Sons.

Meadows, Donella H., Dennis L. Meadows, Jorgen Randers, and William W. Behrens III. 1972. *The Limits to Growth: A Report for the Club of Rome's Project on the Predicament of Mankind*. New York: Universe Books.

Mingers, John 2014. *Systems Thinking, Critical Realism and Philosophy: A Confluence of Ideas*. Oxon: Routledge.

Moallemi, Enayat A., Lu Aye, Fjalar J. de Haan, and John M. Webb. 2017. "A dual narrative-modelling approach for evaluating socio-technical transitions in electricity sectors." *Journal of Cleaner Production* 162: 1210–24.

Moallemi, Enayat A., Fjalar de Haan, and Jonathan Köhler. 2019. "Exploratory modelling of transitions: An emerging approach for coping with uncertainties in transitions research." In *Modelling Transitions – Virtues, Vices, Visions of the Future*, edited by E. Moallemi and F. de Haan. This volume: Routledge.

Moallemi, Enayat A., Fjalar de Haan, Jan Kwakkel, and Lu Aye. 2017. "Narrative informed exploratory analysis of energy transition pathways a case study of India's electricity sector." *Energy Policy* 110: 271–87.

Morrison, J. Bradley, and Rogelio Oliva. 2018. "Integration of behavioral and operational elements through system dynamics." In *The Handbook of Behavioral Operations*, edited by K. Donohue, E. Katok, and S. Leider. New York: Wiley.

Nordhaus, William D. 1973. "World dynamics: Measurement without data." *The Economic Journal* 83 (332): 1156–83.

Papachristos, George. 2011. "A system dynamics model of socio-technical regime transitions." *Environmental Innovation and Societal Transitions* 1: 202–33.

Papachristos, George. 2012. "Case study and system dynamics research: Complementarities, pluralism and evolutionary theory development." *30th International Conference of the System Dynamics Society*, St Gallen, Switzerland online proceedings. ISBN 978-1-935056-10-2.

Papachristos, George. 2014a. "Towards multi-system sociotechnical transitions: Why simulate." *Technology Analysis and Strategic Management* 26 (9): 1037–55.

Papachristos, George. 2014b. "Transition inertia due to competition in supply chains with remanufacturing and recycling: A systems dynamics model." *Environmental Innovation and Societal Transitions* 12: 47–65.

Papachristos, George. 2017. "Diversity in technology competition. The link between platforms and sociotechnical transitions." *Renewable and Sustainable Energy Reviews* 73: 291–306.

Papachristos, George. 2018a. "A mechanism based transition research methodology: Bridging analytical approaches." *Futures* 98: 57–71.

Papachristos, George. 2018b. "System dynamics modelling and simulation for sustainability transition research." *Environmental Innovation and Societal Transitions* 31: 248–261. https://doi.org/10.1016/j.eist.2018.10.001.

Papachristos, George, and Emmanuel. Adamides. 2014. "Internal supply-chain competition in remanufacturing: Operations strategies, performance and environmental effects." *International Journal of Logistics Systems and Management* 19 (2): 187–211.

Papachristos, George, and Emmanuel. Adamides. 2016. "A retroductive systems-based methodology for socio-technical transitions research." *Technological Forecasting and Social Change* 108: 1–14.

Papachristos, George, Aristotelis Sofianos, and Emmanuel Adamides. 2013. "System interactions in socio-technical transitions: Extending the multi-level perspective." *Environmental Innovation and Societal Transitions* 7: 53–69.

Papachristos, George, and Geerten Van de Kaa. 2018. "Understanding platform competition through simulation: A research outline." *Technology Analysis and Strategic Management* 30 (12): 1409–21.

Rahmandad, Hazhir. 2012. "Impact of growth opportunities and competition on dynamics of capability development." *Organization Science* 23 (1): 138–54.

Rahmandad, Hazhir, and John D. Sterman. 2008. "Heterogeneity and network structure in the dynamics of diffusion: Comparing agent-based and differential equation models." *Management Science* 54 (5): 998–1014.

Repenning, Nelson 2001. "Understanding fire-fighting in new product development." *Journal of Product Innovation Management* 18 (5): 285–300.

Repenning, Nelson, and John D. Sterman. 2002. "Capability traps and self-confirming attribution errors in the dynamics of process improvement." *Administrative Science Quarterly* 47 (2): 265–95.

Richardson, George P. 2011. "Reflections on the foundations of system dynamics." *System Dynamic Review* 27 (3): 219–43.

Rotmans, Jan, Rene Kemp, and Marjolein van Asselt. 2001. "More evolution than revolution: Transition management in public policy." *Foresight* 3: 15–31.

Rouwette, Etiënne, Inge Bleijenbergh, and Jac Vennix. 2016. "Group model-building to support public policy: Addressing a conflicted situation in a problem neighborhood." *Systems Research and Behavioral Science* 33 (1): 64–78.

Sayer, Andrew. 2000. *Realism and Social Science*. London: SAGE Publications.

Smith, Adrian, and Rob Raven. 2012. "What is protective space? Reconsidering niches in transitions to sustainability." *Research Policy* 41: 1025–36.

Smith, Adrian, Jan-Peter, Voß, and John Grin. 2010. "Innovation studies and sustainability transitions: The allure of the multi-level perspective and its challenges." *Research Policy* 39: 435–48.

Sorrell, Steven. 2018. "Explaining sociotechnical transitions: A critical realist perspective." *Research Policy* 47: 1267–82.

Sterman, John D. 1982. *The Energy Transition and the Economy: A System Dynamics Approach* (Doctoral diss.), Massachusetts Institute of Technology.

Sterman, John D. 1989. "Misperceptions of feedback in dynamic decision making." *Organizational Behaviour and Human Decision Processes* 43 (3): 301–35.

Sterman, John D. 1994. "Learning in and about complex systems." *System Dynamics Review* 10 (2–3): 291–330.

Sterman, John D. 2000. *Business Dynamics: Systems Thinking and Modelling for a Complex World*. New York: McGraw Hill.

Sterman, John D. 2002. "All models are wrong: Reflections on becoming a systems scientist." *System Dynamics Review* 18 (4): 501–31.

Sterman, John D. 2008. "Risk communication on climate: Mental models and mass balance." *Science* 322: 532–3.

Sterman, John D. 2011. "Communicating climate change risks in a skeptical world." *Climatic Change* 108: 811–26.

Sterman, John D. 2015a. "Stumbling towards sustainability: Why organizational learning and radical innovation are necessary to build a more sustainable world – But not sufficient." In *Organizational and Strategic Change and the Challenge of Sustainability*, edited by R. Henderson, M. Tushman, and R. Gulati. Oxford: Oxford University Press.

Sterman, John D. 2015b. "Learning for ourselves: Interactive simulations to catalyze science-based environmental activism." In *Science-Based Activism*, edited by P.E. Stoknes and K.A. Eliassen, 253–79. Bergen: Fagbokfolaget.

Sterman, John D. 2018. "System dynamics at sixty: The path forward." *System Dynamics Review* 34 (1–2): 5–47.

Sterman, John D., Thomas Fiddaman, Travis Franck, Andrew Jones, Stephanie McCauley, Philip Rice, Elizabeth Sawin, and Lori Siegel. 2013. "Management flight simulators to support climate negotiations." *Environmental Modelling and Software* 44: 122–35.

Sterman, John D., Rogelio Oliva, Kevin Linderman, and Elliott Bendoly. 2015. "System dynamics perspectives and modeling opportunities for research in operations management." *Journal of Operations Management* 39–40: 1–5.

Struben, Jeroen, Derek Chan, and Laurette Dubé. 2014. "Policy insights from the nutritional food market transformation model: The case of obesity prevention." *Annals of the New York Academy of Sciences* 1331 (1): 57–75.

Struben, Jeroen, and John D. Sterman. 2008. "Transition challenges for alternative fuel vehicle and transportation systems." *Environment and Planning B: Urban Analytics and City Science* 35: 1070–97.

Teece, David. 2010. "Business model, business strategy and innovation." *Long Range Planning* 43 (2–3): 172–94.

Turner, Graham M. 2014. "Is global collapse imminent? An updated comparison of the limits to growth with historical data." *MSSI Research Paper No. 4*, Melbourne Sustainable Society Institute, The University of Melbourne.

Turnheim, Bruno, and Frank W. Geels. 2012. "Regime destabilisation as the flipside of energy transitions: Lessons from the history of the British coal industry (1913–1997)." *Energy Policy* 50: 35–49.

Turnheim, Bruno, Frans Berkhout, Frank W. Geels, Andries HOf, Andy McMeekin, Björn Nykvist, and Detlef van Vuuren. 2015. "Evaluating sustainability transition pathways: Bridging analytical approaches to address governance challenges." *Global Environmental Change* 35: 239–253.

Van de Kaa, Geerten, George Papachristos, and Hans. de Bruijn. 2019. "The governance of platform development processes: A metaphor and a simulation model." *Technology Forecasting and Social Change* 138: 190–203.

Van den Bergh, Jeroen C.J.M., Bernhard Truffer, and Giorgos Kallis. 2011. "Environmental innovation and societal transitions: Introduction and overview." *Environmental Innovation and Societal Transitions* 1: 1–23.

Vennix, Jac A.M. 1996. *Group Model-building: Facilitating Team Learning Using System Dynamics.* Chichester: Wiley.

Voss, Jan-Peter, Dierk Bauknecht, and Rene. Kemp. 2006. *Reflexive Governance for Sustainable Development.* Cheltenham: Edward Elgar.

Wainstein, Martin E., and Adam G. Bumpus. 2016. "Business models as driver of the low carbon power system transition: A multi-level perspective." *Journal of Cleaner Production* 126: 572–85.

Walrave, Bob, and Rob Raven. 2016. "Modelling the dynamics of technological innovation systems." *Research Policy* 45: 1833–44.

9 Socio-technical representation of electricity provision across scales

Angela Rojas and Fjalar J. de Haan

1 Introduction

Systems of service provision, the infrastructure to meet society's needs for energy, food, water and sanitation, must transition to be low-carbon and resilient systems if we are to achieve humanity's climate change mitigation and adaptation objectives. In many of these systems, the transition is happening. We should now focus on where it can take us and what that may mean (e.g. in sustainability terms).

These systems of service provision have been often described as socio-technical systems. However, these descriptions usually suffer from a lack of detailed representation of the system's components. To understand how these systems work, it is useful to break them down into their constituent parts and analyse their interactions.

Current theories and frameworks like the multi-level perspective, for example in Geels (2002) rendition, has significantly contributed to advance the understanding of socio-technical change. The multi-level perspective discusses socio-technical system's technology and organisations but only in the broadest of terms. While this may work well for historical narratives, we need an analytical framework for a precise, formal representation of socio-technical components and relationships. This provides not only a basis for software implementations, with which systems can be interrogated against various metrics, but also as an analytical tool to understand how the system works.

The analysis of socio-technical systems in transition also poses its challenges as it involves technical and societal re-configurations. On the technological side, the analysis may become scattered across scales. In the electricity sector, electricity is generated and used on-site, while the regional and national grids remain key components of the system. The increased penetration of distributed generation leads to voltage-rise effects and power-quality issues that compromise the long-term stability of the system (Lopes et al. 2007). However, if managed properly within the existing infrastructure, penetration of distributed energy resources may as well provide significant services to the grid.

On the societal side, analytical challenges are associated with emerging organisational forms (e.g. energy hubs, community energy groups, prosumer groups), novel ownership models, institutional structures and resources that may constrain or enable sustainability transitions. Also, ownership and management of co-production and other alternative models for electricity provision shift the balance of production and consumption, leading to broader changes on how socio-technical systems for electricity provision are shaped and their sustainability impacts in the future.

Integrating these understandings enables analyses across scales and of the inter-relations between contractual (e.g. financial) and energy exchanges. An example of the former would be analysing local, distributed solutions in the context of a grid-based centralised system, while the latter could be an enquiry into their effects on prices and capacity. Recognising these aspects in a socio-technical system at different points in time (i.e. snapshots) contributes to a precise, quantifiable understanding of the dynamics of the transition. This is where a formalised analytical framework to query the socio-technical system at different points in time is needed.

Some transition modellers have recognised this need and have proposed to formalise socio-technical systems using computational ontology or relational diagrams. Van Dam, Nikolic, and Lukszo (2012) propose a formalisation of socio-technical systems to describe the key concepts in terms of actors and technical components, as well as the physical and economical properties of the system. In this instance, the authors propose a system formalisation with the purpose of using it in agent-based model simulations.

In the energy sector, (Chappin 2011, 80) represented the power production system in socio-technical terms, highlighting the technical network with the physical flows between assets, as well as the social network and markets. We would like to contribute to this body of work. Our aim is to provide a formalised way to represent and visualise socio-technical systems, with a focus on the electricity system. The representation handles (re)configurations of infrastructures across scales (e.g. spatial, temporal, technological, functional and organisational) as well as the various social configurations that make them work.

This analytical framework is being developed in the context of a computational modelling effort. It provides the formal basis for software implementations. A static implementation – that is, representing system states at particular times – is being undertaken at the time of writing. Additionally, the framework provides a basis for a simulation model. However, the key contribution is provided by the static implementation already: a model representation that can be interrogated programmatically (i.e. with software) of the social and technical aspects of electricity systems across scales and, moreover, the interactions between both aspects. The greater definitional rigour demanded by the analytical representation facilitates subsequent comparability of cases and a cumulative body of knowledge. If it is used consistently, it will contribute to a systematic and comparable empirical dataset of socio-technical configurations that will help to advance the understanding of transitions, a significant need recognised by other transitions researchers in the area (see Chapter 4 by deHaan and Chapter 12 by deHaan, Martínez and Spekkink in this volume).

We consider this framework as a complementary visual and analytical tool that binds traditional transitions frameworks and computational modelling together.

We believe a transition analysis needs to be based on a representation of socio-technical systems that does justice to its configuration, including infrastructures on and across scales, and the organisational networks that enable them to perform their societal function – a representation based on a networked understanding of the organisational (i.e. 'socio') aspects and a nested-systems understanding of the infrastructural (i.e. 'technical') aspects of socio-technical systems.

The chapter is organised as follows. Firstly, we introduce the analytical framework for the electricity system in detail (Section 2), starting with the elements of socio-technical systems to establish the vocabulary (Section 2.1), followed by the relationships the framework conceptualises (Section 2.2) and finishing with the integrated picture (Section 2.3). We then illustrate the framework's usage and usefulness, applying it to three cases, using residential demand in the electricity sector in Victoria, Australia (Section 3). The first case (Section 3.1) is a snapshot of the electricity system in Victoria when it was a public monopoly restructured into three statutory companies operating the electricity system (1993–1994). The second case is a snapshot of the Eastern Australian system, the so-called National Energy Market (Section 3.2) in which the electricity system of Victoria is part of. This case is a grid-based system with a centralised distribution and governance structure, which is a 'business-as-usual' case for many developed nations. The third is a distributed system configuration. The case includes the explanation of this alternative and a case of a neighbourhood microgrid currently under development in Victoria (Section 3.3). We make some suggestions on how the analytical framework could be used and some limitations in Section 4. Finally, we present the conclusions in Section 5.

2 Analytical framework

The main task of the analytical framework we present in this chapter is to enable a formalised *representation* of an electricity system to capture its configuration at any point in time and to explore how a socio-technical system could look in the future.

Systems of service provision have been described as socio-technical systems, which are (networks of) actors, institutions, material artefacts and knowledge constantly interacting to provide specific services to society (Geels 2004; Markard, Raven, and Truffer 2012; Weber 2003). Most recently, Geels et al. 2017 defined socio-technical systems as 'the interlinked mix of technologies, infrastructures, organisations, markets, regulations and user practices that together deliver social function'. Thus, a socio-technical transition should exhibit a profound alteration of the way these socio-technical components interlink to deliver social functions.

Therefore, a representation of socio-technical systems should strive to include these definitional components, including those in the social (i.e. organisational, demand) and in the technical (i.e. infrastructures, and technologies) dimensions. As the exploration of alternative ways of electricity provision is one of the key envisioned uses of the framework, it needs to be able to represent both conventional layouts of a socio-technical electricity system and a broad spectrum of alternatives. What components does our framework need to represent all these systems? What are the common elements?

2.1 Socio-technical components of electricity systems

Currently, a conventional system for electricity provision uses centralised power plants, usually of a large size. Most of the time, this infrastructure is energy, carbon and capital intensive. Electricity is transported using long-distance grids, known as transmission and distribution networks, from the power station to the place where it is going to be used. This infrastructure was designed to allow the unidirectional flow of electricity.

In a conventional configuration, end-users receive electricity and pay to their retailer of choice. Retailers and other aggregators distribute the payments to grid operators and generators. These transactions go 'up the chain' via markets and other regulated transactions. The system is designed to provide electricity as a service; for this, it needs to balance supply and demand constantly, and to do so it uses automatic controls, reserves and market-based mechanisms.

An alternative configuration would, for example, use on-site generation and possibly some type of storage. It would take advantage of a microgrid configuration and information and communication technologies (ICT) to manage the system more efficiently. The technologies that facilitate this configuration may already co-exist, though to a limited extent, with conventional systems.

For example, solar photovoltaic (PV) systems are integrated with current infrastructure. However, high penetration of intermittent generation, if not well configured and managed, poses challenges to conventional infrastructure. The lack of preparedness (e.g. bidirectional flows, limited visibility of new generation, lack of technical controls and organisational coordination mechanisms) of current infrastructure to manage a higher share of variable capacity increases the complexity of the supply–demand balancing tasks.

Similarly, the flow of monetary transactions also changes. Current mechanisms such as markets specialised on balancing services will need to accommodate aggregated small-scale generation that reaches the conventional grid. Also, instruments such as feed-in-tariffs enable the users with on-site generation to participate in the system as small service providers. The pace of on-site generation uptake, mainly of renewable technologies and storage, poses significant challenges not only on the technical management of current grid infrastructure but also to its governance arrangements.

On the technical side, one observes that the system is naturally represented in terms of *assets*. The relationships between assets are their physical connections over which the electricity flows. More specifically, we identify the broad classes of assets for generation, transportation, storage and consumption (see Table 9.1).

Table 9.1 Main technical components of the electricity system

Assets	*Examples*
Generation	Coal-fired or otherwise fossil-fuelled power plants, nuclear, PV panels, hydroelectric dams and wind farms.
Transportation	All types of grids and related assets used to transport electricity: (inter) national, regional, local grids. Also, microgrids, grids in buildings, as well as terminal stations, substations and feeders.
Storage	Gravitational/pumped hydro, household or neighbourhood electrochemical batteries, and molten salt.
Consumption	Anything connected to the electricity system that consumes energy. Household appliances are aggregated into 'dwellings'.

It is safe to assume that each asset has an inbuilt control feature; some assets are able to provide balancing services to the system through different devices (e.g. shunt capacitors, inductors, batteries) or has manual or automatic control systems and market mechanisms. Whether an asset provides such a service or not depends on the functional role the asset has within the system. For example, a generator or a consumption asset can be configured to provide some balancing mechanism (e.g. frequency control, demand response) and participate in the market under specific rules.

On the social side, the entities are naturally represented in terms of *organisations*. Though here broad classes can also be identified, like retailers, operators, and regulators, organisations are more variegated than assets. It is common in the socio-technical literature to refer to these entities as *actors* (Geels 2004; Markard, Raven, and Truffer 2012; Weber 2003; see also Chapter 7 by Holtz and Chappin in this volume).

An actor can be an individual or a group of individuals that are part of an organisation; therefore, we will use 'actor' and 'organisation' interchangeably. Actors partake in two kinds of relationships, namely those with other (actors) organisations, which we will call actor–actor relationships, and those between organisations and assets, which we will call actor–asset relationships. Looking at how actor–actor relations are organised will help to clarify both actor–actor and actor–asset relationships.

The relationships on the social side of an electricity provision system are ultimately geared at connecting energy production with energy consumption – this is the societal need the system is there for. On the technical side, the assets provide the electricity production. To turn electricity production into provision, actors engage in *transactions* and *contracts*, of which the latter can be thought of as a continuing or repeated transaction.

This means that another important class of entities in this analytical framework is that of *markets*. Beckert (2009, 248) defines *markets* as 'arenas of social interaction. They provide a social structure and institutional order for the voluntary exchange of rights in goods and services, which allows actors to evaluate, purchase, and sell these rights'.[1]

Table 9.2 Actor roles based on the relation to market institutions (rules)

	Actors	*Description*
Rule makers	Governmental institutions, other authorities	Governance: providing laws and regulation, standards.
Rule implementers	Regulatory bodies, market operators	Implementation of rules: coordinating markets, compliance enforcement.
Rule followers	Electricity generators, grid operators, retailers, households, banks, brokers	Comply with rules: owning, leasing, exploiting assets. May occasionally participation in rule making.

In our framework, a market will be a context (arena) in which a particular set of rules (institutions, regulations) applies to interactions (transactions, contracts) between actors. However, it can also be regarded as an operational arena, where standards and regulations are designed to coordinate technical features of the service. Whether the arena is used for coordinating the technical operation of the system, the monetary transactions or both, we call it market in our framework.

On the technical side, an operator has to balance electricity supply and demand at all times. Usually, the operator first approves the interested participants in the system. This is done according to technical standards embedded in regulatory frameworks. Then, the operator coordinates the physical system and the participants using control (e.g. frequency signals and SCADAS) and market mechanisms (e.g. bidding processes). These (arenas) markets are designed and regulated (i.e. institutions) by institutional actors. Also, they are operated and supervised by organisations.

From the above, we find that with reference to actor–actor relations, the following broad classes provide useful distinctions in terms of the overall governance of the system: *rule makers*, *rule implementers* and *rule followers* (see Table 9.2), where 'rules' refers to market institutions. As assets are used to generate electricity for *some* market, organisations' relationships with assets are also delimited by market and other institutional rules.

Consequently, one finds actors who directly own or exploit assets amongst the rule followers.[2] Separate actors take responsibility of rule making and rule implementing. A description of these categories and example of actors can be seen in Table 9.2.

2.2 Socio-technical layout: networks and recursion

With the main components and their relationships outlined, we can now put these together for a representation of the system and its structure. We call this the *socio-technical layout*. Significantly, the socio-technical layout has the characteristics of a recursive and networked system, which we explain in this section.

From the perspective of an electricity consumer, the 'end-user', the electricity system might as well be a black box. The consumer pays for the service (i.e. bill) and electricity comes out of an *interface*, which, for simplicity, we can think of as a wall outlet at the end side of the system. Behind this interface, there are the generating, transporting and storage assets that allow electricity to be consumed.

Usually the user remains unaware of how the system behind the interface is wired to accomplish this feat. Notably, this does not only hold from the perspective of an end-user such as a household. Up the chain, a distributor, say, may obtain electricity from 'further up' and likewise receive it from some interface without necessarily having to know exactly how the system is wired and controlled behind it.

The concept of the interface helps to distinguish one system from the other and its components. It serves for instance to define *system boundaries*. At each such interface, the electricity system behind it could be considered a black box providing electricity to an organisation, whether that is a household or some intermediate organisation like a distribution business. That organisation would engage in a monetary transaction for the service, indicating that the flow of electricity has a monetary flow counterpart. Analysing how the money flows relate to the electricity flows is one of the key aims of this framework.

Thus, from the perspective of organisations at every interface from generation to consumption, the electricity system could be considered a black box. This must mean that each such box contains other ones, and we essentially have a recursive system. We will call the boxes *service provision modules*, abbreviated as SPMs.

SPMs contain generation and transportation assets and possibly storage assets. The recursivity comes in via an SPM's generation assets – each of which may be an asset *or* the interface of a nested SPM. An example would be a national electricity system (the perspective of organisations 'high up the chain'), with several generation assets, (e.g. coal-fired and gas power plants), as well as the facility to import electricity from neighbouring regions. Figure 9.1a schematically depicts the main components of an SPM and Figure 9.1b shows a linear recursion.

Assets are connected to actors; they are owned, leased and exploited up to where one or several organisations (e.g. distribution business) provide an interface. These actors engage in contracts and transactions to recompense for the electricity services provided to each other.

In other words, each SPM has an associated network of actors with links formed by transactions and contracts in the context of a market. These are again the rule-following actors of Table 9.2. The market (there may be several) often has an associated organisation, such as a market operator, responsible for its functioning and to enforce compliance with its rules. These are Table 9.2's rule implementers.

Rule makers are typically governments or government agencies. Other entities provide standards and protocols that might end up being part of regulations. Though other actors are at times part of the rule-making processes, this depends on the availability of regulatory avenues to do so or the actors' market power.

A minimal socio-technical layout is depicted in Figure 9.2. Here, rule-follower actors (a) and (b) interact through the market. Most of the time (a) actors hold a direct relationship with the assets, whereas (b) actors can be considered assetless actors. The dashed lines represent the relationships between assets and (a) actors. The straight lines represent the relationships between

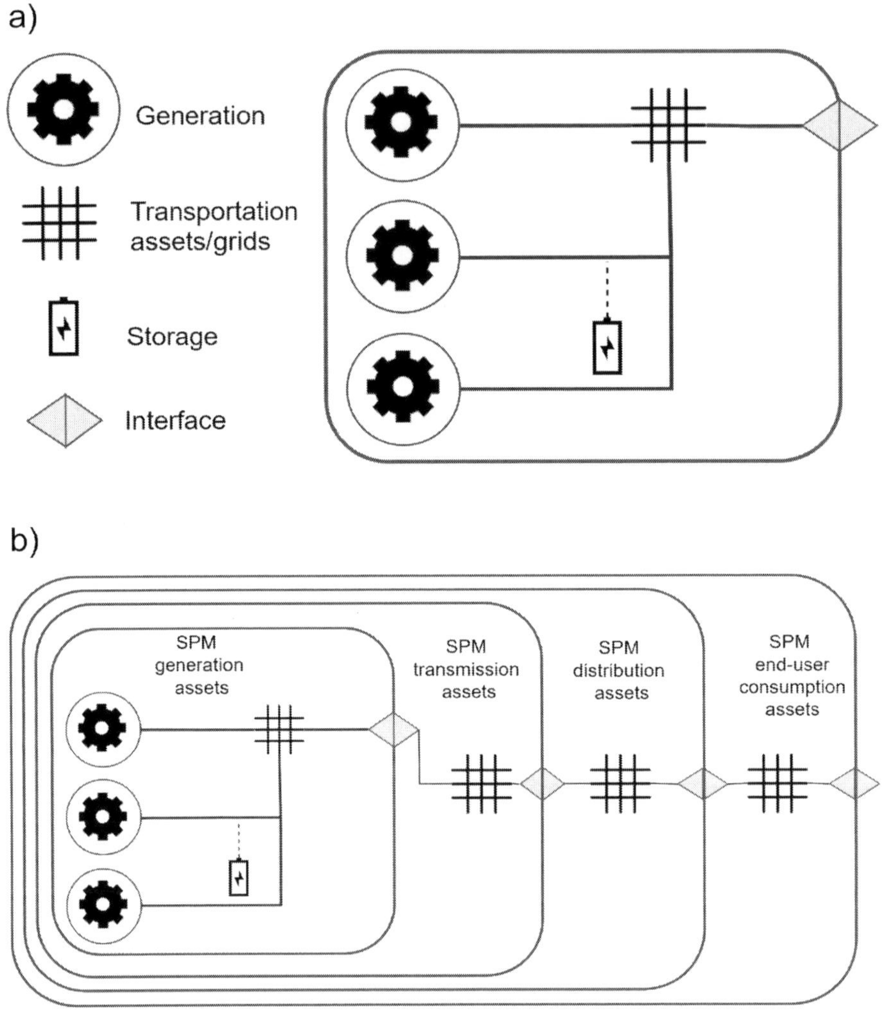

Figure 9.1 Service provision module (SPMs). Top: SPM components. Bottom: Linearly recursing SPMs, from the household perspective up.

(a) and (b) actors, while the dotted lines represent the relationship (c) actors (e.g. regulators) and (d) actors (e.g. operators) hold via markets.

We use arrows to highlight the directionality of the relationships. Bidirectional arrows are mainly buy-and-sell transactions under some type of contractual mechanism set by the market institution; unidirectional arrows signal the actor's role over the arena. For instance, (c) actors are the rule makers in the system. They are the governmental entities regulating the markets and are connected with a dotted line to the rules box located within the market.

In some markets, actors know who they are making transactions with (e.g. a household knows who its electricity retailer is). However, there are other types of markets, such as the wholesale market, where participants engage in transactions but do not necessarily know who their counterpart is. For instance, in the wholesale market, the operator is the intermediary actor coordinating the transactions between retailers and electricity producers. This type of market is the one depicted in the market 'box' of Figure 9.2 with the multiple directional arrows between the actors (small circles in the box) engaging

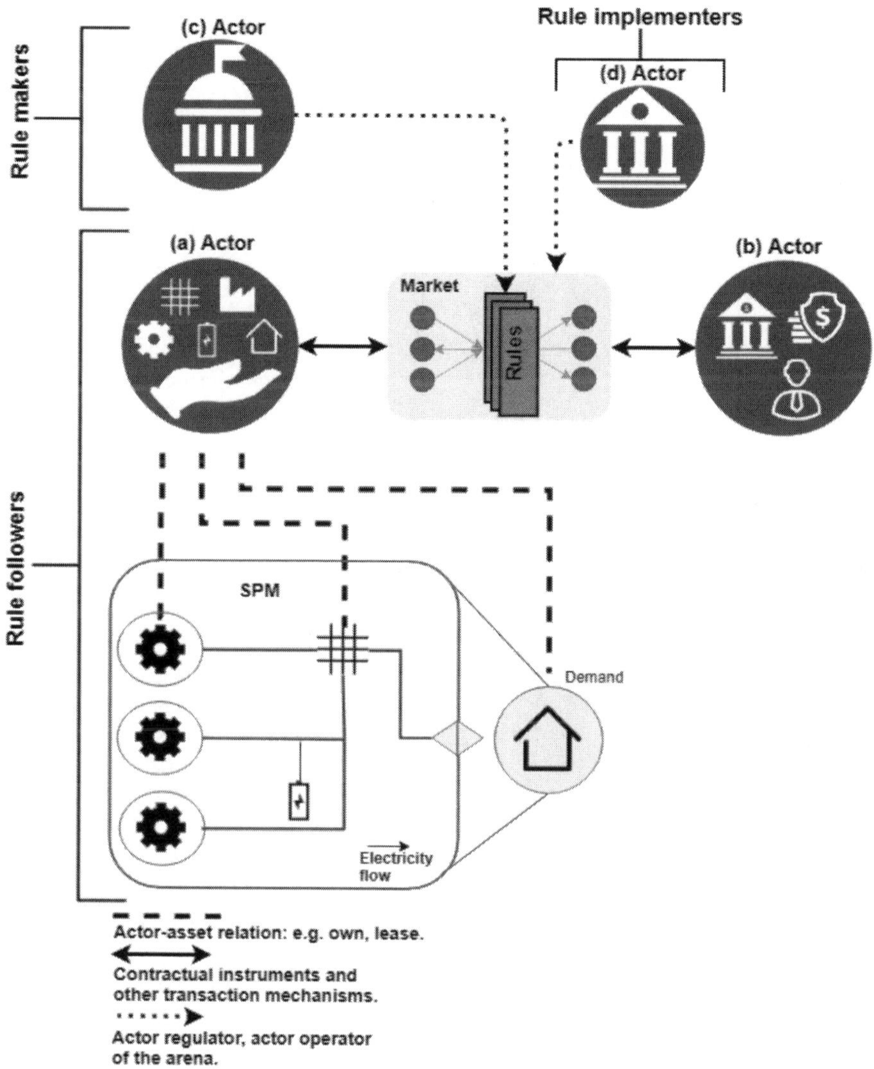

Figure 9.2 Socio-technical layout with SPM and actors transacting in a market context.

in transactions framed by regulations and standards (inner rectangular box). Also, a wholesale market may be operated by a (d) actor, which can be one (or more) organisation (s).

2.3 Preparing for change

Though we will not be dealing with socio-technical dynamics in detail in this chapter, the analytical framework needs to be prepared to deal with change. Changes in a socio-technical layout can be represented by adding or subtracting components (i.e. organisations, assets and arenas) or relations amongst them. For assets and matters like ownership and perhaps most transactions, this seems sufficient.

A complicating factor is that organisations can change considerably over time, with significant consequences for the layout and dynamics of the system. For example, a community energy group may start out as an unofficial organisation sharing the power from their communally owned solar panels. They may subsequently form a cooperative, grow and produce a commercial proprietary-limited company offshoot. What we mean is that organisations can change their *organisational form*. This has consequences for the kind of assets organisations may own or exploit and how they engage in markets.

Organisational form is what Romanelli (1991, pp. 81–82) refers to as 'those characteristics of an organisation that identify it as a distinct entity and, at the same time, classify it as a member of a group of similar organizations'. This is still a very broad delineation of the concept. Practical proxies for the organisational form in the context of this framework are the business structure and the ownership model. However, these can be extended to other characteristics depending on the analysis, which may include dates of the organisation's registration, the date and type of status change (e.g. name, merging, dissolution, disaggregation, etc.) and the type of economic activity.

In general, business structures are categories like incorporated associations, proprietary and limited companies, public companies, trusts, sole traders, limited partnerships, cooperatives and so on and so forth. We also allow what we call informal and proto business structures, which are either not or not yet recognised by the relevant legal system (e.g. grassroots movements). Ownership model refers to how the organisation is owned (e.g. donation based, community investment, community–developer partnership, community–council partnership, and multi-household models of community energy). These models vary depending on different commercial, financial and legal aspects. Thus, dynamics are also characterised by changes in organisational forms.

The way to identify the organisational forms within the framework is by using the snapshot of the socio-technical layout to map out the organisations, their business structures and (if possible) the ownership model for each group of rule makers, implementers and followers. In general terms, we understand an organisation to consist of a group of persons with a shared goal. Identifying the organisations interacting in the system helps to highlight their goals, motivations or incentives behind each relationship.

For instance, within the rule-follower group, one could identify an organisation that owns generation assets, is a cooperative and has a partnership with a local council as ownership model. Perhaps the main goal of the cooperative is to own their electricity from renewable resources. There might be an environmental motivation behind it as well as a common sense of belonging. This type of actors participate in the system perhaps with different objectives and motivations than the ones from a proprietary and limited company.

3 Application to three cases: Victoria, Australia

In this section, we use the analytical framework to represent three configurations associated to the electricity system in Victoria, Australia. The cases are: (a) a Victorian residential neighbourhood as part of the 1993–1994 configuration. Just before the electricity system was fully privatised; (b) a grid-connected residential neighbourhood as part of the Eastern-National electricity system in Australia; and (c) a residential neighbourhood developing a microgrid.

Finally, we briefly discuss some socio-technical challenges with the change of configuration (i.e. multiple microgrids enter the system).

3.1 Case a: Victoria electricity system (1993–1994)

The electricity sector in Victoria has considerably changed its organisational forms and overall structure since it started to operate. Before the 1910s, electricity was generated and distributed by municipalities, by private companies under franchise to the councils or by joint private–public bodies (Abbott 2006).

In 1910, brown coal reserves in the La Trobe Valley (Victoria) started to be developed close to the source as a result of technological advancements in electricity transmission. During this time and after different commissions and legislation changes, the electricity system started to be transferred to public hands. The system was then part of a statutory authority company managed by a Board of Commissioners who were part of the State Electricity Commission of Victoria (SECV or SEC). SEC still depended on parliament and reported to a minister.

SEC was a vertically integrated monopoly model in operation in Victoria until mid to late 1990s. Even though the electricity sector remained a monopoly under SEC, the organisation was disaggregated into three statutory businesses in 1993–1994: (1) Generation Victoria, in charge of electricity production; (2) National Electricity, responsible of the transmission assets, balancing supply and demand, and security of supply; and (3) Electricity Services Victoria, responsible of the distribution assets and retail business (Victoria State Government 2013).

The socio-technical layout for the Victorian electricity system of 1993–1994 is depicted in Figure 9.3. The Victorian government was the sole rule maker, while the rule implementers were the Board of Commissioners of SEC. Most of the participants at the service supply side (i.e. excluding households) were statutory companies. However, these participants with asset–actor relationships

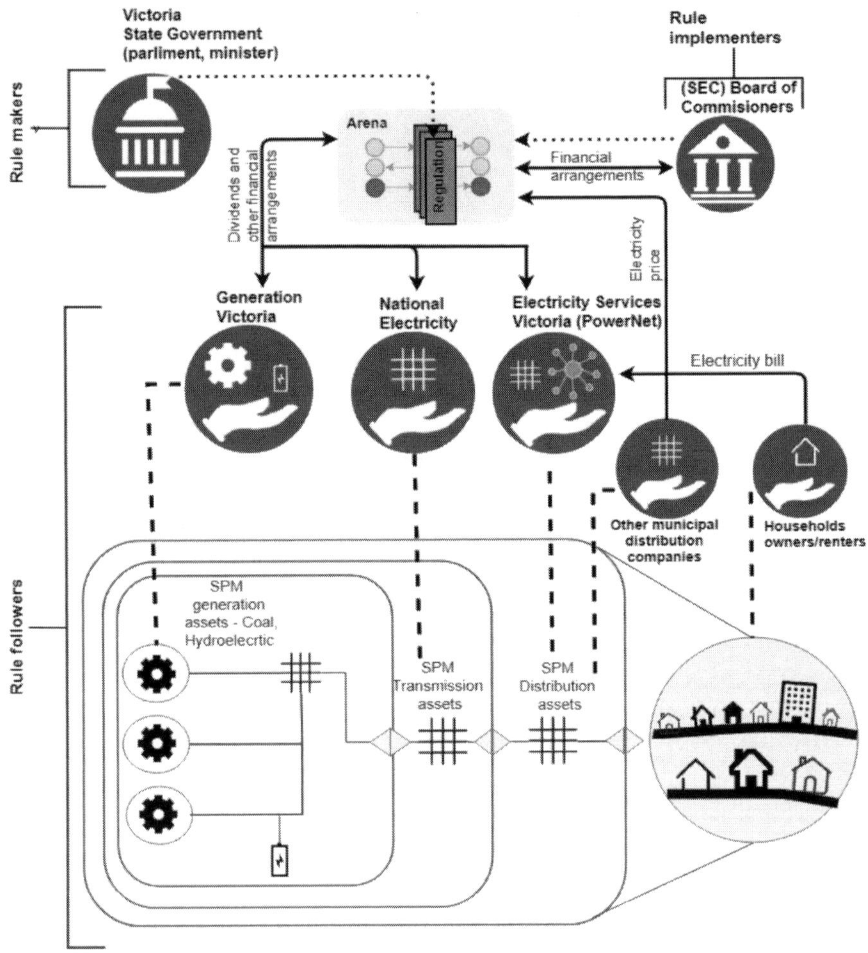

Figure 9.3 Socio-technical layout of a Victorian electricity system, 1993–1994.

acted as separated organisations from the State Government, and can therefore be considered rule followers.

In this configuration, the market is a regulatory framework composed by different acts (e.g. Electricity Industry Act and the Corporations Act). Public companies in charge of generation, transmission, distribution and retail pay dividends to SEC. They also receive loans and other operational monetary allowances via financial agreements from SEC.

Few private and municipal distribution companies were still operating at that time. They also participated in the market, using different financial agreements with SEC. Every company in the system has to comply with operational rules for the service provision (e.g. mechanisms for balancing supply and demand) included in regulatory frameworks. Finally, end-users pay their bills to Electricity Services Victoria.

3.2 Case b: business as usual, the Victorian electricity system and the Eastern-Australian NEM

After an enquiry into the electricity generation and distribution companies on the basis of inefficient resource use under the influence of governmental authorities (McConnell et al. 2012), the electricity system started an institutional and regulatory reform that led to the corporatisation of the system in the mid to late 1990s. The responsibilities of electricity generation, transmission and distribution assets fell to private organisations. One of the main objectives of this reform was to bring competition to the market to avoid misconduct by preventing misuse of market power.

The Victorian actor–asset and actor–actor relationships for this case are represented in the socio-technical layout in Figure 9.4. At the bottom of the socio-technical layout are the actor–asset relationships (connected with dashed lines), followed by the actor–actor relationships at the top (connected with straight lines and the name of transaction). We explain in general terms what these relationships are and then specify the actors and relationships for our case in Victoria, Australia.

The Eastern and Southern states in Australia are interconnected via high-voltage girds (i.e. transmission lines) in the National Electricity Market (NEM). There are five regional market jurisdictions in the NEM. The state of Victoria is one of these jurisdictions. Through the NEM, states import and export electricity by means of 40,000 kilometres of transmission and distribution grids. For our case in Victoria, without loss of generality, we assume that the demand in the neighbourhood is completely supplied by Victoria's regional generation; thus, neighbouring states are not considered in this case.

It is important to highlight that the current electricity system has increased in complexity with respect to the 1993–1994 configuration; thus, for visualisation purposes, all the actors shown in cases b and c fall into the rule-followers category. However, we mention the rule makers and implementers in the text.

Amongst the actor–asset relationships, from left to right, we have the:

- Electricity producer associated with the generation SPM
- Transmission operator who owns and operates the transmission SPM
- Distribution business who owns and operates the distribution SPM
- Households of the neighbourhood under study who own or lease the consumption assets

The SPMs are linearly recursive reflecting the black-box characteristics of the system, where each actor is only interested on the assets contained in their SPM but not what is beyond its interface. For instance, the transmission SPM has the grid assets. The operation of this SPM depends on the voltage and current at

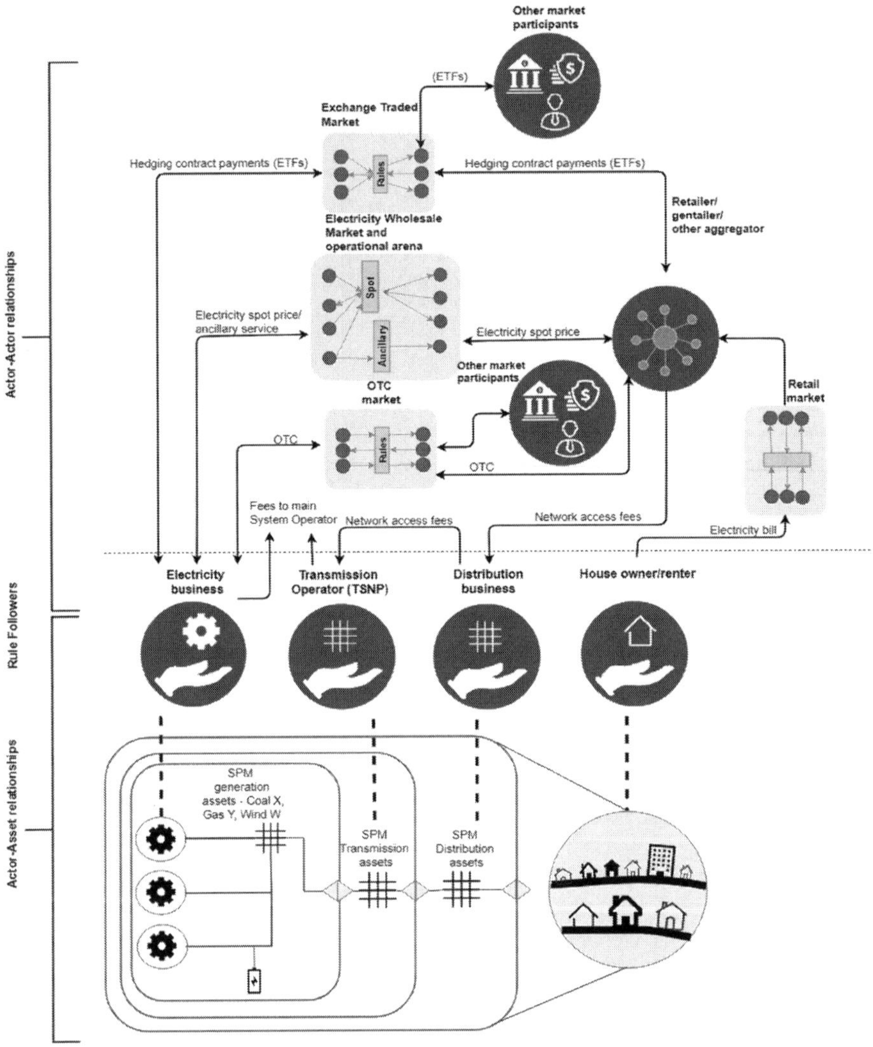

Figure 9.4 Socio-technical layout of a conventional electricity system.

the sending and receiving ends, not on knowing exactly which generator units are feeding the electricity into the grid.

Starting from the generation SPM, examples of actor–asset relationships in Victoria are the subsidiary of EnergyAustralia which owns and operates the Yallourn coal-fired power station; AGL Energy owns and operates the Loy Yang 'A' coal-fired power plant; or H.R.L Morrison & Co and Malakoff, owners of the Macarthur wind farm with AGL, the wind farm's operator. These actors 'plan, build, own and operate electricity generation facilities and new power lines to connect generation to the transmission network' (AEMO 2012).

The transmission SPM is owned and operated by organisations known in the Australian system as Transmission System Network Providers (TSNPs). TSNPs build, own and operate the transmission networks in Australia. AusNet is the only TSNP in Victoria. This company is owned by different organisations. However, most of the company's ownership is shared by Singapore Power and State Grid Corporation of China. AusNet shares can be traded in the Australian Securities Exchange or stock exchange market.

The distribution SPM ('poles and wires', in the argot of the sector) in Victoria are owned and operated by different companies overlooking specific geographical areas. Depending on the area under study, this SPM is owned and operated by Powercore & Citipower Australia, Ausnet services, United Energy Distribution or Jemena. All these organisations are subsidiaries of one or more organisations.

In all three cases analysed in this chapter, households are the actors related to consumption assets. They occupy and own or lease these assets, which are the loads in the electricity system. We do not detail the characteristics of the residential demand in this chapter. However, we suggest that an analysis on the consumption side of neighbourhood SPM is essential to evaluate the changes in the system from energy-efficiency measures or grid-balancing mechanisms, including demand response and more broadly demand management measures.

An example of the type of demand analysis that can be done is one based on specific building characteristics (physical assets) and key demographic data (actors). Also, one can think of an analysis of organisational forms related to demographic data; for instance, the relationship between community energy groups born as grassroots movements and the demographic characteristics of an area under study.

Going back to this case, the conventional configuration, actors associated with the consumption assets do not have onsite generation or storage nor a defined business structure, as they hold a fixed role in the system as residential electricity consumers. All the actors introduced so far are rule followers in the system.

Moving up in Figure 9.4, we find the actor–actor relationships as well as the different markets via which these actors interact. To clarify how the Australian energy system works, we broadly identify four markets according to the distinct rules and rule implementers associated to each: (1) the exchange-traded market, (2) the wholesale or spot market, (3) the over-the-counter (OTC) market and (4) the retail market.

Different actors, depending on their organisational form, may participate in these markets. Prices in one market can affect prices in the other. For instance, electricity prices in the spot market affect those in the exchange-traded market, as prices of transaction instruments (e.g. derivatives), depend on the average spot prices over a determined period of time.

Electricity businesses, retailers, energy brokers, banks and speculators (e.g. hedge funds) can participate in the exchange-traded market. These participants use different instruments to hedge risks of electricity price volatility and signal future investments. The main instrument used is exchange traded funds (ETFs), which are investment vehicles listed on the Australian stock exchange.

The wholesale market is the arena where electricity-registered generators are required to sell their production to supply the state's demand. This market is managed by the Australian Energy Market Operator (AEMO) (i.e. rule implementer), who is responsible for national planning and security of the national electricity grid (AER 2009). AEMO decides which generators will be deployed first to produce the cheapest electricity following a 'merit order approach'. Also, AEMO is the responsible of balancing the grid through control and market mechanisms. These are the regulation, contingency and emergency services. Regulation services are constant automated controls to keep the grid frequency at 50Hz; contingency services are called in case of major disturbances in the system.

Some market participants (electricity producers and big consumers) are part of this operational role, and they can register to provide these services. These participants have contracts in the ancillary markets (e.g. frequency-control and Ancillary Systems [FCAS]). The electricity wholesale market and operational arena are in the same box in Figure 9.4 since they are operated by the same rule implementer; however, the rules though which participants operate can be slightly different (e.g. different types of contracts). Finally, retail organisations also participate in the wholesale market, buying the electricity for their own use and their portfolio of customers.

The OTC market is an arena where participants enter into confidential bilateral contracts to manage risk (AER 2017, 59). These markets comprise direct transactions between actors, mainly generators and retailers, often with the assistance of an energy broker. The main transactions settled in this type of market are the power purchase agreements with different characteristics. One such agreement can be a fixed-price long-term contract between an energy-intense consumer (e.g. retailer, industry actor) with an electricity generator. Finally, the retail markets comprise the organisations and rules that allow a retailer actor to sell electricity to commercial and residential consumers.

These markets are overseen by specific organisations that act as rule implementers or rule makers. For instance, the rule implementer for the spot market and the retail market is the Australian Energy Regulator (AER),[3] while the principal rule maker of these markets is the Australian Energy Market Commission (AEMC). Similarly, the Australian Securities and Investment Commission is the rule maker of the OTC and exchange traded markets (ASX).

As can be seen in Figure 9.4, actors are represented with dark circles and markets with shaded boxes. Both the exchange traded market and the wholesale market are arenas where many market players participate in many transactions (e.g. bidding processes) and standardised contracts. These transactions are represented with dashed straight lines. The names of the main transactions are on top of these lines.

Conversely, the retail and OTC markets are represented with straight lines between the dark circles in the market box from Figure 8.4. These are the bilateral transactions between actors following certain institutional settings. For instance, the households of the neighbourhood pay directly the retailer of their choice, and, in the OTC market, the participants of the transactions are known by all parties and might be mediated by energy brokers.

3.3 Case c: neighbourhood microgrid project: socio-technical potential and the Yackandandah case

As mentioned earlier, we use the neighbourhood microgrid as an example of an alternative physical (i.e. technical) configuration. In a microgrid, distributed energy resources can be integrated in a coordinated way with the use of optimisation algorithms, communication devices and digital platforms.

In this configuration, electricity is generated as close as possible to the point of demand. Microgrids are also regarded as flexible architectures for the deployment of distributed energy resources (Hirsch, Parag, and Guerrero 2018).

With a microgrid, generation and conversion of energy can be achieved by small generation units, using solar energy, though other types such as diesel generators, small hydro generators (mostly for rural areas) and small wind turbines can also be used onsite to generate electricity. This technical microgrid SPM configuration is shown at the bottom left of Figure 9.5.

In contrast to the conventional system, mostly dominated by an oligopoly, assets in a microgrid can be owned and operated by individual end-users and new organisations (i.e. social). Incumbent actors from conventional electricity configurations may as well participate in these new arrangements. All these actors can collectively organise to own, operate, share or profit from the electricity system.

Microgrids open the possibility to participant actors for sharing their surplus of energy locally, stabilising the grid and increasing their resilience to energy outages. Since microgrids entail the connection of different participants, this technical configuration could start with a community energy (CE) project. CE projects usually involve individuals willing to have control over their own generation and, if possible, share the surplus with the community. These communities are usually environmentally aware and might be incentivised by the opportunity of sourcing their electricity from renewable sources, the decreasing costs of technologies, feed-in-tariffs or other future financial opportunities.

In terms of organisational forms, a CE group may have a not-for-profit business structure, such as non-trading cooperative, incorporated association or public company by guarantee. And, depending on different commercial, financial and legal aspects, they might choose a specific ownership model such as donation/philanthropic, community investment, community–developer partnership or community–council partnership (Lane et al. 2015).

In urban areas, unless the microgrid is completely isolated from the main grid (i.e. off-grid), current owners and operators of distribution assets are considered key participants in this alternative setting. Therefore, another potential ownership model will be a community–distribution business partnership.

Microgrid cases in urbanised regional areas in Victoria are emerging as demonstration initiatives. They usually count with the participation of a distribution business subsidiary and use some type of government incentive. One of these initiatives is being developed in Yackandandah, located at 300 kilometres northeast from Melbourne. AusNet is the transmission and distribution operator in this area of Victoria.

In the Yackandandah case, the community came together and pledged to be 100% powered by renewable energy by 2022.[4] They are a fully voluntary community energy group. This type of structure allowed them to follow a donation-based ownership model. Using this model they ran a fund, called the 'perpetual energy fund', and used it to invest in renewable energy generation, energy efficiency and storage in the community.[5] The loans for the projects are then repaid from the electricity bills savings and reinvested in the community.

The community energy group participated in a bulk buy of more than 500kW of solar and 100KWh of batteries for 14 households. At this point, the community could be considered an embedded generator, shown at the bottom right of Figure 9.5. The community also partnered with Mondo Power, a subsidiary of AusNet, who developed the R&D to monitor and control the future microgrid.

At the time of writing, the community was in the process of creating a community energy retailer that will track the electricity supplied back to the grid from their embedded generation and will pay the corresponding rate to the homeowner. The surplus of energy will either be paid with feed-in-tariffs (FiT) or as electricity savings. This potential relationship is shown at the top right of Figure 9.5. The community retailer will be another actor in the retail market, operating under the same market institutions and regulations as other retailers. At the same time, this community retailer will have a direct relationship with the community members.

This type of partnership makes Mondo Power the operator of the microgrid. The community retailer will most likely be paying Mondopower a service fee for managing the grid, which will be included in each household electricity bill. The new microgrid operator actor opens the possibility to incumbent distribution businesses, aggregators, retailers and emerging organisations to have a role in this type of configuration.

Although the microgrid configuration in Figure 9.5 did not significantly change the actor–actor relationships (top part) with respect to the conventional configuration shown in Figure 9.4, it enabled new relationships that did not exist in the conventional setting. However, the technical SPM (bottom left) did change significantly.

The recursion is still linear, but several new assets such as solar PV and storage are visible at the microgrid SPM. Thus, even if the neighbours do not know exactly where the energy generated comes from when they use their consumption assets, the microgrid operator has visibility over the local generators as well as the energy coming from the distribution SPM. Note that the visibility of the microgrid operator is limited by the interface at the Distribution SPM. This demonstrates the flexibility of the proposed analytical framework to represent different socio-technical layouts.

The SPMs can be used to analyse the reconfiguration changes when multiple microgrids that manage distributed energy resources at larger scales enter the

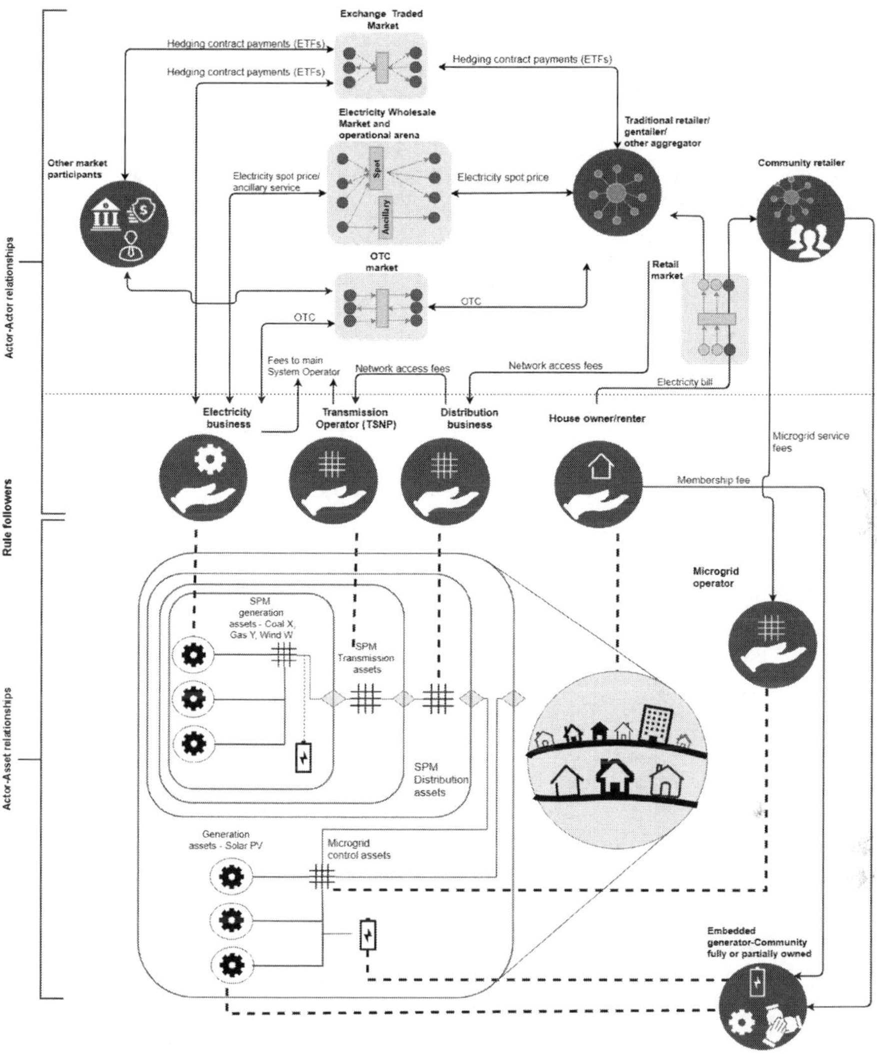

Figure 9.5 Socio-technical layout of an alternative electricity system.

system. Not only different technologies and aggregators will need to be coordinated by an overarching organisation, but also new markets might emerge. An example of such a body can be a Distributor System Operator (DSO), who could act as the equivalent of a Transmission System Operator but managing all the distributed energy resources at the distribution level (medium- and low-voltage grids) in a centralised manner.

Depending on the relationships between a DSO and other actors in the system, the DSO could have different responsibilities. For instance, if the DSO acts as an intermediary between aggregators and the transmission system operators, it will be fully in charge for the procurement of the system flexibility and coordination. This will be a highly complex and challenging task for this actor and brings uncertainty on who could be the most appropriate player in this space.

Moreover, the appearance of vacant roles within the system bring along questions that challenge not only governance structures but also market settings. To give an example, what if a new market is implemented at a local geographical level (e.g. local government area)? This market will fall under a local jurisdiction and at the distribution grid. Therefore, how does this market need to be managed to balance grid issues and electricity prices? Who will manage this market? Which regulation should be in place to make sure the correct operation of the grid? How will balancing mechanisms such as demand response be regulated and operated at different levels? Our analytical framework can play a role in mapping such potential future configurations and their various consequences.

4 Potential applications

In this section, we discuss how we envision the analytical framework to be applied for the study of transitions.

Notably, sustainability questions over certain periods of time can be validated and explored with the socio-technical layouts of the system. The study of the snapshots enhanced with an analysis of transitions pathway typologies can well be used to explore future transition pathways. The framework is precise enough to capture some core aspects of the electricity system, while it is general and flexible enough to provide a common view on different system states over time while avoiding getting lost in the details.

Another avenue for the utilisation of the analytical framework presented in this chapter is as the basis of a simulation model which will contribute to the body of work on socio-technical energy transitions models (STET) (Li, Trutnevyte, and Strachan 2015). The framework can served as basis of an agent-based modelling (ABM) endeavour, as ABMs especially suited to answer questions related to the creation of collective phenomena from the interaction of independent agents, identify influential agents on the generated behaviour and identify breakpoints in time at which change occur (Lorenz 2009; Pyka and Grebel 2006). Using the analytical framework as a complementary tool, a detailed actor analysis can be further formalised using, for instance, the ontology proposed by Van Dam, Nikolic, and Lukszo (2012) towards the development of an ABM. A straightforward ABM analysis using the framework may be using actors' decision making in different settings, such as different markets. These decisions will depend on the characteristics of each actor in a socio-technical configuration – actors with and without assets, types of assets, actors' relations and arena's rules. More on how transition modellers use ABM in the field can be found in Hansen, Liu, and Morrison (2019), Chappin (2011) and Chapter 7 by Holtz and Chappin in this volume.

System dynamics (SD) modelling (Forrester 1968, 1973; Sterman 2001) is another suitable modelling approach to be used with the socio-technical layouts and SPMs. SD has already been used in electricity sector modelling, mainly as a policy assessment tool (Ahmad et al. 2016). The definition in SD of stocks and flows is suitable to represent different electricity and monetary flows of a socio-technical layout. Similarly, the feedback concepts from SD can be used to understand the multiple loops in the system and represent its dynamics. An in-depth analysis of such models used for transitions research can be found in Chapter 8 by Papachristos and Struben in this volume.

5 Conclusion

We proposed an analytical framework for the socio-technical representation of electricity systems and argued how it can be considered as a complementary tool for the study of transitions. More specifically, the analytical framework aims to contribute to the body of work on formalisation of socio-technical system analysis, including the work of the transition modellers and ontologists (Van Dam, Nikolic, and Lukszo 2012). However, it complements more broadly other areas of transitions research. In this sense, the framework is *agnostic* towards the use of specific transition theories, as well as computational modelling and simulation techniques. The determination of the underpinning theoretical framework or computer modelling technique will depend on the aim of the research.

Therefore, the analytical framework facilitates software implementation and quantitative analysis, while allowing also a qualitative analysis, as illustrated by the cases in Section 3. The application of the framework to the cases demonstrated its flexibility without losing precision. Although the framework applied to the cases strove to represent some of the core elements of a socio-technical system, it did not pretend to be exhaustive in the definition of matters such as demand analysis, actors' motivations, behaviours, or the nature behind actors' relationships.

We offered an analysis using the SPMs and highlighted the networked understanding of the organisational (i.e. 'socio') aspects and a nested system understanding of the infrastructural (i.e. 'technical') aspects of socio-technical systems. Finally, we encourage future work to be done in other systems of service provision with similar characteristics to the electricity system to be analysed using the same (more, or improved) components defined in the analytical framework presented in this chapter.

Acknowledgements

The CRC for Low Carbon Living provided funding for this research. The authors would also like to acknowledge the valuable input of Lu Aye, Greg Foliente, Seona Candy and Dylan McConnell. Thanks to all the chapter reviewers who kindly provided exhaustive and highly constructive feedback.

Notes

1 Also from work (in German) by Aspers and Beckert (2008).
2 Even the exception of a fully vertically integrated and socialised electricity provision system fits this scheme. In such a case, the government may seem to be simultaneously rule maker, implementer and follower. However, usually such a government would set up separate organisations for the various aspects. Though these organisations may participate in rule making, our framework separates the implementer and follower roles within the system.
3 www.aer.gov.au/about-us/our-role
4 http://totallyrenewableyack.org.au/about/community-energy/
5 http://totallyrenewableyack.org.au/perpetual-energy-fund/

References

Abbott, Malcolm. 2006. "The performance of an electricity utility: The case of the State Electricity Commission of Victoria, 1925–93." *Australian Economic History Review* 46 (1): 23–44.

AEMO. 2012. "Energy infrastructure in your community." *Technical Report*, Australian Energy Market Operator (AEMO). supporthub@aemo.com.au.

AER. 2009. "Electricity transmission." *Technical Report 5*, Australian Energy Regulator. www.aer.gov.au/system/files/Chapter.

AER. 2017. "State of the energy market." *Technical Report*, Australian Energy Regulator. www.aer.gov.au/publications/state-of-the-energy-market-reports/state-of-the-energy-market-may-2017.

Ahmad, Salman, Razman Mat Tahar, Firdaus Muhammad-Sukki, Abu Bakar Munir, and Ruzairi Abdul Rahim. 2016. "Application of system dynamics approach in electricity sector modelling: A review." *Renewable and Sustainable Energy Reviews* 56: 29–37.

Aspers, Patrik, and Jens Beckert. 2008. "Märkte". In *Mrkte, Wirtschaft + Gesellschaft*, edited by A. Maurer, 225–46. Wiesbaden: VS Verlag für Sozialwissenschaften.

Beckert, Jens. 2009. "The social order of markets." *Theory and Society* 38 (3): 245–69.

Chappin, Émile Jean Louis. 2011. *Simulating Energy Transitions*. Delft: Next Generation Infrastructures Foundation.

de Haan, Fjalar J. 2019. "Making it a science – Aspirations and hesitations of transitions research." In *Modelling Transitions – Virtues, Vices, Visions of the Future*, edited by E. Moallemi and F. De Haan. This volume: Routledge.

de Haan, Fjalar J., Alfonso Martinez Aranz, and W. Spekkink. 2019. "Data-driven transitions research – Methodological considerations for event-based analysis. In *Modelling Transitions – Virtues, Vices, Visions of the Future*, edited by E. Moallemi and F. De Haan. This volume: Routledge.

Forrester, Jay W. 1968. "Industrial dynamics – After the first decade." *Management Science* 14 (7): 398–415.

Forrester, Jay W. 1973. *Principles of Systems: Text and Workbook*. Vol. 2. Cambridge, MA: Wright-Allen Press.

Geels, Frank W. 2002. "Technological transitions as evolutionary reconfiguration processes: A multi-level perspective and a case-study." *Research Policy* 31 (8–9): 1257–74.

Geels, Frank W. 2004. "From sectoral systems of innovation to socio-technical systems: Insights about dynamics and change from sociology and institutional theory." *Research Policy* 33 (6–7): 897–920.

Geels, Frank W., Benjamin K. Sovacool, Tim Schwanen, and Steve Sorrell. 2017. "Socio-technical transitions for deep decarbonization." *Science* 357 (6357): 1242–4.

Hansen, Paula, Xin Liu, and Gregory M. Morrison. 2019. "Agent-based modelling and socio-technical energy transitions: A systematic literature review." *Energy Research & Social Science* 49: 41–52.

Hirsch, Adam, Yael Parag, and Josep Guerrero. 2018. "Microgrids: A review of technologies, key drivers, and outstanding issues." *Renewable and Sustainable Energy Reviews* 90: 402–11.

Holtz, Georg, and Emmile J.L. Chappin. 2019. "Considering actor behaviour: Agent-based modelling of transitions." In *Modelling Transitions – Virtues, Vices, Visions of the Future*, edited by Enayat Moallemi and Fjalar De Haan. This volume: Routledge.

Li, Francis GN, Evelina Trutnevyte, and Neil Strachan. 2015. "A review of socio-technical energy transition (STET) models." *Technological Forecasting and Social Change* 100: 290–305.

Lopes, JA Pecas, N. Hatziargyriou, J. Mutale, P. Djapic, and N. Jenkins. 2007. "Integrating distributed generation into electric power systems: A review of drivers, challenges and opportunities." *Electric Power Systems Research* 77 (9): 1189–203.

Lorenz, Tobias. 2006. "Abductive fallacies with agent-based modeling and system dynamics." In *International Workshop on Epistemological Aspects of Computer Simulation in the Social Sciences*, pp. 141–52. Berlin, Heidelberg: Springer.

Markard, Jochen, Rob Raven, and Bernhard Truffer. 2012. "Sustainability transitions: An emerging field of research and its prospects." *Research Policy* (6): 955–67.

McConnell, Dylan. 2012. *Regulation in the NEM, Facilitating the PV Revolution: Mapping the Regulatory Frameworks in the National Electricity Market*. Report Melbourne, Australia: Melbourne Energy Institute.

Papachristos, G., and J. Struben. 2019. "System dynamics methodology and research: Opportunities for transitions research." In *Modelling Transitions – Virtues, Vices, Visions of the Future*, edited by E. Moallemi and F. De Haan. This volume: Routledge.

Pyka, Andreas, and Thomas Grebel. 2006. "Agent-based modelling – A methodology for the analysis of qualitative development processes." In *Agent-based Computational Modelling*, pp. 17–35. Physica-Verlag HD.

Romanelli, Elaine. 1991. "The evolution of new organizational forms." *Annual Review of Sociology* 17 (1): 79–103.

Sterman, John D. 2001. "System dynamics modeling: Tools for learning in a complex world." *California Management Review* 43 (4): 8–25.

Van Dam, Koen H., Igor Nikolic, and Zofia Lukszo, eds. 2012. *Agent-based modelling of socio-technical systems*. Vol. 9. Dordrecht: Springer Science & Business Media.

Victoria State Government. 2015. "Guide to community-owned renewable energy for Victorians." *Technical Report*, State of Victoria. www.ecodev.vic.gov.au.

Victoria State Government, A. 2013. *Electricity Industry (Residual Provisions) Act 1993*. www.legislation.vic.gov.au (Accessed 15 January 2017).

Weber, K. Matthias. 2003. "Transforming large socio-technical systems towards sustainability: on the role of users and future visions for the uptake of city logistics and combined heat and power generation." *Innovation: The European Journal of Social Science Research* 16 (2): 155–75.

Part 3

The future of modelling transitions

10 Models as scenario tools for developing robust transformative plans

Shirin Malekpour

1 Introduction

This chapter focuses on long-term planning under uncertainty as a practical backdrop within which transition modelling could be used. Sustainability transitions are long-term processes that often take many years to unfold (Markard, Raven, and Truffer 2012). It takes time for innovative solutions and practices to develop and take off, and it also takes time to work against the inertia of well-established systems and unlock them (Köhler et al. 2019). The future of transitions is open-ended as there are often multiple potential solutions, multiple transition pathways, multiple actors and multiple factors in play that can shape and re-shape the trajectories of change.

Given the highly complex and uncertain nature of sustainability transitions, long-term planning in this context faces significant challenges associated with disrupting the status quo and big systemic change. While long-term planning can draw lessons from descriptive case studies of past and emerging transitions available to us (Köhler et al. 2019), various kinds of computational models have been used to structurally analyse complexity and uncertainty through various degrees of abstraction (Holtz et al. 2015). Models can represent systems in a clear and coherent way, facilitate inferences about factors and processes underlying a phenomenon and help with systemic experiments to anticipate the outcomes of any given action (Köhler et al. 2019). Despite their potential, models have not been a prominent tool for investigating and planning of sustainability transitions (Holtz et al. 2015). This chapter tries to address this gap by unpacking long-term planning and the challenge of dealing with uncertain futures. The research question guiding this chapter is: *how can modelling address significant uncertainties and add value to processes of long-term planning in sustainability transitions?*

2 Long-term planning under uncertainty

Long-term planning in all sectors of the economy – such as water, energy, food, transport – is surrounded by uncertainties. Uncertainty exists in various aspects of planning decisions (Haasnoot et al. 2011; Störmer et al. 2009), including in:

- Context conditions: the environment within which planning decisions have to be made is highly uncertain. Issues such as population growth, climate change, changes in policy and institutional landscape have significant impacts on the long-term success or failure of our strategies, and how they will unfold in the future is uncertain.

- Societal values: how society, including customers and clients, will define goals and evaluate the success or failure of our strategies is fluid, disputed and subject to change.
- Emerging solutions: innovative technologies, products and practices can fundamentally change the effectiveness of our strategies. Their emergence, as well as their uptake by society, is uncertain.

The notion of uncertainty has been interpreted differently in different fields. Most traditional definitions of uncertainty associate the concept with the lack of knowledge or inadequate information (van Dorsser et al. 2018). In the 1920s, Knight made a distinction between *uncertainty* and *risk* (Knight 1921). According to him, risk is the calculable and controllable part of what is unknown (probability of an unfortunate event multiplied by the damage incurred), whereas uncertainty is the incalculable and uncontrollable part. In other words, Knightian uncertainty acknowledges an essential unpredictability and immeasurability of some of the future events, whereas risk is a lower level of knowledge inadequacy which can be dealt with through quantifiable probabilities.

In planning and decision-making, uncertainty often refers to the gap between available knowledge, and the knowledge needed for making the best decision (Marchau et al. 2019). However, as Walker and his colleague argue, uncertainty can be present even in the presence of information, particularly when this information reveals complexities that were previously unknown or underestimated (Walker et al. 2003). They provide an all-encompassing definition of uncertainty as: *any departure from the ideal of complete determinism* (Walker et al. 2003).

Such an inclusive definition of uncertainty, however, does not imply a blanket view of different types or levels of uncertainty. In fact, a spectrum of different levels of uncertainty exists, ranging from the ideal of complete determinism (or complete knowledge) at one end, to complete ignorance at the other end (Walker et al. 2003). Below, I describe the different levels/zones on the uncertainty spectrum. I also explain that when we are in different zones, we need to adopt different approaches to planning and decision making, to be able to cope with the level of uncertainty that exists in any particular zone. A summary of the following discussion is presented in Figure 10.1, which is inspired by the work of Walker, Marchau, and Kwakkel (2013) in conceptualising uncertainty in policy making.

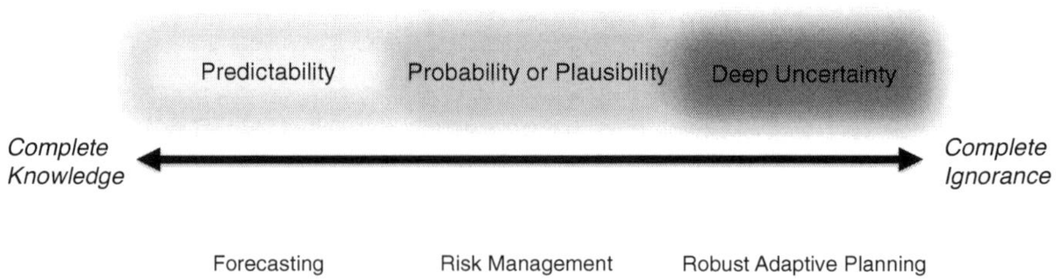

Figure 10.1 The uncertainty spectrum and planning approaches for each zone.

2.1 Predictability

When there is close-to-complete knowledge about an issue or a system, we are in the *predictability zone* (or Level 1 uncertainty as characterised by Walker and his colleagues). In predictable systems, historical data can be used to predict future outcomes. Any uncertainty that exists does not overwhelm our knowledge of the system (Bankes, Lempert, and Popper 2002). In the predictability zone, we may have a single deterministic model of the system, and we can estimate the outcomes of any action reasonably well, within an error margin or using sensitivity analysis. Controlled engineering systems, or closed physical systems, are often subject to this level of uncertainty.

When the issue we face is located within the predictability zone, forecasting may be used as a reasonable planning approach (Malekpour, de Haan, and Brown 2016). A forecast is a single projection, or, the best estimate of the future, often derived from an extrapolation of past trends, or expert judgement (Goodwin and Wright 2010; van Dorsser et al. 2018). A solution, or strategy, could then be devised to fit the forecasted future conditions.

2.2 Probability or plausibility

When uncertainty grows further, we get into the *probability* or *plausibility zone*, where a few alternative futures can be imagined. In this zone, we can study the system using a single probabilistic/stochastic model (Level 2 uncertainty according to Walker and his colleagues), or a few plausible models (Level 3 uncertainty according to Walker and colleagues). This level of uncertainty exists in various fields, such as engineering, business, finance, epidemiology, etc. (Rotmans 1998).

In planning and decision-making, probability is often dealt with through risk management and probabilistic approaches (van Dorsser et al. 2018), and plausibility is dealt with through developing a limited number of future scenarios. The majority of these approaches are based on Bayesian decision analysis, which assumes that the future can be described from a system model that can relate current and past actions to future outcomes, and that uncertainty can be captured using probability distributions for model inputs, or using a few alternative models (Lempert, Popper, and Bankes 2003). In using probabilistic/stochastic models, what a decision maker needs to do is to understand risks and the relationship between inputs and outputs, and to take mitigation actions to minimise negative outcomes. In using alternative models, one can examine different outcomes resulting from different models, and choose the strategy that produces the best outcomes across different models or different scenarios.

2.3 Deep uncertainty

Uncertainties surrounding social phenomena or long-term policies can be best characterised as *deep uncertainty*. Deep uncertainty, a term coined by Lempert and his colleagues, refers to a situation where we do not know how the future will unfold (Lempert, Popper, and Bankes 2003). In deep uncertainty, decision makers do not know, or cannot agree on, the appropriate conceptual models that describe the system, and cannot quantify the likelihood of occurrence for different futures (Level 4 uncertainty according to Walker and his colleagues). In simple words, in deep uncertainty anything can happen, and it is no longer possible to assign probabilities or characterise the future with a limited set of scenarios. Issues related to social policy or long-term strategy are often subject to deep uncertainty, as they have open boundaries with complex underlying mechanisms and processes that are not well understood. Therefore, any attempt to characterise them through probabilities and risk assessment will only be misleading (Malekpour et al. 2017; Rotmans 1998).

When the issue we face is located in the deep uncertainty zone, conventional planning and decision-making based on forecasting and risk management cannot be used. The reason is simple: a solution/strategy may be optimal under a defined set of circumstances, but will not be optimal, or may even fail, if those circumstances change (Malekpour, de Haan, and Brown 2016). Given the extent of uncertainty in this situation, all we can do is to develop a strategy that is robust across varying conditions (known unknowns), and/or can adapt to new conditions (unknown unknowns), without the need to be explicit about the likelihood of those conditions. Over the past couple of decades, a range of planning approaches has been developed to assist decision makers with devising robust and/or adaptive long-term plans. Examples include: assumption-based planning (Dewar et al. 1993), robust decision-making (Lempert, Popper, and Bankes 2003), adaptive policy making (Walker, Rahman, and Cave 2001) and dynamic adaptive policy pathways (Haasnoot et al. 2013). See Walker, Marchau, and Kwakkel (2013) for a review of these approaches, including their similarities and differences.

3 Long-term planning in sustainability transitions

Long-term planning in the context of sustainability transitions can be characterised as planning under deep uncertainty (Holtz et al. 2015; Walker, Haasnoot, and Kwakkel 2013). Transitions add extra layers of uncertainty and complexity to the already uncertain and complex planning environments. Transitions involve redirecting the trajectories of development in different sectors (water, energy, transport, etc.), nurturing innovative solutions and enabling systemic change (Markard, Raven, and Truffer 2012; Loorbach 2010; Truffer et al. 2010). Desired outcomes result from a complex interplay between diverse political, economic, social and technological drivers. Steering the transition process requires a departure from business-as-usual optimal plans, towards adaptive strategies that can cope with the uncertainty of uncharted transition pathways. Long-term planning in this context not only has to deal with uncertainty and complexity along the way, but also has to make sure that all efforts are channelled towards a desired future state – that is, sustainability – which is in itself an ambiguous, subjective and normative concept (Rotmans 2005; de Haan et al. 2014).

The majority of transitions scholars refrain from using the term 'planning' in their conceptual frameworks and instead use 'management', 'steering', 'navigating' or 'policy design' (Voß, Smith, and Grin 2009). Despite this avoidance, empirical evidence suggests that long-term strategic planning (e.g. of infrastructure systems) is in fact a major driver for sustainability transitions (Carroli 2018; Truffer et al. 2010; Lutz et al. 2017; Bulkeley, Castán Broto, and Maassen 2014). In avoiding the language of planning, transitions literature tries to dissociate itself from the pitfalls, and the ill-repute, of an earlier generation of planning approaches, which were focused on command and control (Voß, Smith, and Grin 2009).

First-generation planning approaches were underpinned by the rationality paradigm, and concerned with building up large infrastructure and setting up bureaucracies, in order to establish the welfare state (Malekpour, Brown, and de Haan 2015; Forester 1982). This was done through highly technocratic processes, using forecasting and predictive measures to understand future change (Malekpour, Brown, and de Haan 2015). After the 1970s, with the rise of the neo-liberal ideology and the concept of 'bounded rationality' (Simon 1956), the command-and-control approach fell into disrepute (Voß, Smith, and Grin 2009). It was criticised for its positivistic and non-democratic approach, and its inability to address macroeconomic challenges of its time (Malekpour, Brown, and de Haan 2015). Long-term planning was linked with long-range, wide-scale and highly interventionist societal planning, which was no longer popular (Voß, Smith, and Grin 2009).

The new generation of long-term planning approaches, however, is more concerned with nurturing, rather than commanding, societal change (Malekpour, Brown, and de Haan 2015; Dominguez, Truffer, and Gujer 2011; Carroli 2018). These approaches acknowledge the unintended consequences of previous development plans, as well as the presence of different worldviews, incomplete information and increasing uncertainties in contemporary planning and policy contexts. As Norgaard argues, good planners and policy makers in the context of sustainable development are those who take no prior stance on potential solutions or appropriate frameworks, but enable conversations between different groups of actors from different backgrounds, cultures and scientific disciplines to gain a shared understanding of the complex issues they face (Norgaard 1988). Voβ and his colleagues also recognise the relationship between sustainability transitions and long-term planning, and conceptualise transition management – the most notable meta-governance framework in transitions literature – as an example of the new generation planning/policy design approaches (Voß, Smith, and Grin 2009).

Dealing with deep uncertainty is at the heart of the transition management framework (see also Chapter 13 by Moallemi, de Haan and Köhler in this volume). According to Derk Loorbach, one of the pioneers of transition management,

> transitions to a sustainable society are like discovery journeys into the unknown; they are about exploration, learning, discovery and change.

> Since the destination (what is a sustainable society) is unclear and the roads towards it highly uncertain, the only way forward is to take small steps and regularly evaluate whether we are coming closer to or drifting away from our ideal destination.
>
> (Loorbach 2007, 3)

This is very much aligned with the approach taken by adaptive planning scholarship (Walker, Haasnoot, and Kwakkel 2013), except that in adaptive planning there is little emphasis on a normative direction at the societal level, for example a sustainable society.

4 Scenario planning as a way of informing long-term plans

Scenario planning has been developed as a way of assisting planning and decision-making in dealing with future issues. It emerged in the aftermath of World War II as a method for military planning (Varum and Melo 2010). It soon expanded to the public policy realm (mainly through the work of the American think-tank RAND Corporation) and corporate strategy (mainly through the work of Royal Dutch Shell) in response to increased uncertainty and complexity in planning contexts, such as the 1970s oil shocks, widespread deregulation, globalisation and the emergence of disruptive technologies (Schoemaker 1993).

Scenarios are descriptions of how a system (or an event) may develop (or unfold) into the future (Schoemaker 1993; Walker 2000). Scenario planning, also referred to as scenario building or scenario analysis, is the construction of descriptions about how the future could unfold, to help decision makers recognise, consider and reflect on uncertainties that surround planning decisions (Goodwin and Wright 2010).

Scenarios have been used in various sectors and various disciplines in completely different ways, employing diverse techniques and methodological approaches (Hajer and Pelzer 2018; Malekpour, de Haan, and Brown 2016; Rogers and Gunn 2015; Hunt et al. 2012). One of the attempts to organise the diverse world of scenarios and scenario-planning approaches has been by Börjeson and colleagues who provide a typology of scenarios from the decision-makers' perspective (Börjeson et al. 2006). According to them, three types of scenarios exist, which correspond to three types of questions a user may want to ask about the future. The three types are:

1 Predictive scenarios, which address the question *What will happen in the future?*
2 Explorative scenarios, which address the question *What can happen in the future?*
3 Normative scenarios, which address the question *What should happen in the future?*

This typology is particularly useful in the context of long-term strategic planning as it provides a structure for thinking about scenarios from a very practical perspective, which is often required by planners and decision makers. Below, I elaborate on the scenario types and link each of them with another typology of scenarios by Sondeijker and her colleagues (Sondeijker et al. 2006), which is based on the historical development of scenarios. I also position each type of scenario within the zones on the uncertainty spectrum described earlier (Figure 10.1), as well as within the strategic-planning context it may best fit.

4.1 *Predictive scenarios*

While the majority of the scenario-planning literature emphasises that scenarios are not predictions of the future (Schoemaker 1993; Goodwin and Wright 2010; Hunt et al. 2012), in practice, many estimates and forecasts of the future have been developed with a scenario name tag attached to them. A notable example is population growth projections, which are often referred to as population scenarios (see, for example, Rhodes 2003). Predictive scenarios try to describe *what is likely to happen* in the future. This is often based on past trends and extrapolation into the future, or, through probabilistic assessment of the likelihood of certain circumstances and their outcomes (Börjeson et al. 2006). Sondeijker and her colleagues call this group of scenarios 'the first generation of scenarios', which are not more than statistical predictions of future conditions (Sondeijker et al. 2006).

Predictive scenarios are often developed to plan for futures that are expected to occur. Voros calls them probable futures (Voros 2003). They are foreseeable futures that are almost free from surprises. Predictive scenarios are therefore useful when uncertainties surrounding future issues are low, when driving factors influencing the system are more or less known or when we are dealing with a short-term future. In other words, predictive scenarios are useful when we are within either the predictability zone or the probability/plausibility zone on the uncertainty spectrum. In this situation, we can develop an optimal plan that performs well under reasonably well-understood circumstances.

4.2 *Explorative scenarios*

Unlike predictive scenarios which start from the past or from the present and develop into the future, explorative scenarios start from the future (Börjeson et al. 2006). They are the second generation of scenarios which came about when the economic growth trends broke following the oil crisis in the 1970s, resulting in a loss of confidence in trend extrapolation and forecasting approaches (Sondeijker et al. 2006). Explorative scenarios acknowledge discontinuities, novelty, irregular and radical change; what we refer to today as disruptions (Malekpour et al. 2017). They try to understand *what can/might happen*, irrespective of current trends. Often a range of scenarios is developed to cover a wide span of possible developments (Börjeson et al. 2006) or possible futures (Voros 2003).

In other words, explorative scenarios provide a set of alternative futures, which could be totally different from each other.

Explorative scenarios are useful when the underlying mechanisms that drive future conditions are not fully understood or when irregular and radical change can disrupt the system (Börjeson et al. 2006). This is when we are in the deep uncertainty zone, where the future cannot be estimated or quantified, but rather imagined and monitored along the way. In this situation, scenarios are used as a basis for discussing different strategies, in order to find the strategy that is the most robust across those alternative futures. This is the approach used by robust decision-making (Lempert, Popper, and Bankes 2003) and most adaptive planning approaches in the policy analysis literature (see, for example, Lawrence et al. 2013; Haasnoot et al. 2013).

4.3 Normative scenarios

With the rise of normative concepts such as sustainability, a third generation of scenarios was developed to allow decision makers and their clients to describe a desirable future and think about how to achieve it (Sondeijker et al. 2006). In addressing the question of *what should happen*, normative scenarios create conditions for proactive and emancipatory actions that can shape the future. The construction of preferred futures (Voros 2003) has been referred to differently in different disciplines. Urban studies call them imaginaries, while the sustainability transitions literature calls them visions. In transition management, visions are qualitative societal ambitions and goals that provide a basis for transformative actions (Loorbach 2007). They are not a fixed end state; they are moving targets that can evolve over time. A vision provides a reference point for actors who collaborate on its realisation, and a narrative for mapping the possibilities and mobilising resources (Hajer and Pelzer 2018).

Normative scenarios are useful in developing transformative plans that go beyond complex and uncertain forces of the future, to actively shape and reshape it. Through the process of backcasting from a future vision, different transition pathways can be developed to define long-, medium- and short-term outcomes and strategies. Given the highly uncertain nature of transformation processes, and the uncharted pathways ahead, normative scenarios of transitions to sustainability can be positioned within the deep uncertainty zone.

4.4 An emerging generation of scenarios: normative + explorative

While the transitions literature deals with the dynamic and transient nature of change processes, when it comes to using scenarios, it has been primarily focused on normative scenarios (see, for example, Sondeijker 2009; Ferguson, Frantzeskaki, and Brown 2013; Hajer and Pelzer 2018). The literature on policy analysis and planning under deep uncertainty, on the other hand, uses exploratory scenarios as a fundamental tool for dealing with change (Haasnoot et al. 2011; Haasnoot et al. 2013). Long-term planning in the context

Table 10.1 Typology of scenarios, and the strategic-planning context in which they fit.

Scenario Generation	*Scenario Type*	*Future Type*	*Uncertainty Zone*	*Question of Interest*	*Plan Type*
1st	Predictive	Probable or plausible	Predictability, probability or plausibility	What will happen?	Optimal
2nd	Explorative	Possible	Deep uncertainty	What can happen?	Robust
3rd	Normative	Preferred	Deep uncertainty	What should happen?	Transformative
4th (Emerging)	Normative + explorative	Preferred and possible	Deep uncertainty	How to achieve a desired future and cope with what can happen along the way?	Robust transformative

of sustainability transitions could benefit from a more explicit combination of normative and explorative scenarios. The normative aspect will enable actors to develop a shared vision of a desired future, identify interim outcomes in the medium and short term and develop proactive strategies to achieve them over time. The explorative aspect will enable actors to stress-test their strategies across a range of possible disruptions and reveal their vulnerabilities. These disruptions could be positive and facilitate progress towards the future vision, or they could be negative and hinder, or derail, progress. Positive disruptions could be anticipated and utilised as windows of opportunity, while negative disruptions could be mitigated, or their impacts could be minimised through developing appropriate coping strategies. A more explicit integration of normative and exploratory scenarios could constitute the fourth generation of scenario-planning approaches focused on developing robust transformative plans – plans that can proactively shape the future (not fix it), while coping with various disruptive circumstances along the way (see Moallemi and Malekpour 2018 and Kok et al. 2011 for examples of such an integration).

The typology of scenarios as discussed in this section is presented in Table 10.1.

5 Transition models as normative and explorative scenario tools

Having discussed long-term planning under uncertainty, and the use of scenario planning in developing robust transformative plans, I now arrive at the explicit focus of this volume and pose the question: *Can modelling, as a scenario tool, add value to long-term planning in sustainability transitions?* The relationship between computational modelling and transitions is an interesting one, since transitions are characterised as social processes of a complex nature, often described through social science theories in the form of heuristics that do not easily translate into computational elements (Holtz et al. 2015). Here, I will address the question from the perspective of long-term planning to discuss how models could contribute to challenges in the planning process. How social science theories can be translated into computational descriptors and the methodological challenges and technical limitations of modelling transitions is not the focus of this discussion and have been extensively discussed elsewhere (see, for example, Halbe et al. 2015; Holtz et al. 2015; de Haan et al. 2016; as well as Chapter 12 by de Haan et al., Chapter 13 by Moallemi et al., and Chapter 7 by Holtz and Chapin in this volume).

Halbe and his colleagues in a study of model use in sustainability transitions have identified three roles for modelling: models for understanding transitions, models for providing policy advice and models for facilitating stakeholder processes (Halbe et al. 2015, Chapter 11 by Halbe in this volume). Modelling for understanding transitions is curiosity-driven. It is part of the fundamental transitions science, which aims at understanding change processes and theory development (see, for example, de Haan et al. 2016). Modelling for policy advice is problem-driven. It is related to the applied transitions science, aimed at developing practical solutions to case-specific problems (see, for example, Yücel 2010). Modelling for facilitating stakeholder processes is engagement-driven. It is related to post-normal science (Funtowicz and Ravetz 1993) aimed at enabling the participation of stakeholders for strategic activities in transitions management (see, for example, Trutnevyte, Stauffacher, and Scholz 2011). Out of these three categories, the last two are most relevant in the context of long-term planning, given the practical and multi-stakeholder nature of planning processes. Hence, I use those two categories to structure the arguments in the rest of this section.

5.1 Models as scenario tools to inform policy

As discussed in the previous section, explorative scenarios try to capture the breadth of the possibility space of the future under deep uncertainty, while normative scenarios try to channel all actions and strategies towards a desired future. Explorative scenarios often involve a range or a portfolio of scenarios to represent alternative possibilities. A policy option could then be viewed against those scenarios, to assess its performance in achieving the desired outcomes (the normative aspect), as well as its robustness in the face of changing circumstances (the explorative aspect).

Qualitative scenario-planning approaches, which solely rely on the intuitive knowledge and expertise of their participants, rather than computational tools, may fall short of their full exploratory potential (Bryant and Lempert 2010; Kwakkel, Auping, and Pruyt 2013). Due to their qualitative nature, qualitative approaches can only handle a limited number of scenarios, and therefore often summarise the breadth of the future into a small number of working scenarios (see, for example, Kok et al. 2011; Makropoulos et al. 2008).

Computational modelling tools, on the other hand, enable us to produce a virtually infinite number of scenarios. In addition, models allow us to analyse the implications of different scenarios in a systematic way. They can capture a diverse set of causes and effects, pieces of knowledge and uncertainties in a coherent and logical representation (Holtz et al. 2015). This improves our limited ability to follow and link complex chains of inference, which often lead to over-reliance on past experience and observed realities in qualitative scenario planning (Wright and Goodwin 2009).

Exploratory modelling and analysis (Bankes 1993) is an approach to modelling that allows the construction of a portfolio of futures, and experimentation with different parameters, or policy options. Each single model run is one computational experiment (Bankes, Lempert, and Popper 2002) that reveals how a system would behave if the various assumptions made about the parameters in that particular run were true. In reality, such experiments might be impossible (e.g. when a system does not exist yet) or impractical (e.g. when the economic or socio-political costs of tampering with parameters are high) or unethical (e.g. when the side-effects of any tampering are not well understood) (Holtz et al. 2015). Exploratory models allow us to conduct many such experiments in a safe space, and unpack the consequences of different strategies. We can check which strategies may achieve the desired transformations and under what conditions, and also under what conditions they may fail. This could assist with developing robust transformative strategies; that is, strategies that can achieve the desired transformations, while their robustness in the face of numerous futures has been tested.

While modelling is a niche area, or sub-field, in transitions studies, within that niche, exploratory modelling has been the dominant scenario modelling approach to address policy questions (see, for example, Yücel 2010; Moallemi et al. 2017; Kwakkel and Pruyt 2013).

5.2 Models as scenario tools to facilitate stakeholder processes

Combining qualitative participatory processes with quantitative computational modelling can facilitate stakeholder processes in normative and explorative scenario planning (Moallemi and Malekpour 2018). In scenario processes that solely rely on qualitative methods to construct a limited number of scenarios (often 3 to 10), the reductionist approach can become arbitrary and raise conflict among stakeholders with diverse, sometimes contradictory, perspectives and interests (Bryant and Lempert 2010). Groves and Lempert (2007), for instance, explain the tensions in a scenario planning exercise for developing a 25-year water resources strategy in California. Stakeholders expressed their concerns that a few scenarios would not represent all future uncertainties, and focusing on a few futures among many possible futures could result in the exclusion of some water management strategies at the expense of others.

Exploratory modelling of transitions can shift our attention from identifying the relevant future scenarios, to identifying the vulnerabilities of our strategies and policy options. Stakeholders will no longer need to build consensus around future scenarios, as numerous future scenarios could be developed using the computational power of models. Instead, they can test the implications of different actions in achieving the desired transformations, and in dealing with uncertainty along the way.

Models, when used in participatory settings, can also help with developing a shared vision and a common understanding of issues and strategies among stakeholders. Visions in the context of sustainability transitions are often owned by some key actors, and not necessarily shared among all stakeholders (Trutnevyte, Stauffacher, and Scholz 2011). Modelling allows us to test different visions against quantitative future scenarios in order to unpack their implications (Trutnevyte, Stauffacher, and Scholz 2011). It can create a virtual experience of long-term issues, which are often beyond our direct experience. The capability of models in demonstrating the implications of different assumptions can make the complexity of transition strategies more accessible, and the consequences of policy options more explicit (Holtz et al. 2015). A critical issue here is the transparency of the model structure and its underlying assumptions. A model that is not a 'black box' and its structure is accessible for stakeholders can assist them to arrive at a shared long-term view of the future, and a uniform translation of the required actions to achieve it (Malekpour, de Haan, and Brown 2013). This could result in better coordination among different actors in driving transformative actions.

6 Conclusion

This chapter described the practical context within which modelling of transitions would be used. I defined this context as long-term planning, acknowledging that much of the transitions literature avoids using the word 'planning'. This avoidance may be due to the negative image of traditional planning, or planning history, as a command-and-control approach that tries to engineer societal change. The new generation of long-term planning approaches, however, has moved far beyond the traditional positivistic thinking and acknowledges the deeply uncertain and highly complex environment of actions and strategies. They recognise the need for exploratory analysis of multiple future pathways, and for robust and adaptive measures that can cope with changing circumstances.

Sustainability transitions, with their normative agenda, add extra layers of complexity to long-term planning endeavours. Long-term planning in this context has to cope with deep uncertainties and disruptions along the way, while making sure all short-, medium- and long-term strategies are channelled towards a desired future – a destination which may itself evolve over time. To support long-term planning in sustainability transitions, I proposed a combination of normative and explorative scenario planning. This combination will envision a desired future and the pathways to realisation, while testing the robustness of those pathways against changing circumstances.

Computational models can add value to scenario planning exercises in different ways. They can generate numerous future possibilities, not just a few, making them a strong exploratory tool under deeply uncertain futures. When used in a participatory setting, models can become a platform for negotiation, reflection and learning among stakeholders with diverse worldviews, backgrounds and interests (Pahl-Wostl and Hare 2004). This could increase the likelihood of achieving a shared vision among different actors in sustainability transitions, and of better coordination for implementing transformative strategies.

While there have been a few attempts to integrate normative and explorative scenario planning in sustainability transitions (e.g. Kok et al. 2011), and fewer that use computational models for producing scenarios (e.g. Moallemi et al. 2017; Moallemi and Malekpour 2018), the field is in infancy and requires further conceptual and methodological developments. Moreover, there is limited – if any – empirical data available to provide insights on the effectiveness of such an integration in sustainability transitions. Future research could provide a better understanding of:

1 How to develop models as scenario tools in long-term planning of sustainability transitions: this includes theoretical and methodological issues, as well as technical limitations in modelling complex social processes.
2 How to use those models in a participatory way and in combination with qualitative techniques.
3 The effectiveness of models as normative and explorative scenario tools in improving long-term planning in sustainability transitions.

There are examples of trying to address some of the above in the policy analysis literature (especially in the decision-making under deep uncertainty stream) which can provide insights for the sustainability transitions research community.

References

Bankes, Steven. 1993. "Exploratory modeling for policy analysis." *Operations Research* 41 (3): 435–49.

Bankes, Steven, Robert Lempert, and Steven Popper. 2002. "Making computational social science effective: Epistemology, methodology, and technology." *Social Science Computer Review* 20 (4): 377–88. http://doi.org/10.1177/089443902237317.

Börjeson, Lena, Mattias Höjer, Karl-Henrik Dreborg, Tomas Ekvall, and Göran Finnveden. 2006. "Scenario types and techniques: Towards a user's guide." *Futures* 38 (7): 723–39. http://doi.org/10.1016/j.futures.2005.12.002.

Bryant, Benjamin P., and Robert J. Lempert. 2010. "Thinking inside the box: A participatory, computer-assisted approach to scenario discovery." *Technological Forecasting and Social Change* 77 (1). Elsevier Inc.: 34–49. http://doi.org/10.1016/j.techfore.2009.08.002.

Bulkeley, Harriet, Vanesa Castán Broto, and Anne Maassen. 2014. "Low-carbon transitions and the reconfiguration of urban infrastructure." *Urban Studies* 51 (7): 1471–86. http://doi.org/10.1177/0042098013500089.

Carroli, Linda. 2018. "Planning roles in infrastructure system transitions: a review of research bridging socio-technical transitions and planning." *Environmental Innovation and Societal Transitions*. 29. Elsevier: 81–9. http://doi.org/10.1016/j.eist.2018.06.001.

de Haan, Fjalar J., Briony C. Ferguson, Rachelle C. Adamowicz, Phillip Johnstone, Rebekah R. Brown, and Tony H.F. Wong. 2014. "The needs of society: A new understanding of transitions, sustainability and liveability." *Technological Forecasting and Social Change* 85 (June). Elsevier Inc.: 121–32. http://doi.org/10.1016/j.techfore.2013.09.005.

de Haan, Fjalar J., Alfonso Martinez Aranz, and Wouter Spekkink. 2019. "Data-driven transitions research – Methodological considerations for event-based analysis." In *Modelling Transitions – Virtues, Vices, Visions of the Future*, edited by Enayat Moallemi and Fjalar de Haan. This volume: Routledge.

de Haan, Fjalar J., Briony C. Rogers, Rebekah R. Brown, and Ana Deletic. 2016. "Many roads to Rome: The emergence of pathways from patterns of change through exploratory modelling of sustainability transitions." *Environmental Modelling and Software* 85. Elsevier Ltd: 279–92. http://doi.org/10.1016/j.envsoft.2016.05.019.

Dewar, James A., Carl H. Builder, William M. Hix, and Morlie H. Levin. 1993. *Assumption-Based Planning: A Planning Tool for Very Uncertain Times*. Santa Monica: RAND.

Dominguez, Domain, Bernhard Truffer, and Willi Gujer. 2011. "Tackling uncertainties in infrastructure sectors through strategic planning: The contribution of discursive approaches in the urban water sector." *Water Policy* 13 (3): 299–316. http://doi.org/10.2166/wp.2010.109.

Ferguson, Briony C., Niki Frantzeskaki, and Rebekah R. Brown. 2013. "A strategic program for transitioning to a water sensitive city." *Landscape and Urban Planning* 117 (September). Elsevier B.V.: 32–45. http://doi.org/10.1016/j.landurbplan.2013.04.016.

Forester, John. 1982. "Planning in the face of power." *Journal of the American Planning Association* 48 (1): 67–80.

Funtowicz, Silvio O., and Jerome R. Ravetz. 1993. "Science for the post-normal age." *Futures* 25 (7): 739–55. http://doi.org/10.1016/0016-3287(93)90022-L.

Goodwin, Paul, and George Wright. 2010. "The limits of forecasting methods in anticipating rare events." *Technological Forecasting and Social Change* 77 (3). Elsevier Inc.: 355–68. http://doi.org/10.1016/j.techfore.2009.10.008.

Groves, David G., and Robert J. Lempert. 2007. "A new analytic method for finding policy-relevant scenarios." *Global Environmental Change* 17 (1): 73–85. http://doi.org/10.1016/j.gloenvcha.2006.11.006.

Haasnoot, Marjolijn, Jan H. Kwakkel, Warren E. Walker, and Judith ter Maat. 2013. "Dynamic adaptive policy pathways: A method for crafting robust decisions for a deeply uncertain world." *Global Environmental Change* 23 (2). Elsevier Ltd: 485–98. http://doi.org/10.1016/j.gloenvcha.2012.12.006.

Haasnoot, Marjolijn, H. Middelkoop, E. van Beek, and W.P.A. van Deursen. 2011. "A method to develop sustainable water management strategies for an uncertain future." *Sustainable Development* 19 (6). John Wiley & Sons, Ltd.: 369–81. http://doi.org/10.1002/sd.438.

Hajer, Maarten A., and Peter Pelzer. 2018. "2050 – An energetic odyssey: Understanding 'techniques of futuring' in the transition towards renewable energy." *Energy Research and Social Science* 44 (January). Elsevier: 222–31. http://doi.org/10.1016/j.erss.2018.01.013.

Halbe, Johannes. 2019. "Participatory modelling in sustainability transitions research." In *Modelling Transitions – Virtues, Vices, Visions of the Future*, edited by Enayat Moallemi and Fjalar de Haan. This volume: Routledge.

Halbe, Johannes, Dominik E. Reusser, Georg Holtz, Marjolijn Haasnoot, Annette Stosius, Wibke Avenhaus, and Jan H. Kwakkel. 2015. "Lessons for model use in transition research: A survey and comparison with other research areas." *Environmental Innovation and Societal Transitions* 15. Elsevier B.V.: 194–210. http://doi.org/10.1016/j.eist.2014.10.001.

Holtz, Georg, Floortje Alkemade, Fjalar de Haan, Jonathan Köhler, Evelina Trutnevyte, Tobias Luthe, Johannes Halbe, et al. 2015. "Prospects of modelling societal transitions: Position paper of an emerging community." *Environmental Innovation and Societal Transitions* 17. Elsevier B.V.: 41–58. http://doi.org/10.1016/j.eist.2015.05.006.

Holtz, Georg, and Emile J.L. Chapin. 2019. "Considering actor behaviour: Agent-based modelling of transitions." In *Modelling Transitions – Virtues, Vices, Visions of the Future*, edited by Enayat Moallemi and Fjalar de Haan. This volume: Routledge.

Hunt, Dexter V.L., D. Rachel Lombardi, Stuart Atkinson, Austin R.G. Barber, Matthew Barnes, Christopher T. Boyko, Julie Brown, et al. 2012. "Scenario archetypes: Converging rather than diverging themes." *Sustainability* 4 (4): 740–72. http://doi.org/10.3390/su4040740.

Knight, Frank H. 1921. *Risk, Uncertainty and Profit*. Boston and New York: Houghton Mifflin Company.

Köhler, Jonathan, Frank W. Geels, Florian Kern, Jochen Markard, Anna Wieczorek, Floortje Alkemade, Flor Avelino, et al. 2019. "An agenda for sustainability transitions research: State of the art and future directions." *Environmental Innovation and Societal Transitions* 31. Elsevier: 1–32. http://doi.org/10.1016/j.eist.2019.01.004.

Kok, Kasper, M. van Vliet Mathijs, I. Bärlund Ilona, Anna Dubel, and Jan Sendzimir. 2011. "Combining participative backcasting and exploratory scenario development: Experiences from the SCENES project." *Technological Forecasting and Social Change* 78 (5): 835–51. http://doi.org/10.1016/j.techfore.2011.01.004.

Kwakkel, Jan H., Willem L. Auping, and Erik Pruyt. 2013. "Dynamic scenario discovery under deep uncertainty: The future of copper." *Technological Forecasting and Social Change* 80 (4). Elsevier Inc.: 789–800. http://doi.org/10.1016/j.techfore.2012.09.012.

Kwakkel, Jan H., and Erik Pruyt. 2013. "Exploratory modeling and analysis, an approach for model-based foresight under deep uncertainty." *Technological Forecasting and Social Change* 80 (3). Elsevier Inc.: 419–31. http://doi.org/10.1016/j.techfore.2012.10.005.

Lawrence, Judy, Andy Reisinger, Brett Mullan, and Bethanna Jackson. 2013. "exploring climate change uncertainties to support adaptive management of changing flood-risk." *Environmental Science & Policy* 33 (November). Elsevier Ltd: 133–42. http://doi.org/10.1016/j.envsci.2013.05.008.

Lempert, Robert J., Steven W. Popper, and Steven C. Bankes. 2003. *Shaping the Next One Hundred Years: New Methods for Quantitative, Long-Term Policy Analysis*. Santa Monica: RAND Corporation.

Loorbach, Derk Albert. 2007. *Transition Management New Mode of Governance for Sustainable Development*. Rotterdam: Erasmus University of Rotterdam.

Loorbach, Derk Albert. 2010. "Transition management for sustainable development: A prescriptive, complexity-based governance framework." *Governance* 23 (1): 161–83. http://doi.org/10.1111/j.1468-0491.2009.01471.x.

Lutz, Lotte Marie, Lisa Britt Fischer, Jens Newig, and Daniel Johannes Lang. 2017. "Driving factors for the regional implementation of renewable energy – A multiple case study on the German energy transition." *Energy Policy* 105 (February). Elsevier Ltd: 136–47. http://doi.org/10.1016/j.enpol.2017.02.019.

Christos K. Makropoulos, Fayyaz Ali Memon, Chris Shirley-Smith, and David Butler. 2008. "Futures: An exploration of scenarios for sustainable urban water management." *Water Policy* 10 (4): 345–73. http://doi.org/10.2166/wp.2008.014.

Malekpour, Shirin, Rebekah R. Brown, and Fjalar J. de Haan. 2015. "Strategic planning of urban infrastructure for environmental sustainability: Understanding the past to intervene for the future." *Cities* 46. Elsevier Ltd: 67–75. http://doi.org/10.1016/j.cities.2015.05.003.

Malekpour, Shirin, Rebekah R. Brown, Fjalar J. de Haan, and Tony H.F. Wong. 2017. "Preparing for disruptions: A diagnostic strategic planning intervention for sustainable development." *Cities* 63. Elsevier Ltd: 58–69. http://doi.org/10.1016/j.cities.2016.12.016.

Malekpour, Shirin, Fjalar J. de Haan, and Rebekah R. Brown. 2013. "Marrying exploratory modelling to strategic planning: Towards participatory model use." In *MODSIM2013, 20th International Congress on Modelling and Simulation*, 2248–54. Adelaide: Modelling and Simulation Society of Australia and New Zealand.

Malekpour, Shirin, Fjalar J. de Haan, and Rebekah R. Brown. 2016. "A methodology to enable exploratory thinking in strategic planning." *Technological Forecasting and Social Change* 105. Elsevier Inc.: 192–202. http://doi.org/10.1016/j.techfore.2016.01.012.

Marchau, Vincent A.W.J., Warren E. Walker, Pieter J.T.M. Bloemen, and Steven W. Popper. 2019. *Decision Making Under Deep Uncertainty: From Theory to Practice*. Springer. https://doi.org/10.1007/978-3-030-05252-2.

Markard, Jochen, Rob Raven, and Bernhard Truffer. 2012. "Sustainability transitions: An emerging field of research and its prospects." *Research Policy* 41 (6). Elsevier B.V.: 955–67. http://doi.org/10.1016/j.respol.2012.02.013.

Moallemi, Enayat A., Fjalar J. de Haan, and Jonathan Köhler. 2019. "Exploratory modelling of transitions: An emerging approach for coping with uncertainties in transitions research." In *Modelling Transitions – Virtues, Vices, Visions of the Future*, edited by Enayat Moallemi and Fjalar de Haan. This volume: Routledge.

Moallemi, Enayat A., Fjalar de Haan, Jan Kwakkel, and Lu Aye. 2017. "Narrative-informed exploratory analysis of energy transition pathways: A case study of India's electricity sector." *Energy Policy* 110 (August). Elsevier Ltd: 271–87. http://doi.org/10.1016/j.enpol.2017.08.019.

Moallemi, Enayat A., and Shirin Malekpour. 2018. "A participatory exploratory modelling approach for long-term planning in energy transitions." *Energy Research and Social Science* 35 (February 2017). Elsevier: 205–16. http://doi.org/10.1016/j.erss.2017.10.022.

Norgaard, Richard B. 1988. "Sustainable development: A co-evolutionary view." *Futures* 20 (6): 606–20. http://doi.org/10.1016/0016-3287(88)90003-1.

Pahl-Wostl, Claudia, and Matt Hare. 2004. "Processes of social learning in integrated resources management." *Journal of Community & Applied Social Psychology* 14 (3). John Wiley & Sons, Ltd.: 193–206. http://doi.org/10.1002/casp.774.

Rhodes, Bruce G. 2003. "Uncertainties in water resources planning." In *28th International Hydrology and Water Resources Symposium*. Wollongong, Australia: The Institution of Engineers.

Rogers, Briony C., and Alexander W. Gunn. 2015. *Towards a Water Sensitive Elwood*. Melbourne: Cooperative Research Centre for Water Sensitive Cities.

Rotmans, Jan. 1998. "Methods for IA: The challenges and opportunities ahead." *Environmental Modeling and Assessment* 3: 155–79.

Rotmans, Jan. 2005. *Societal Innovation: Between Dream and Reality Lies Complexity*. Rotterdam: Erasmus University.

Schoemaker, Paul J.H. 1993. "Multiple scenario development: Its conceptual and behavioral foundation." *Strategic Management Journal* 14: 193–213.

Simon, Herbert A. 1956. "Rational choice and the structure of the environment." *Psychological Review* 63 (2): 129–38.

Sondeijker, Saartje. 2009. *Imagining Sustainability: Methodological Building Blocks for Transition Scenarios*. Rotterdam: Erasmus University of Rotterdam.

Sondeijker, Saartje, Jac Geurts, Jan Rotmans, and Arnold Tukker. 2006. "Imagining sustainability: The added value of transition scenarios in transition management." *Foresight* 8 (5): 15–30. http://doi.org/10.1108/14636680610703063.

Störmer, Eckhard, Bernhard Truffer, Damian Dominguez, Willi Gujer, Anja Herlyn, Harald Hiessl, Hans Kastenholz, et al. 2009. "The exploratory analysis of trade-offs in strategic planning: Lessons from regional infrastructure foresight." *Technological Forecasting and Social Change* 76 (9). Elsevier Inc.: 1150–62. http://doi.org/10.1016/j.techfore.2009.07.008.

Truffer, Bernhard, Eckhard Störmer, Max Maurer, and Annette Ruef. 2010. "Local strategic planning processes and sustainability transitions in infrastructure sectors." *Environmental Policy and Governance* 20: 258–69. http://doi.org/10.1002/eet.550.

Trutnevyte, Evelina, Michael Stauffacher, and Roland W. Scholz. 2011. "Supporting energy initiatives in small communities by linking visions with energy scenarios and multi-criteria assessment." *Energy Policy* 39 (12). Elsevier: 7884–95. http://doi.org/10.1016/j.enpol.2011.09.038.

van Dorsser, Cornelis, Warren E. Walker, Poonam Taneja, and Vincent A.W.J. Marchau. 2018. "Improving the link between the futures field and policymaking." *Futures*. Elsevier Ltd: 1–10. http://doi.org/10.1016/j.futures.2018.05.004.

Varum, Celeste Amorim, and Carla Melo. 2010. "Directions in scenario planning literature – A review of the past decades." *Futures* 42 (4). Elsevier Ltd: 355–69. http://doi.org/10.1016/j.futures.2009.11.021.

Voros, Joseph. 2003. "A generic foresight process framework." *Foresight* 5 (3): 10–21. http://doi.org/10.1108/14636680310698379.

Voß, Jan-Peter, Adrian Smith, and John Grin. 2009. "Designing long-term policy: Rethinking transition management." *Policy Sciences* 42 (4): 275–302. http://doi.org/10.1007/s11077-009-9103-5.

Walker, Warren E. 2000. "Policy analysis: A systematic approach to supporting policymaking in the public sector." *Journal of Multi-Criteria Decision Analysis* 9: 11–27.

Walker, Warren E., Marjolijn Haasnoot, and Jan Kwakkel. 2013. "Adapt or perish: A review of planning approaches for adaptation under deep uncertainty." *Sustainability* 5 (3): 955–79. http://doi.org/10.3390/su5030955.

Walker, Warren E., Paul Harremoes, Jan Rotmans, Jeroen P. van der Sluijs, Marjolein B.A. van Asselt, Peter Janssen, and Mafrtin P. Krayer von Krauss. 2003. "Defining uncertainty a conceptual basis for uncertainty management." *Integrated Assessment* 4 (1): 5–17.

Walker, Warren E., Vincent A.W.J. Marchau, and Jan H. Kwakkel. 2013. "Uncertainty in the framework of policy analysis." In *Public Policy Analysis: International Series in Operations Research & Management Science*, edited by W. Thissen and W. Walker. Vol 179. Boston, MA: Springer.

Walker, Warren E., S. Adnan Rahman, and Jonathan Cave. 2001. "Adaptive policies, policy analysis, and policy-making." *European Journal of Operational Research* 128 (2): 282–9. http://doi.org/10.1016/S0377-2217(00)00071-0.

Wright, George, and Paul Goodwin. 2009. "Decision making and planning under low levels of predictability: Enhancing the scenario method." *International Journal of Forecasting* 25 (October). Elsevier B.V.: 813–25. http://doi.org/10.1016/j.ijforecast.2009.05.019.

Yücel, Gönenç. 2010. *Analyzing Transition Dynamics: The Actor-Option Framework for Modelling Socio-Technical Systems*. Delft: Delft University of Technology.

11 Participatory modelling in sustainability transitions research

Johannes Halbe

1 Introduction

Participatory modelling has been identified as a promising method to support understanding of transitions, provide policy advice and facilitate stakeholder processes (e.g. Holtz et al. 2015; Halbe et al. 2015). For instance, participatory modelling can reveal stakeholders' mental models and preferences which can help to parameterise simulation models and design effective policies (e.g. Tàbara et al. 2008). In addition, modelling with stakeholders can support social learning and collaboration (e.g. Halbe, Pahl-Wostl, and Adamowski 2018).

Participatory modelling has a long history in the fields of environmental management (e.g. Voinov et al. 2016), operational research (e.g. Vennix 1996) and public policy (e.g. Rouwette, Bleijenbergh, and Vennix 2016). However, the term 'participatory modelling' is a broad term that includes various methodological frameworks, such as group model building (Vennix 1996), mediated modelling (van den Belt 2004) or companion modelling (Barreteau et al. 2003). In addition, the term covers various forms of participation ranging from stakeholders being actively involved in model building (e.g. Videira et al. 2012) to stakeholders using a model in form of a serious game (e.g. Wiese et al. 2014). Furthermore, various methods for participatory modelling exist ranging from qualitative to semi-quantitative and quantitative modelling methods (Beall and Ford 2010).

This chapter aims at reviewing the state of the art of participatory modelling in transitions research and identifying promising future research directions. Given the aforementioned diverse connotations of participatory modelling, a review of participatory modelling applications either has to focus on a specific type of participatory modelling by providing a precise definition, or to explicitly account for the various types of participatory modelling. In this chapter, the second route is chosen by specifying the various dimensions of participatory modelling through a framework which was built based upon a review of literature from research fields in which participatory modelling has a longer tradition (e.g. environmental management). The framework defines key dimensions of participatory modelling, including different model uses, modes of knowledge capture and exchange, timing of stakeholder involvement and methods applied. The framework is applied to structure a systematic literature review on participatory modelling in the field of sustainability transitions research. Search terms are specified to identify publications in the field that combine modelling and stakeholder engagement. Each participatory modelling exercise identified in the literature will be analysed to eventually provide an overview on emphases and gaps in the current state of the art of participatory modelling in sustainability transitions.

This chapter is structured as follows. In Section 2, a framework is presented that defines key dimensions of participatory modelling. In Section 3, the literature review method is provided. Section 4 presents the review results. Section 5 offers a discussion of review results and proposes a number of future research directions. Finally, conclusions are provided in Section 6.

2 Different forms and methods of participatory modelling

This section presents a framework which includes central dimensions of participatory modelling. Four dimensions of participatory modelling are distinguished: (1) the *model use* comprising distinct purposes, application contexts and epistemological foundations of modelling; (2) the various *modes of knowledge capture and exchange* between modelers, stakeholders and decision makers; (3) the *timing* of stakeholder involvement in the modelling process; and (4) the different general approaches and *modelling methods* that allow for an involvement of stakeholders. In the following, each dimension is presented in more detail.

2.1 Model uses

Halbe et al. (2015) distinguish between three model uses based upon different epistemological foundations of modelling, which are linked to certain model purposes and application contexts. The model use of *understanding transitions* has the purpose to develop general insights into transition processes, which is related to traditional core science (Funtowicz and Ravetz 1993). Hence, model development is more curiosity-driven by tailoring the model according to a specific research question or a phenomenon of interest. Participatory modelling for this model use can take the form of involving experts in model development and parameterisation. The model use of *providing case-specific decision-support* has the purpose to examine practical solutions to case-specific problems. The modelling process is more problem-driven and has an applied character in the sense that model results are intended to have a high practical value, for instance by proposing case-specific policies, strategies and measures. A special challenge of this model use is the handling of uncertainties, which can be accomplished by advanced modelling methods, such as exploratory modelling (Kwakkel and Yücel 2012, see Chapter 10 by Malekpour and Chapter 13 by Moallemi et al. in this volume). Participatory modelling could take the form of involving model users, such as managers or policy makers, in model design to increase their understanding and trust in the model. Finally, model use for *facilitating stakeholder processes* is stakeholder-driven and has the purpose of supporting stakeholder processes, for example through problem framing, vision development or assessment of strategies. This model use is linked to a post-normal type of science (Funtowicz and Ravetz 1993), as stakeholders with various backgrounds need to be involved to address issues with epistemological and ethical uncertainties. For example, participatory modelling processes can support learning about complex systems and foster effective communication between stakeholders. A central challenge of this model use is the choice of appropriate modelling methods that remain understandable even for stakeholders with limited mathematical knowledge (cf. Halbe, Pahl-Wostl, and Adamowski 2018).

Of course, the distinct model uses are of ideal-typical nature, and specific modelling applications can relate to two or even three model uses. Nevertheless, the classification can help to identify the primary model use of practical modelling applications, which allows for a targeted comparison of experiences gained.

2.2 Modes of knowledge capture and exchange

Participatory modelling can serve knowledge exchange between decision makers and stakeholders, which can be accomplished through different types of process designs. Lynam et al. (2007) identify three different modes of knowledge capture and exchange[1]. Participatory modelling processes can be designed in an *extractive* mode in which knowledge, preferences and values of individual stakeholders are analysed by researchers who later pass the synthesised knowledge to subsequent decision-making processes (i.e. stakeholders do not exchange knowledge and are not directly involved in decision-making). Examples for this extractive mode are modelling exercises to elicit individual mental models (e.g. Inam et al. 2015 and Halbe, Pahl-Wostl, and Adamowski 2018) or to improve the quality or acceptance of an existing model (cf. Hare 2011). Participatory processes can be also design in a *co-learning* mode, which implies that multiple stakeholders exchange and synthesise knowledge. However, stakeholders are not involved in decision-making, as information from the participatory processes is only passed down to the decision-making process (Lynam et al. 2007). In a participatory research context co-learning can be achieved through role-playing games (e.g. Gurung, Bousquet, and Trébuil 2006), collaborative serious games (e.g. Learmonth et al. 2011) or group modelling (e.g. Penn et al. 2013). Processes designed in a *co-management* mode finally enable an active involvement of stakeholders in knowledge synthesis and the decision-making process (Lynam et al. 2007). For example, this can be accomplished in transdisciplinary research projects (cf. Brandt et al. 2013) or participatory processes that have a clear decision-making mandate from policy makers (e.g. Palmer et al. 2013).

2.3 Timing

Participatory modelling exercises can span long periods of time (up to years), in particular if quantitative simulation models are developed. Therefore, stakeholders may not be involved throughout the whole process, but, depending on the purpose of their involvement, their participation can be limited to particular times. Hare (2011) distinguishes the following timing schemes:

- Stakeholders are involved at the beginning of the modelling process (*front-end participatory modelling)* in order to provide input on the model definition (e.g. Knoeri, Binder, and Althaus 2011).
- Stakeholders are involved at the end of the modelling process (*back-end participatory modelling*), which could involve the testing and use of the model (e.g. Sessa and Ricci 2014).
- Stakeholders are involved at the beginning AND end of the modelling process, but stakeholders are not included in the actual building of the model (*front- and back-end participatory modelling*). For example, stakeholders are involved in defining the model (front-end) and participate in its validation (back-end) (e.g. Rosales-Carreón and García-Díaz 2015).
- Stakeholders are involved during the whole modelling process (*co-construction participatory modelling*), which is mostly applied for conceptual models rather than simulation models (e.g. Videira et al. 2014).

2.4 Applied methods

Participatory modelling methods are usually applied in a broader involvement process. For example, causal loop diagrams, a qualitative modelling method, can be applied in the course of an individual interview with stakeholders or in stakeholder workshops (Halbe, Pahl-Wostl, and Adamowski 2018). Thus, conducting participatory modelling processes also requires expertise in organizing stakeholder processes and applying further research and engagement methods. This section presents a brief overview of general approaches for stakeholder involvement in research and some widely applied participatory modelling methods.

General approaches for stakeholder involvement

Surveys are widely applied to interrogate knowledge and perceptions of stakeholders through questionnaires. Several factors should be considered in questionnaire design, such as the choice for open or closed questions, or design of rating scales (Krosnick 2018). Testing of questionnaires is another critical step for which several methods are available (e.g. Brancato et al. 2006).

The *Delphi method* is a particular survey method aiming at soliciting information from a group of experts. A group communication process is designed with the aim to achieve a convergence of opinions on a certain topic (Hsu and Sandford 2010). Therefore, a series of surveys is prepared that interrogate the opinion of individual experts (Okoli and Pawlowski 2004). After each round the surveys are designed in such a way that each expert has to comment on the group opinion. This feedback structure is the peculiar element of the Delphi method (Ludwig 1997).

Interviews should also be carefully designed (e.g. clear objectives, selecting data analysis methods) and pre-tested (Guthrie 2010). Three different types of interviews can be distinguished (see Guthrie 2010). First, there are *unstructured interviews* where questions are not prepared in advance but are intuitively raised in the course of a conversation. Unstructured interviews generate qualitative data and allow for in-depth inquiry of a topic. Second, *semi-structured interviews* are more standardised (i.e. structured by a set of topics) but still allow for some flexibility. Semi-structured interviews usually generate a combination of qualitative and quantitative data. Third, *structured interviews* follow a formal standardised questionnaire using set questions and set response codes. Thereby, structured interviews mainly provide quantitative data that can be analysed by statistical methods.

Focus groups are a form of group interviews that can be used to generate insights into perceptions regarding a certain topic or product (Kitzinger 1995). The facilitation of communication and interaction between participants is a central part of the method. About 5 to 10 people participate in a carefully planned series of discussions, which is guided by an experienced facilitator, to receive their perceptions on a tropic of interest (Krueger and Casey 2014). Participants can ask questions, exchange anecdotes or comment on each other (Kitzinger 1995). The particular design of focus groups is influenced by their purpose, which can be market research, scientific research or participatory processes (Krueger and Casey 2014).

Stakeholder workshops are a widely used context for participatory modelling. The organisation of such workshops is not trivial but requires consideration of various factors. Various methods are available to support an effective organisation of stakeholder workshops, such as stakeholder analysis (e.g. Prell, Reed, and Hubacek 2011) or process analysis and evaluation (e.g. Jones et al. 2009; Forrest and Wiek 2014). Practical experiences have been often collected in the form of handbooks (cf. Nielsen, Bryndum, and Bedsted 2017). For example, the research project HarmoniCOP developed a comprehensive handbook that addresses several aspects of the design of stakeholder workshops (HarmoniCOP 2005). An important element is the setting of ground rules, such as asking for commitment to the process (HarmoniCOP 2005). The organisation of the actual workshop requires the selection of a facilitator, stakeholders and a venue. Various challenges can emerge in the course of a workshop, such as dominant stakeholders or emerging conflicts.

Participatory modelling methods

A broad range of qualitative and quantitative methods can be used for participatory modelling. In the following, an overview of some well-known participatory modelling methods is provided. These modelling methods can be combined with further non-technical approaches, such as *narrative storylines* or *collages* (e.g. Kok et al. 2006).

A widely applied method for qualitative, conceptual participatory modelling is systems thinking using *causal loop diagrams* (see Chapter 9 by Rojas and de Haan in this volume). Causal loop diagrams support the visualisation and analysis of multi-causal relationships and feedback processes (e.g. Vennix 1996; Inam et al. 2015). The method can be applied in the setting of an individual interview or a group meeting (e.g. Halbe, Pahl-Wostl, and Adamowski 2018). A stepwise approach is usually followed to develop a causal loop diagram of a problem situation (e.g. Inam et al. 2015) or the barriers and drivers of a sustainability innovation (Halbe et al. 2015; Halbe 2016; Halbe and Pahl-Wostl 2019). The process starts by defining a start variable (e.g. a problem variable or an innovation variable) before variables representing influencing factors and consequences are added and connected to the start variable via causal links. In subsequent steps, feedback loops, solutions and barriers are added to the model (see Halbe, Pahl-Wostl, and Adamowski 2018 for a detailed presentation of the method). Another qualitative modelling method is *participatory mapping* which focuses on spatial relationships by animating participants to draw maps of a particular problem situation. Participatory rapid appraisal is such a mapping approach in which stakeholders draw maps, diagrams and timelines in a creative process to express their point of view in an integrated way (Chambers 1994). Such conceptual modelling and mapping methods can also be combined by a diagnostic scoring procedure. Stakeholders can prioritise the importance of different aspects of a problem by using simple voting techniques (cf., Sheil and Liswanti 2006).

Semi-quantitative modelling methods produce quantitative results, which however need to be interpreted qualitatively. A prominent semi-quantitative method is *fuzzy cognitive mapping*, which bases upon a conceptual model in which relationships are weighted using numerical values in the range of −1 and 1. In participatory modelling processes, the weights can be also set by using an ordinal scale, such as +++ for strong positive links, ++ for positive links with medium strength and + for weak positive links (an analogous scale is used for negative links) (Jetter and Kok 2014). Such a fuzzy cognitive map can be interpreted as a type of a recursive neural network (Kosko 1993) in which impulses pass through the network until a stable state (i.e. variable take constant values) or a stable limit cycle (i.e. variable values oscillate) is reached. Scenarios can be simulated by fixing the value of a scenario variable. In the following, impulses are sent throughout the network, which are normalised through so-called 'squashing function', which restrict variable values to intervals (i.e. in case of a sigmoid squashing function, variable values range between 0 and 1, while a trivalent function produces values between −1 and 1) (see Jetter and Kok 2014 for more details). The results of the scenario analysis using fuzzy cognitive mapping are interpreted by comparing the relative difference between variable values (i.e. variable *X* increases more strongly than variable *Y* in a certain scenario) (see Halbe and Adamowski 2019). Various fuzzy cognitive mapping software tools exist, such as the FCMapper (Wildenberg et al. 2010; Olazabal and Pascual 2016) or Mental Modeler (Gray et al. 2013; Henly-Shepard, Grey, and Cox 2015). *Cross-impact balance analysis* is another semi-quantitative modelling method that can be applied in participatory processes. By using this method, relationships between system descriptors are analysed by quantitatively defining whether a descriptor raises or lowers the occurrence probability of another descriptor (Weimer-Jehle 2006). These cross impacts between descriptors are specified in a matrix by using an ordinal scale, for example ranging from −3 denoting a strongly restricting direct influence to 0 representing no direct influence and +3 for strongly promoting direct influence (Weimer-Jehle 2006). The matrix can be filled out by a group of experts in a group setting or through individual interviews, which are later synthesised by researchers. As a result, cross-impact balance analysis identifies consistent combination of system descriptors and thus can help in the identification of plausible scenarios (Schweizer and Kriegler 2012). *Multi-criteria analysis* is another semi-quantitative approach that can support decision processes by analysing sets of actions that minimise trade-offs and maximise sustainability benefits (Ness et al. 2007). Multi-criteria analysis usually starts with the definition of goals, criteria and alternatives for action (Wang et al. 2009).

Quantitative participatory modelling approaches can impede the involvement of stakeholders with limited mathematical knowledge in model development. Thus, qualitative methods are often combined with quantitative modelling to allow for stakeholder participation. As an example, *narratives* are developed by stakeholders before simulation models are applied by experts to infer consequences and potential incoherencies (e.g. Alcamo 2008; Moallemi and Malekpour 2018). Another example is the *companion modelling* approach in which stakeholders are engaged through *role-playing games* related to an environmental issue (Barreteau et al. 2003). The game results, such as strategies chosen by players, are subsequently utilised in an *agent-based model* to systematically investigate potential outcomes of alternative rules and agent behavior (e.g. Gurung, Bousquet, and Trébuil 2006). *System dynamics modelling* is also a widely applied quantitative participatory modelling method (see Chapter 8 by Papachristos and Struben in this volume), which is often combined with qualitative causal loop diagrams. Graphical user interfaces support a more active participation of stakeholders in the model-building process (e.g. Vennix 1996; van den Belt 2004) or utilisation of the available model (e.g. Stave 2003). *Serious games* are an effective approach to support the playful application of models by stakeholders, which can induce learning about the system's complexity (e.g. Sterman 2014; Dörner et al. 2016). Serious games require an appealing gaming interface that allows for players to set scenarios (e.g. by choosing between alternative strategies) and see results of their actions. Serious games can build upon different modelling methods, such as system dynamics or agent-based models.

3 Literature review methodology

A literature review is an accepted and widely used method to screen available scientific evidence about a certain topic and to summarise and synthesise research findings (cf. Petticrew and Roberts 2006; Liberati et al. 2009). Narrative reviews can be influenced by the researchers' bias, as the selection, interpretation and synthesis of findings are usually accomplished in an implicit way (Tranfield, Denyer, and Smart 2003). Due to this lack of transparency and reproducibility of narrative reviews, guidelines have been developed that support a more systematic review that explicitly explains the rationale, literature selection, risk of bias and synthesis method (cf. Liberati et al. 2009). A systematic review is defined as 'a review of a clearly formulated question that uses systematic and explicit methods to identify, select, and critically appraise relevant research, and to collect and analyze data from the studies that are included in the review' (Moher et al. 2009, 1).

The literature review in this chapter answers the following question: What are the emphases and gaps of participatory modelling in the sustainability transitions research field? Scopus was used as the literature database to identify research articles of participatory modelling applications in sustainability transitions research up to 2017. A query was designed to screen titles, abstracts and key words for the following aspects (search terms are written in italics):[2]

- Sustainability transitions research: *(Transition OR transform*) AND Sustainab**
- Modelling applications: *Modeling OR Modelling*
- Participation: *participat* OR stakeholder OR collaborat**

The query resulted in a list of 105 publications. A limitation of this review is the exclusion of conference papers, reviews, editorials, book chapters, literature without peer review (e.g. grey literature) and literature in a language other than English. In addition, a number of studies might have been excluded in which the participatory element is not the focus of the study, but rather a side topic (i.e. the participatory element has not be mentioned in the title, abstract or key words but in the main text). This might be especially the case for modelling studies that apply an extractive mode of knowledge capture and exchange (see Section 2.2), for example by involving stakeholders in the parameterisation of the model.

In the next step, the abstracts of all 105 publications were screened by the author of this chapter. Articles were included in the review that complied with the three quality criteria: (1) the study topic was linked to sustainability transitions (i.e. broad societal change towards sustainability), (2) a modelling study is provided, and (3) stakeholders participated in the modelling process (statistical modelling studies were excluded as the statistical analysis of survey data was not considered as participatory modelling). Eighty articles were excluded, as they did not comply with the aforementioned quality criteria. The full text of the remaining 25 publications was analysed by the author of this chapter using the dimensions of participatory modelling identified in Section 2.

4 The review of participatory modelling in sustainability transitions research

This section presents the results of the literature review. First, all articles that were included in the review are introduced in Section 4.1 in a tabular form.

Second, the review results are quantitatively analysed in Section 4.2 to provide an overview of emphases and gaps.

4.1 The framework-based analysis of results

For each article, the following information is provided in Table 11.1: (1) the general topic of the study, (2) the particular model used (see Section 2.1), (3) the knowledge capture and exchange mode applied (see Section 2.2), (4) the timing of stakeholder involvement (see Section 2.3), (5) the general involvement approach in which participatory modelling was embedded (see Section 2.4), and (6) the particular modelling method(s) that were applied in the participatory process (see Section 2.4).

4.2 Identifying patterns, emphases and gaps

Figure 11.1a shows the number of articles per year resulting from the Scopus query. While modelling studies in the field of sustainability transitions increased exponentially from 2005 to 2017, the number of participatory modelling studies in the field only increased to a small extent (see Figure 11.1b). The number of articles included in this review shows a slightly increasing trend from 2005 when the first study was detected to the endpoint of the review in 2017 (see Figure 11.1b).

Figure 11.2 shows a detailed analysis of review results. The studies included in this review utilised all model uses with an emphasis on decision-support (64%) and facilitation (60%) (see Figure 11.2a). Knowledge capture and exchange were mostly organised in a co-learning mode (60%) followed by the extractive (40%) and co-management modes (16%) (see Figure 11.2b). The low number of studies in the co-management mode shows a clear gap in the current literature. The timing of involvement has followed a front- and back-end modelling (44%) or co-construction modelling approach (40%) in the most cases (see Figure 11.2c). Front-end (20%) and back-end (12%) approaches were applied to a lesser extent. A co-construction mode allows for a deeper involvement of stakeholders in relation to other knowledge capture modes. Other knowledge capture modes (i.e. front- and/or back-end modes) have been applied in 76% of reviewed studies (i.e. in some studies, co-construction modeling has been combined with other knowledge capture and exchange modes), which points to a current gap in the literature regarding participatory processes following a co-construction mode. Stakeholder workshops were the dominant setting (68%) in which participatory modelling processes were conducted (see Figure 11.2d). Further involvement processes were surveys (32%) and interviews (28%). A broad range of participatory modelling methods has been applied (see Figure 11.2e) ranging from qualitative methods, such as collages or narratives, to quantitative modelling methods, including agent-based models, system dynamics models and expert simulation models. Expert simulation models showed the highest prevalence in 32% of the studies followed by causal loop diagrams (28%), agent-based

Table 11.1 Review results of participatory modelling in sustainability transitions research, using the various dimensions of participatory modelling identified in Section 2: model uses, modes of knowledge capture and exchange, timing (if a study used participatory methods in different timing schemes, methods are specified in brackets), involvement process, participatory modelling methods

Source	*Topic*	*Model Uses*	*Knowledge Capture and Exchange*	*Timing*	*Involvement Process*	*Participatory Modelling Methods*
Magnuszewski, Sendzimir, and Kronenberg (2005)	Adaptive management; sustainable regional development	Facilitation	Co-learning	Co-construction	Stakeholder workshops	Conceptual modelling (causal loop diagrams)
García-Barrios, Speelman, and Pimm (2008)	Natural resource management	Facilitation	Co-learning	Back end	Stakeholder workshops	Role-playing games; simulation models (equation-based; multi-criteria analysis)
Levine et al. (2008)	Sustainable development of villages	Decision-support and facilitation	Extractive	Front and back end	Stakeholder workshops	System dynamics, serious gaming, physical models, drawings
Tàbara et al. (2008)	Water management	Decision-support and facilitation	Co-learning and co-management	All forms: front and back end (role-play + agent-based model), co-construction (collages + matrix)	Stakeholder workshops, focus groups	Collages, cause–effect response matrix, agent-based model, role-playing game
Whitmarsh and Nykvist (2008)	Sustainable mobility	Decision-support and facilitation	Extractive	Front end	Stakeholder workshops	Agent-based model
Gourmelon et al. (2011)	Environmental education	Understanding and facilitation	Co-learning	Front and back end	Stakeholder workshops (development); classroom setting (application)	Conceptual modelling, agent-based model, role-playing game

(*Continued*)

Table 11.1 (Continued)

Source	*Topic*	*Model Uses*	*Knowledge Capture and Exchange*	*Timing*	*Involvement Process*	*Participatory Modelling Methods*
Knoeri, Binder, and Althaus (2011)	Recycling in construction sectors	Understanding and decision-support	Extractive	Front end	Surveys	Multi-criteria decision analysis
Videira et al. (2012)	Integrated maritime policy	Decision-support and facilitation	Co-learning and co-management	Co-construction	Individual Interviews; Survey; Stakeholder workshops	Causal loop diagrams
Halbe et al. (2013)	Integrated water resource management	Decision-support and facilitation	Co-learning	Co-construction	Stakeholder workshops	Causal loop diagrams
Penn et al. (2013)	Bio-based economy	Decision-support and facilitation	Co-learning	Co-construction	Stakeholder workshop	Fuzzy cognitive mapping
Iwaniec et al. (2014)	Urban planning	Decision-support	Co-management	Front and back end (city vision narrative, causal loop diagrams, system dynamics, board game); co-construction (narratives)	Stakeholder workshops	Narratives, conceptual modelling (causal loop diagrams), system dynamics; board game
Burks-Copes and Kiker (2014)	Ecosystem-based management	Decision-support	Co-management	Front and back end (iterative)	Stakeholder workshops	Conceptual modelling (drivers-stressors, responses, indicators)
Sessa and Ricci (2014)	Global future scenarios	Understanding	Extractive	end	Delphi Survey	Narratives
Wiese et al. (2014)	Transition towards a renewable energy system	Facilitation	–	Back end	Open source model	Energy system model (serious game)

Rodriguez et al. (2014)	Adaptation of farming systems to climate change	Understanding, decision-support and facilitation	Extractive, co-learning	Front and back end (parametrisation + scenario analyses)	Interviews and stakeholder workshops	Whole-farm dynamic simulation model
Videira et al. (2014)	Degrowth pathways	Understanding	Co-learning	Co-construction	Survey, Stakeholder workshop	Causal loop diagrams, cross-impact analysis
Rosales-Carreón and García-Díaz (2015)	Sustainable construction	Understanding	Extractive, co-learning	Front and back end	Semi-structured interviews	Agent-based model, systemic diagram
Withycombe Keeler et al. (2015)	Sustainable water governance	Facilitation	Extractive, co-learning	Front and back end	Surveys	Narratives, impact and consistency matrices, dynamic water demand and supply simulation model (WaterSim 5.0)
Moraine et al. (2016)	Crop–livestock integration	Decision-support and facilitation	co-learning	Co-construction	Surveys, stakeholder workshops	Multi-criteria assessment
Li et al. (2016)	Urban transitions towards sustainability	Understanding and Decision-support	Extractive	Front-end	Surveys, semi-structured interviews	Conceptual modelling using DPSIR framework, material flow analysis, causal loop diagram
Ward and Butler (2016)	Rainwater harvesting niche governance	Understanding and Decision-support	Extractive	Front end	Surveys, interviews	Participatory social actor analysis

(Continued)

Table 11.1 (Continued)

Source	*Topic*	*Model Uses*	*Knowledge Capture and Exchange*	*Timing*	*Involvement Process*	*Participatory Modelling Methods*
Zivkovic et al. (2016)	Sustainable urban heating systems	Decision-support and facilitation	Co-learning	Co-construction (qualitative scenarios), front and back end (quantitative scenarios, energy system model)	Stakeholder workshops	Qualitative and quantitative scenarios (participatory backcasting), energy system model (LEAP)
Cruz et al. (2017)	Integrated traffic and parking management	Decision-support	Extractive	Front end	Survey, semi-structured interviews	Life cycle analysis
Delmotte et al. (2017)	Integrated assessment of future agricultural systems	Understanding and decision-support	Co-learning	Co-construction (narratives), front and back end (bio-economic model)	Stakeholder workshops	Narrative scenarios, bio-economic model
Valkering et al. (2017)	Transition dynamics in city regions	Understanding, decision-support and facilitation	Co-learning	Front and back end	Interviews, Stakeholder workshops	Causal loop diagrams, agent-based model (serious game)

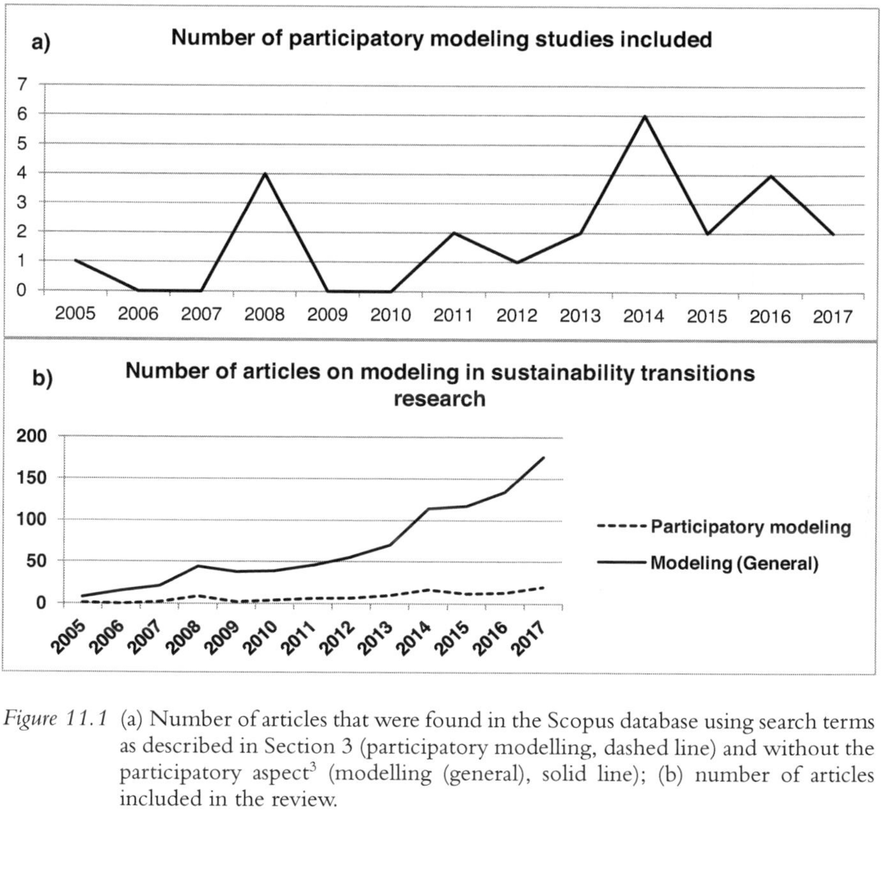

Figure 11.1 (a) Number of articles that were found in the Scopus database using search terms as described in Section 3 (participatory modelling, dashed line) and without the participatory aspect[3] (modelling (general), solid line); (b) number of articles included in the review.

models (20%), and narratives, role-playing games and multi-criteria analysis (each 12%). The following methods were applied to a smaller extent in the reviewed literature: system dynamics models (8%), cross-impact analysis (4%), fuzzy cognitive mapping (4%) and collages (4%). All participatory processes using quantitative simulation models, such as expert models, agent-based models and system dynamics models, were not conducted in a co-construction mode, but used a front-end, back-end or a front-end/back-end mode. This is also the case for role-playing games that were initially developed by researchers and only later applied by stakeholders. Qualitative modelling methods, including collages, narratives and causal loop diagrams, were mostly applied in a co-construction mode. However, there are also two studies in which causal loop diagrams were applied by researches using a front-end mode (Li et al. 2016) or a front- and back-end mode (Valkering et al. 2017). Cross-impact analysis, fuzzy cognitive mapping and multi-criteria assessment are located in the middle of the qualitative–quantitative continuum, and thus are interesting method to provide a bridge between qualitative and quantitative analysis. While multi-criteria analysis was mostly used in a front- and back-end mode

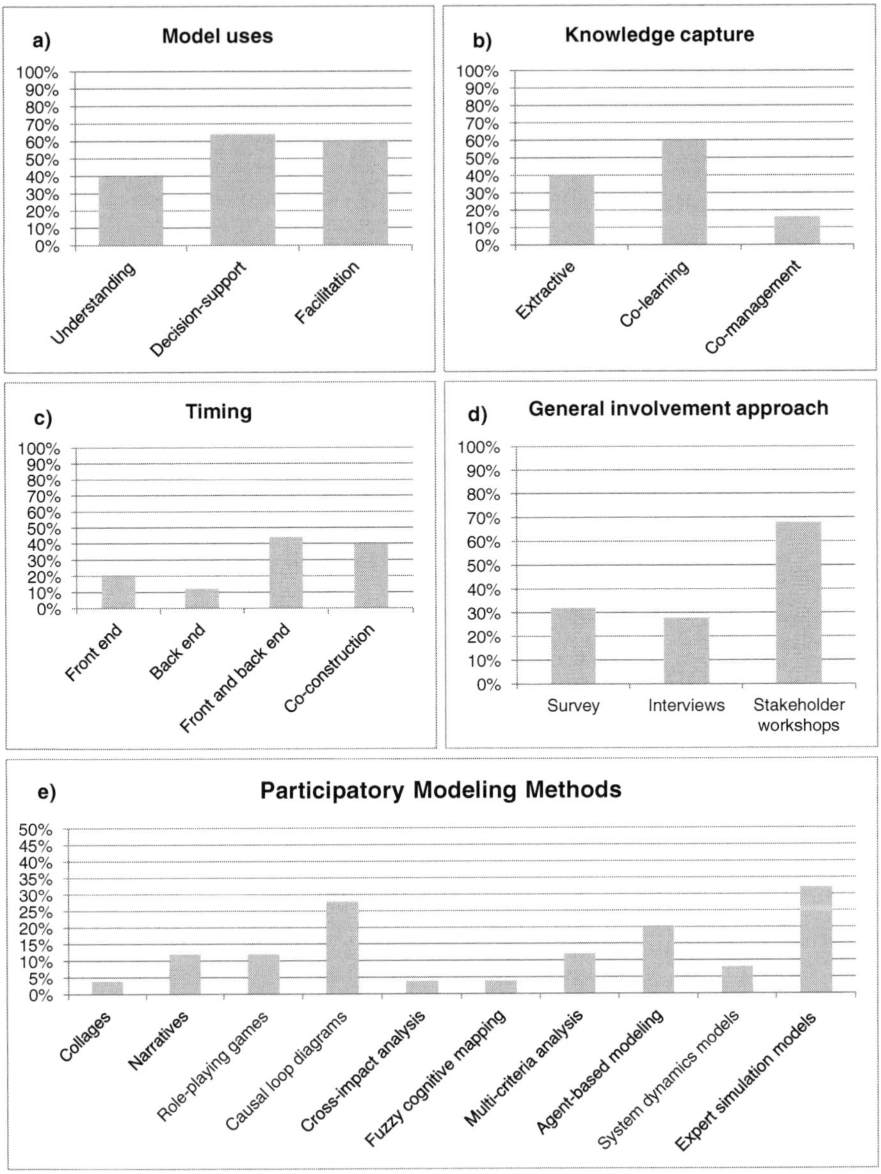

Figure 11.2 Analysis of review results using the various dimensions of participatory modelling, including model uses (a), modes of knowledge capture and exchange (b), timing (c), general involvement approach (d) and modelling methods (e).

(García-Barrios, Speelman, and Pimm 2008; Knoeri, Binder, and Althaus 2011; Moraine et al. (2016) used a co-construction mode though), fuzzy cognitive mapping (Penn et al. 2013) and cross-impact analysis (Videira et al. 2014) have been applied in a co-construction mode.

5 Discussion

The review provides an overview of the various methods and forms of participatory modelling in sustainability transitions research. The participatory modelling framework developed in Section 2 was able to cover the various dimensions of participatory modelling, including different model uses, modes of knowledge capture and exchange, timing of stakeholder involvement and applied methods and involvement contexts. Based upon the framework, the review revealed a number of emphases and gaps in the literature.

One of the gaps is indicated by the relatively low number of participatory modelling studies that were sorted to the model use of understanding sustainability transitions. On the one hand, this might be due to the aim of this model use to generate general knowledge and insight in a more curiosity-driven process of fundamental research, which means that both the model developers and target audience are predominantly researchers. Thus, stakeholders play a minor role in the research process and distribution of results compared to other model uses. Involving stakeholders in this model use is therefore more a side topic and thus might not be explicitly mentioned in the abstract or keywords. On the other hand, modelling for decision-support and facilitation has to closely consider the needs and requirements of stakeholders. Modelling for facilitating stakeholder processes even demands an active involvement of stakeholders in model development. Thus, these model uses rely upon more intense forms of stakeholder participation, which is more likely to be mentioned in the abstract or keywords of related research articles.

Another gap in the literature is linked to the co-management mode of knowledge capture and exchange. An extractive and co-learning mode are more in line with a control paradigm (see Halbe et al. 2018), in which decision makers still want to control the outcome of stakeholder participation by retaining their power of decision. Thus, co-management processes require a paradigm change from an outcome orientation (i.e. outcomes are pre-defined at the beginning of the process) towards a process-orientation (i.e. process characteristics are pre-defined but outcome remains open). Further applications of participatory modelling and process design methods in transitions research can be helpful to gather further experiences that can foster the application of participatory methods in practice. In this respect, process design and evaluation are tremendous challenges due to the complexity of real-world transition processes. Methodological and conceptual frameworks are needed to guide the design and evaluation of participatory processes and support the development of best practices and cross-case comparison (e.g. Forrest and Wiek 2014). An explicit link of participatory modelling studies to the process phases of the transition management framework (Loorbach 2010) can be a helpful approach to reduce complexity of process design and evaluation.

Furthermore, the timing of stakeholder involvement in modelling was mostly following a front- and/or back-end approach. Future research should focus more on co-construction participatory modelling, as this approach allows for most intense form of stakeholder participation. Qualitative model methods are predominantly used for model co-construction. However, fuzzy cognitive mapping and cross-impact analysis are promising semi-quantitative modelling methods for model co-construction. In this review, only two studies were detected that used semi-quantitative modelling methods: Penn et al. (2013) applied fuzzy cognitive mapping; Videira et al. (2014) applied cross-impact analysis. Nevertheless, there is a vast experience of applying fuzzy cognitive mapping and cross-impact analysis in other research fields, which can be used in sustainability transitions research (e.g. Weimer-Jehle 2006; Jetter and Kok 2014; Halbe and Adamowski 2019).

The high share of expert simulation models (32%) in the reviewed modelling studies is a surprising result. On the on hand, the high number can be explained by the broad nature of the 'expert simulation model' category (i.e. various model applications were subsumed). On the other hand, the result still shows the high relevance and usage of expert simulation models in the transitions research field. Established simulation models can have the advantage of being tested in various case studies so that experts have a high trust in modelling results. In addition, expert models can simulate complex processes in a sophisticated way, such as water demand and supply dynamics (e.g. Withycombe Keeler et al. 2015), which would require tremendous time to be built in a participatory setting. Still, expert simulation models have the disadvantage of the black-box character for non-experts, which can limit trust in their output. A potential way to deal with this issue is the combination of expert models with other participatory modelling methods, also called informal or loose coupling. Various studies included in this review have applied such a mixed-method approach. For example, companion modelling (see Section 2.4) is a well-known and widely applied approach that follows this route by combining role-playing games and agent-based modelling. Methodological frameworks can help to specify the contribution of each modelling methods and their linkages (see Liu et al. 2008).

Another interesting option to combine participatory and expert-based simulation models is the use of formal coupling; that is, the dynamical coupling of a model developed by stakeholders with an established expert model. Thereby, the stakeholder process can focus on modelling pressing issues (e.g. social-economic aspects), while well-known aspects, such as hydrological processes, are covered by an expert model. Inam et al. (2017a, 2017b) describes such an approach by using causal loop diagrams and system dynamics modelling to analyze socio-economic issues of soil salinisation, while using an expert model for including physical aspects. Model coupling is a time-intensive task though, as data exchange and iterative execution of models have to be realised (see also Prodanovic and Simonovic (2010) who coupled a system dynamic model with a physical model to analyze flood management strategies). Malard et al. (2017) developed a tool for a more rapid and stakeholder-friendly coupling of system dynamics and physical models, which can support the implementation of formal model coupling in practice.

Tailoring a participatory modelling process to case-specific requirements is a tremendous challenge, which every facilitator of participatory processes has to tackle. Many studies included in this review followed a more customised approach by combining different methods, uses, timings and modes of knowledge capture and exchange. About half of the studies combined different participatory modelling methods, 60% combined various model uses, 25% combined different timings of stakeholder engagement and 20% different modes of knowledge capture and exchange. These results underline the need for methodological and conceptual frameworks to systematically considering the different dimensions of stakeholder engagement in the design of participatory modelling processes. This also includes the conceptualisation of influential factors on a participatory process (e.g. resource availability of level of conflict), process characteristics (e.g. stakeholder selection, applied methods) and expected outcomes (e.g. knowledge or trust). As mentioned before, established conceptual and methodological frameworks in the transitions research fields can support systematic process design. For example, the transition management approach can be used to identify distinct process phases in a participatory process (e.g. problem structuring and vision development), which demand specific methods and stakeholders.

Comparing the results of this review to findings of a more general systematic review of methodological patterns and trends in the transitions research field by Zolfagharian et al. (2019) provide some insights into the potential of participatory modelling. Interestingly, some methodological limitations in transition research identified by Zolfagharian et al. (2019) have been found as strengths of participatory modelling in this review. First, transition research shows generally a strong emphasis on qualitative methods (86%) (Zolfagharian et al. 2019). In contrast, this review showed a majority of the reviewed studies (72%) applying quantitative modelling methods, including expert simulation models, agent-based models, multi-criteria analysis and system dynamics models. This underlines the relevance of modelling in transition research, which allows for quantitative analysis of transition pathways and sustainability visions. Second, Zolfagharian et al. (2019) note that transitions research is strong in explaining past transitions and 'less strong in designing (practical) interventions' (Zolfagharian et al. 2019, 11). In contrast, our review showed a focus of participatory modelling on decision-support (64%) and process facilitation (60%), while the model use of understanding transitions was only applied by 40% of the studies. These results underline that modelling in general and participatory modelling in particular can be important research approaches to address current challenges in the research field.

6 Conclusions

Participatory modelling is a multi-faceted approach that ranges from the involvement of stakeholders in the development of models to the mere usage of simulation models by stakeholders in the form of a serious game. This article provided an overview of the different methods and forms of participatory modelling that have been applied in the sustainability transitions research field. A systematic literature review revealed the multiple dimensions of participatory modelling, including different modelling methods, model uses, timing schemes of stakeholder participation and modes of knowledge capture and exchange. In addition, more general involvement approaches were examined, such as surveys, interviews and workshops. The results show the various emphases and gaps with regard to participatory modelling in the sustainability transitions research field.

Participatory modelling studies included in the review show a focus on the model uses of decision support and facilitation. The model use of understanding sustainability transitions was applied to a smaller extent. Regarding the modes of knowledge capture and exchange, co-learning processes were mostly applied in which stakeholders learn about the system at hand as well as subjective perspectives. Many studies also applied an extractive mode, in which stakeholder knowledge and perspectives were interrogated without enabling a direct exchange between stakeholders. Only a small number of studies implemented a co-management mode in the sense that either decision-makers participate in the process or the stakeholder process has the mandate to take decisions. With regard to the timing of stakeholder involvement, a front- and/or back-end modelling approach was chosen in most of the cases; that is, stakeholders were involved in the beginning and/or end of the modelling process, but did not actively participate in the actual model development process. The general involvement process was in most cases realised through stakeholder workshops followed by interviews and surveys. The usage of modelling methods shows a mixed picture in the sense that various modelling methods have been applied in the reviewed case studies. Expert simulation models have the highest share followed by causal loop diagrams and agent-based models.

Future research should address the gaps identified in this review, which are related to co-construction participatory modelling (i.e. involving stakeholders during the whole duration of the modelling process), the co-management mode of knowledge capture and exchange (i.e. involving stakeholders in knowledge synthesis and decision-making), semi-quantitative modelling methods and process design frameworks. Co-construction participatory modelling allows for the most intense stakeholder participation (compared to the other timing schemes), which can foster understandability and ownership of model results. The co-management mode of knowledge capture and exchange is particularly suitable to foster social learning and effective implementation of model results. Semi-quantitative modelling methods, such as fuzzy cognitive mapping, can furthermore help to bridge qualitative and quantitative modelling and deal with low data availability. In addition, conceptual and methodological frameworks are needed to tailor participatory processes to case-specific challenges by systematically combining different modelling methods, timings and knowledge capture and exchange modes.

Notes

1 Lynam et al. (2007) use the term 'modes of knowledge capture and use'. We use the term 'modes of knowledge capture and exchange' to avoid any confusion with the term 'model use'.

2 Full query: TITLE-ABS-KEY(Transition OR transform*) AND TITLE-ABS-KEY (Modeling) AND TITLE-ABS-KEY(participat* OR Stakeholder OR collaborat*) AND TITLE-ABS-KEY(sustainab*)) AND (LIMIT-TO (DOCTYPE,"ar") OR LIMIT-TO (DOCTYPE,"re")
3 Full query: TITLE-ABS-KEY(Transition OR transform*) AND TITLE-ABS-KEY (Modeling) AND TITLE-ABS-KEY(sustainab*)) AND (LIMIT-TO (DOCTYPE,"ar") OR LIMIT-TO (DOCTYPE,"re")

References

Alcamo, Joseph. 2008. "Chapter six the SAS approach: Combining qualitative and quantitative knowledge in environmental scenarios." *Developments in Integrated Environmental Assessment* 2: 123–50.

Barreteau, Olivier, Martine Antona, Patrick D'Aquino, Sigrid Aubert, Stanislas Boissau, François Bousquet, and William's Daré. 2003. "Our companion modelling approach." *Journal of Artificial Societies and Social Simulation* 6 (1).

Beall, Allyson M., and Andrew Ford. 2010. "Reports from the field: Assessing the art and science of participatory environmental modeling." In *Societal Impacts on Information Systems Development and Applications*, edited by J. Wang, 195–213. Hershey: IGI Global.

Brancato, Giovanna, Stefania Macchia, Manuela Murgia, Marina Signore, Giorgia Simeoni, Karen Blanke, and Jürgen Hoffmeyer-Zlotnik. 2006. *Handbook of Recommended Practices for Questionnaire Development and Testing in the European Statistical System*. Luxembourg: European Statistical System.

Brandt, Patric, Anna Ernst, Fabienne Gralla, Christopher Luederitz, Daniel J. Lang, Jens Newig, Florian Reinert, David J. Abson, and Henrik von Wehrden. 2013. "A review of transdisciplinary research in sustainability science." *Ecological Economics* 92: 1–15.

Burks-Copes, Kelly A., and Gregory A. Kiker. 2014. "Uncovering lines of evidence hidden in complex problems: Using conceptual models to inform ecosystem-based management of the Missouri river cottonwoods." *Environment Systems and Decisions* 34 (3): 425–42. http://doi.org/10.1007/s10669-014-9509-2.

Chambers, Robert. 1994. "Participatory Rural Appraisal (PRA): Challenges, potentials and paradigm." *World Development* 22 (10): 1437–54.

Cruz, Luís, Eduardo Barata, João-Pedro Ferreira, and Fausto Freire. 2017. "Greening transportation and parking at university of Coimbra." *International Journal of Sustainability in Higher Education* 18 (1): 23–38. http://doi.org/10.1108/IJSHE-04-2015-0069.

Delmotte, Sylvestre, Vincent Couderc, Jean-Claude Mouret, Santiago Lopez-Ridaura, Jean-Marc Barbier, and Laure Hossard. 2017. "From stakeholders narratives to modelling plausible future agricultural systems. Integrated assessment of scenarios for Camargue, Southern France." *European Journal of Agronomy* 82: 292–307. http://doi.org/10.1016/j.eja.2016.09.009.

Dörner, Ralf, Stefan Göbel, Wolfgang Effelsberg, and Josef Wiemeyer (Eds.). 2016. *Serious Games*. Basel, Switzerland: Springer International Publishing.

Forrest, Nigel, and Arnim Wiek. 2014. "Learning from success – Toward evidence-informed sustainability transitions in communities." *Environmental Innovation and Societal Transitions* 12: 66–88. http://doi.org/10.1016/j.eist.2014.01.003.

Funtowicz, Silvio O., and Jerome R. Ravetz. 1993. "Science for the post-normal age." *Futures* 25 (7): 739–55.

García-Barrios, Luis. E., Erika N. Speelman, and Max S. Pimm. 2008. "An educational simulation tool for negotiating sustainable natural resource management strategies among stakeholders with conflicting interests." *Ecological Modelling* 210 (1–2): 115–26. http://doi.org/10.1016/j.ecolmodel.2007.07.009.

Gourmelon, Francoise, Mathias Rouan, Jean F. Lefevre, and Anne Rognant. 2011. "Role-playing game and learning for young people about sustainable development stakes: An experiment in transferring and adapting interdisciplinary scientific knowledge." *Journal of Artificial Societies and Social Simulation* 14 (4): 21.

Gray, Steven A., Stefan Gray, Linda Cox, and Sarah Henly-Shepard. 2013. "Mental modeler: A fuzzy-logic cognitive mapping modeling tool for adaptive environmental management." *Proceedings of the 46th Hawaii International Conference on Complex Systems*: 963–73. https://ieeexplore.ieee.org/abstract/document/6479949

Gurung, Tayan R., Francois Bousquet, and Guy Trébuil. 2006. "Companion modeling, conflict resolution, and institution building: Sharing irrigation water in the Lingmuteychu Watershed, Bhutan." *Ecology and Society* 11 (2).

Guthrie, Gerard. 2010. *Basic Research Methods: An Entry to Social Science Research*. New Delhi: SAGE Publications.

Halbe, Johannes. 2016. *Governance of Transformations Towards Sustainable Development - Facilitating Multi-level Learning Processes for Water, Food and Energy Supply Dissertation*. (PhD diss.), University of Osnabruck.

Halbe, Johannes, and Claudia Pahl-Wostl. 2019. "A methodological framework to initiate and design transition governance processes". *Sustainability* 11(3), 844.

Halbe, Johannes, and Jan Adamowski. 2019. "Modeling sustainability visions: A case study of multi-scale food systems in Southwestern Ontario." *Journal of Environmental Management* 231: 1028–47.

Halbe, Johannes, Kathrin Knüppe, Christian Knieper, and Claudia Pahl-Wostl. 2018. "Towards an integrated flood management approach to address trade-offs between ecosystem services: Insights from the Dutch and German Rhine, Hungarian Tisza, and Chinese Yangtze basins." *Journal of Hydrology* 559: 984–94.

Halbe, Johannes, Claudia Pahl-Wostl, and Jan Adamowski. 2018. "A methodological framework to support the initiation, design and institutionalization of participatory modeling processes in water resources management." *Journal of Hydrology* 556: 701–16.

Halbe, Johannes, Claudia Pahl-Wostl, Jan Sendzimir, and Jan Adamowski. 2013. "Towards adaptive and integrated management paradigms to meet the challenges of water governance." *Water Science and Technology* 67 (11): 2651–60. http://doi.org/10.2166/wst.2013.146.

Halbe, Johannes, Dominik E. Reusser, Georg Holtz, Marjolijn Haasnoot, Annette Stosius, Wibke Avenhaus, and Jan H. Kwakkel. 2015. "Lessons for model use in transition research: A survey and comparison with other research areas." *Environmental Innovation and Societal Transitions* 15: 194–210.

Hare, Matt. 2011. "Forms of participatory modelling and its potential for widespread adoption in the water sector." *Environmental Policy and Governance* 21: 386–402.

HarmoniCOP. 2005. *Learning Together to Manage Together: Improving Participation in Water Management*. Osnabrueck: Druckhaus Bergmann. www.harmonicop.uni-osnabrueck.de/HarmoniCOPHandbook.pdf (retrieved: 12 February 2019).

Henly-Shepard, Sarah, Steven Gray, and Linda J. Cox. 2015. "The use of participatory modeling to promote social learning and facilitate community disaster planning." *Environmental Science & Policy* 45: 109–22.

Holtz, Georg, Floortje Alkemade, Fjalar de Haan, Jonathan Köhler, Evelina Trutnevyte, Tobias Luthe, Johannes Halbe, George Papachristos, Emile Chappin, Jan Kwakkel, and Sampsa Ruutu. 2015. "Prospects of modelling societal transitions: Position paper of an emerging community." *Environmental Innovation and Societal Transitions* 17: 41–58.

Hsu, Chia-Chien, and Brian A. Sandford. 2010. "Delphi technique." *In Encyclopedia of Research Design*, edited by N.J. Salkind, 344–7. Thousand Oaks, CA: SAGE Publications.

Inam, Azhar, Jan Adamowski, Johannes Halbe, and Shiv Prasher. 2015. "Using causal loop diagrams for the initialization of stakeholder engagement in soil salinity management in agricultural watersheds in developing countries: A case study in the Rechna Doab watershed, Pakistan." *Journal of Environmental Management* 152: 251–67. http://doi.org/10.1016/j.jenvman.2015.01.052.

Inam, Azhar, Jan Adamowski, Shiv Prasher, Johannes Halbe, Julien Malard, and Raffaele Albano. 2017a. "Coupling of a distributed stakeholder-built system dynamics socio-economic model with SAHYSMOD for sustainable soil salinity management – Part 1: Model development." *Journal of Hydrology* 551: 596–618.

Inam, Azhar, Jan Adamowski, Shiv Prasher, Johannes Halbe, Julien Malard, and Raffaele Albano. 2017b. "Coupling of a distributed stakeholder-built system dynamics socio-economic model with SAHYSMOD for sustainable soil salinity management. Part 2: Model coupling and application." *Journal of Hydrology* 551: 278–99.

Iwaniec, David M., Daniel L. Childers, Kurt VanLehn, and Arnim Wiek. 2014." Studying, teaching and applying sustainability visions using systems modeling." *Sustainability* 6 (7): 4452–69. http://doi.org/10.3390/su6074452.

Jetter, Antonie J., and Kasper Kok. 2014. "Fuzzy Cognitive Maps for future studies – A methodological assessment of concepts and methods." *Futures* 61 (5): 45–57.

Jones, Natalie A., Pascal Perez, Thomas G. Measham, Gail J. Kelly, Patrick d'Aquino, Katherine A. Daniell, Anne Dray, and Nils Ferrand. 2009. "Evaluating Participatory Modeling: Developing a Framework for Cross-Case Analysis." *Environmental Management* 44: 1180–1195.

Kitzinger, Jenny. 1995. "Qualitative research: Introducing focus groups." *BMJ* 311 (7000): 299–302.

Knoeri, Christof, Claudia R. Binder, and Hans J. Althaus. 2011. "Decisions on recycling: Construction stakeholders' decisions regarding recycled mineral construction materials." *Resources, Conservation and Recycling* 55 (11): 1039–50.

Kok, Kasper, Mita Patel, Dale S. Rothman, and Giovanni Quaranta. 2006. "Multi-scale narratives from an IA perspective: Part II. Participatory local scenario development." *Futures* 38 (3): 285–311. http://doi.org/10.1016/j.futures.2005.07.006.

Kosko, Bart. 1993. "Adaptive inference in fuzzy knowledge networks." In *Readings in Fuzzy Sets for Intelligent Systems*, edited by D. Dubois, H. Prade, and R.R. Yager. San Mateo: Morgan Kaufman.

Krosnick, Jon A. 2018. "Questionnaire design. "In *The Palgrave Handbook of Survey Research*, edited by D.L. Vannette and J.A. Krosnick, 439–55. London: Palgrave Macmillan.

Krueger, Richard A., and Mary A. Casey. 2014. *Focus Groups: A Practical Guide for Applied Research*. Thousand Oaks: SAGE Publications.

Kwakkel, Jan H., and Gönenç Yücel. 2012. "An exploratory analysis of the Dutch electricity system in transition." *Journal of the Knowledge Economy* 5 (4): 670–85.

Learmonth, Gerard P., David E. Smith, William H. Sherman, Mark A. White, and Jeffrey Plank. 2011. "A practical approach to the complex problem of environmental sustainability: The UVa Bay Game." *Innovation Journal: The Public Sector Innovation Journal* 16 (1): 1–8.

Levine, Richard S., Michael T. Hughes, Casey Ryan Mather, and Ernest J. Yanarella. 2008. "Generating sustainable towns from Chinese villages: A system modeling approach." *Journal of Environmental Management* 87 (2): 305–16. http://doi.org/10.1016/j.jenvman.2006.10.026.

Li, Ying, Robert J.S. Beeton, Thomas Sigler, and Anthony Halog. 2016. "Modelling the transition toward urban sustainability: A case study of the industrial city of Jinchang, China." *Journal of Cleaner Production* 134: 22–30.

Liberati, Alessandro, Douglas G. Altman, Jennifer Tetzlaff, Cynthia Mulrow, Peter C. Gøtzsche, John P. Ioannidis, Mike Clarke, P.J. Devereaux, Jos Kleijnen, and David Moher. 2009. "The PRISMA statement for reporting systematic reviews and meta-analyses of studies that evaluate healthcare interventions: Explanation and elaboration." *BMJ* 339: b270.

Liu, Yuqiong, Hoshin V. Gupta, Everett Springer, and Thorsten Wagener. 2008. "Linking science with environmental decision making: Experiences from an integrated modeling approach to supporting sustainable water resources management." *Environmental Modeling and Software* 23 (7): 846–58.

Loorbach, Derk. 2010. "Transition management for sustainable development: A prescriptive, complexity-based governance framework." *Governance* 23 (1): 161–83.

Ludwig, Barbara. 1997. "Predicting the future: Have you considered using the Delphi methodology?" *Journal of Extension* 35 (5): 1–4.

Lynam, Timothy, Wil de Jong, Douglas Shell, Trikurnianti Kusumanto, and Kirsten Evans. 2007. "A review of tools for incorporating community knowledge, preferences, and values into decision making in natural resources management." *Ecology and Society* 12 (1): 5.

Magnuszewski, Piotr, Jan Sendzimir, and Jakub Kronenberg. 2005. "Conceptual modeling for adaptive environmental assessment and management in the Barycz Valley, Lower Silesia, Poland." *International Journal of Environmental Research and Public Health* 2 (2): 194–203.

Malard, Julien J., Azhar Inam, Elmira Hassanzadeh, Jan Adamowski, Héctor A. Tuy, and Hugo Melgar-Quiñonez. 2017. "Development of a software tool for rapid, reproducible, and stakeholder-friendly dynamic coupling of system dynamics and physically-based models." *Environmental Modelling & Software* 96: 410–20.

Moallemi, Enayat A., and Shirin Malekpour. 2018. "A participatory exploratory modelling approach for long-term planning in energy transitions." *Energy Research & Social Science* 35: 205–16.

Moher, David, Alessandro Liberati, Jennifer Tetzlaff, and Douglas G. Altman. 2009. "Preferred reporting items for systematic reviews and meta-analyses: The PRISMA statement." *PLoS Medicine 21* 6 (7): e1000097.

Moraine, Marc, Juliette Grimaldi, Clément Murgue, Michel Duru, and Olivier Therond. 2016. "Co-design and assessment of cropping systems for developing crop-livestock integration at the territory level." *Agricultural Systems* 147: 87–97. http://doi.org/10.1016/j.agsy.2016.06.002.

Ness, Barry, Evelin Urbel-Piirsalu, Stefan Anderberg, and Lennart Olsson. 2007. "Categorising tools for sustainability assessment." *Ecological Economics* 60 (3): 498–508.

Nielsen, Morten V., Nina Bryndum, and Bjørn Bedsted. 2017. "Organising stakeholder workshops in research and innovation – Between theory and practice." *Journal of Public Deliberation* 13 (2): 9.

Okoli, Chitu, and Suzanne D. Pawlowski. 2004. "The Delphi method as a research tool: An example, design considerations and applications." *Information & Management* 42 (1): 15–29.

Olazabal, Marta, and Unai Pascual. 2016. "Use of fuzzy cognitive maps to study urban resilience and transformation." *Environmental Innovation and Societal Transitions* 18: 18–40. http://doi.org/10.1016/j.eist.2015.06.006.

Palmer, Richard N., Hal E. Cardwell, Mark A. Lorie, and William Werick. 2013. "Disciplined planning, structured participation, and collaborative modeling – Applying shared vision planning to water resources." *Journal of the American Water Resources Association* 49 (3): 614–28.

Penn, Alexandra S., Christopher J.K. Knight, David J.B. Lloyd, Daniele Avitabile, Kasper Kok, Frank Schiller, Amy Woodward, Aangela Druckman, and Lauren Basson. 2013. "Participatory development and analysis of a fuzzy cognitive map of the establishment of a bio-based economy in the Humber region." *PloS One* 8 (11): e78319.

Petticrew, Mark, and Helen Roberts. 2006. *Systematic Reviews in the Social Sciences: A Practical Guide*. Malden: Blackwell.

Prell, Christina, Mark S. Reed, and Klaus Hubacek. 2011. "Social network analysis for stakeholder selection and the links to social learning and adaptive co-management." In *Social Networks and Natural Resource Management – Uncovering the Social Fabric of Environmental Governance*, edited by Ö. Bodin and C. Prell, 95–118. Cambridge: Cambridge University Press.

Prodanovic, Predrag, and Slobodan P. Simonovic. 2010. "An operational model for support of integrated watershed management." *Water Resources Management* 24 (6): 1161–94.

Rodriguez, Daniel, Howard Cox, Peter deVoil, and Brendan Power. 2014. "A participatory whole farm modelling approach to understand impacts and increase preparedness to climate change in Australia." *Agricultural Systems* 126: 50–61. http://doi.org/10.1016/j.agsy.2013.04.003.

Rosales-Carreón, Jesús, and César García-Díaz. 2015. "Exploring transitions towards sustainable construction: The case of near-zero energy buildings in the Netherlands." *JASSS* 18 (1). http://doi.org/10.18564/jasss.2625.

Rouwette, Etiënne, Iinge Bleijenbergh, and Jac Vennix. 2016. "Group model-building to support public policy: Addressing a conflicted situation in a problem neighbourhood." *Systems Research and Behavioral Science* 33 (1): 64–78.

Schweizer, Vanessa J., and Elmar Kriegler. 2012. "Improving environmental change research with systematic techniques for qualitative scenarios." *Environmental Research Letters* 7: 044011.

Sessa, Carlo, and Andrea Ricci. 2014. "The world in 2050 and the New Welfare scenario." *Futures* 58: 77–90.

Sheil, Douglas, and Nining Liswanti. 2006. "Scoring the importance of tropical forest landscapes with local people: Patterns and insight." *Environmental Management* 38 (1): 126–36.

Stave, Krystyna A. 2003. "A system dynamics model to facilitate public understanding of water management options in Las Vegas, Nevada." *Journal of Environmental Management* 67 (4): 303–13.

Sterman, John. 2014. "Interactive web-based simulations for strategy and sustainability: The MIT Sloan LearningEdge management flight simulators, Part I". *System Dynamics Review* 30 (1–2), 89–121.

Tàbara, David J., Elisabet Roca, Cristina Madrid, Peter Valkering, Patrik Wallman, and Paul M. Weaver. 2008. "Integrated sustainability assessment of water systems: Lessons from the Ebro river basin." *International Journal of Innovation and Sustainable Development* 3 (1–2): 48–69. http://doi.org/10.1504/IJISD.2008.018193.

Tranfield, David, David Denyer, and Palminder Smart. 2003. "Towards a methodology for developing evidence-informed management knowledge by means of systematic review." *British Journal of Management* 14: 207–22.

Valkering, Pieter, Gönenç Yücel, Ernst Gebetsroither-Geringer, Karina Markvica, Erika Meynaerts, and Niki Frantzeskaki. 2017. "Accelerating transition dynamics in city regions: A qualitative modeling perspective." *Sustainability* 9 (7): 1254. http://doi.org/10.3390/su9071254.

van den Belt, Marjan. 2004. *Mediated Modeling – A System Dynamics Approach to Environmental Consensus Building*. Washington, DC: Island Press.

Vennix, Jac. 1996. *Group Model Building – Facilitating Team Learning Using System Dynamics*. New York: Wiley & Sons.

Videira, Nuno, Rita Lopes, Paula Antunes, Rui Santos, and José L. Casanova. 2012. "Mapping maritime sustainability issues with stakeholder groups." *Systems Research and Behavioral Science* 29 (6): 596–619.

Videira, Nuno, François Schneider, Filka Sekulova, and Giorgos Kallis. 2014. "Improving understanding on degrowth pathways: An exploratory study using collaborative causal models." *Futures* 55: 58–77.

Voinov, Alexey, Nagesh Kolagani, Michael K. McCall, Pierre D. Glynn, Marit E. Kragt, Frank O. Ostermann, Suzanne A. Pierce, and Palaniappan Ramu. 2016. "Modelling with stakeholders – Next generation." *Environmental Modelling & Software* 77: 196–220. http://doi.org/10.1016/j.envsoft.2015.11.016.

Wang, Jiang-J., You-Y. Jing, Chun-F. Zhang, and Jun-H. Zhao. 2009. "Review on multi-criteria decision analysis aid in sustainable energy decision-making." *Renewable and Sustainable Energy Reviews* 13 (9): 2263–78. http://doi.org/10.1016/j.rser.2009.06.021.

Ward, Sarah, and David Butler. 2016. "Rainwater harvesting and social networks: Visualising interactions for niche governance, resilience and sustainability." *Water (Switzerland)* 8 (11): 526. http://doi.org/10.3390/w8110526.

Weimer-Jehle, Wolfgang. 2006. "Cross-impact balances: A system-theoretical approach to cross-impact analysis." *Technological Forecasting and Social Change* 73 (4): 334–61.

Whitmarsh, Lorraine, and Björn Nykvist. 2008. "Integrated sustainability assessment of mobility transitions: Simulating stakeholders' visions of and pathways to sustainable land-based mobility." *International Journal of Innovation and Sustainable Development* 3 (1–2): 115–27. http://doi.org/10.1504/IJISD.2008.018196.

Wiese, Frauke, Gesine Bökenkamp, Clemens Wingenbach, and Olav Hohmeyer. 2014. "An open source energy system simulation model as an instrument for public participation in the development of strategies for a sustainable future." *Wiley Interdisciplinary Reviews: Energy and Environment* 3 (5): 490–504.

Wildenberg, Martin, Michael Bachhofer, Christian M. Adamescu, Geert. De Blust, Ricardo Diaz-Delgadod, Kirsten G. Q. Isak, Flemming Skov, and Varjopuro Riku. 2010. "Linking thoughts to flows-Fuzzy cognitive mapping as tool for integrated landscape modelling." *Proceedings of the 2010 International Conference on Integrative Landscape Modelling* 3: 5.

Withycombe Keeler, Lauren, Arnim Wiek, Dave D. White, and David A. Sampson. 2015. "Linking stakeholder survey, scenario analysis, and simulation modeling to explore the long-term impacts of regional water governance regimes." *Environmental Science and Policy* 48: 237–49. http://doi.org/10.1016/j.envsci.2015.01.006.

Zivkovic, Marija, Kateryna Pereverza, Ooleksii Pasichnyi, Aleksandar Madzarevic, Dejan Ivezic, and Olga Kordas. 2016. "Exploring scenarios for more sustainable heating: The case of niš, serbia." *Energy* 115: 1758–70. http://doi.org/10.1016/j.energy.2016.06.034.

Zolfagharian, Mohammadreza, Bob Walrave, Rob Raven, and Georges Romme. 2019. "Studying transitions: Past, present, and future." *Research Policy* 48(9): 103788. http://doi.org/10.1016/j.respol.2019.04.012.

12 Data-driven transitions research

Methodological considerations for event-based analysis

Fjalar J. de Haan, Alfonso Martínez Arranz, and Wouter Spekkink

1 The case for data-driven transitions research

We argue that serious progress can be made in transitions studies if a more data-driven approach to empirical analysis is adopted. By 'data-driven' we mean a thorough-going empiricism, where data, as many as can be mustered, lead rather than just support inference and hypotheses. Obviously, all empirical research worth its salt will relate its hypotheses to empirical data. What we suggest is a thoroughly systematic attitude towards this. An approach that enables anyone to verify exactly which data are adduced to corroborate what hypothesis – and to perhaps disagree on the basis of those data and propose rival hypotheses. We argue for an approach that facilitates dealing with many data, across cases and using statistical methods where appropriate, an approach amenable to meta-analyses. These are all issues of *methodology* and this chapter is explicitly about methodological considerations. We are, in this regard, theoretically agnostic.

This, of course, does not mean we do not have theoretical and philosophical convictions and it certainly does not mean we think transitions research should be theory averse at all. What we are after is an intensely empiricist methodology, that *postpones* theoretical framing and interpretation as much and as long as possible. The outcome of applying such a methodology should be a practical, shareable and cumulative collection of data that can then be used to test theories and hypotheses of whatever ilk. A sort of 'separation of powers' between the empirical and the theoretical, as it were. Whether one thinks such a separation is possible in principle or not, we think it is something to strive for.

What would such a data-driven approach practically look like? We envision that transition analysts would query a large database of empirical data coming from many cases. These data would be recorded in a standardised way, whilst retaining the detail of the original sources and providing references to them. The analyst would try to identify patterns and the conditions under which they occur ('do all backlashes look the same?', 'is government action relevant for niche formation?'). Hypotheses would be tested straightforwardly by searching for corroborating, contradicting and correlating data ('how often do we find scaling up following legislative change?'). This is clearly not the way transitions research is commonly carried out. Not only is there no such database but moreover it seems contrary to the way data are used in transitions research.

There is, however, no shortage of empirical work done in this field, which further convinces us of the feasibility of an approach such as we propose. For instance, transitions research relies heavily on case study narratives for its underpinning and the field has amassed a considerable body of empirical data in this form. Meta-analysis of the existing empirical literature may be an appropriate way to bootstrap data-driven transitions research, whilst enabling a much-needed re-examination of basic assumptions. A meta-analytic study of this kind was carried out by one of us (Martínez Arranz, 2017), more on this in Section 4.2. Another way is to enhance a common form of empirical research in the field: the study of innovation projects. One of us has been building on an approach where qualitative data on such projects are bracketed and recorded in standardised data sets (cf. Van de Ven and Poole, 1990), after which they are coded to make them amenable to visualisation and analysis with, for example, graph-theoretical tools (Spekkink, 2015). This approach will be discussed further in Section 4.1.

We are not just preaching what we practice though. We observe that data-driven methods have been successfully adopted in areas of research that share several challenges sometimes thought of as typical for transitions, e.g., complex chains of causes and effects, qualitative data, historical and geographical contingency and so forth. In this regard, we think of fields like medicine and psychology, which also share the ubiquity of case research but with an emphasis on replicable conclusions and where resolving conflicting conclusions is a key concern. Meta-analysis, one of the approaches we advocate, is a common data-driven method in these fields.

Data-driven methods necessitate a different way of engaging with empirical material, as well as thinking differently about explanation and theory building. Both, in a fashion, need to be disaggregated. From cases to their component facts and from theoretical frames to mechanisms[1]. We are not simply advocating for comparative case research or $N > 1$ case studies. We are pointing out that case studies may suggest promising generalisations and hypotheses, but that any *particular* case exemplifying them is nothing but precisely that: an example — which may well be unique. The phenomenon that 'theoretical frameworks' accrue more and more 'nuance' as they are 'tested' on more and more case studies is therefore a clear indicator of a problematic relationship with data (as well as questionable theorising[2]).

Data-driven research is not new, nor is it confined to particular areas of inquiry. However, of recent, the term 'data-driven' has gained popularity and is being used in much the same context as the term 'big data'. While our conception is certainly akin to this popular usage, in particular in its emphasis on statistics and postponing of theorising, we are not suggesting an 'algorithms-to-the-rescue' attitude but a rather more old-fashioned and common-sense approach: the Baconian method[3]. What is clear, at least to us, is that computational tools are opening up new ways of doing research on complex societal issues such as transitions. This is not just a matter of increased raw computing power, it is also the potential of new ways to organise research and sharing data inspired by the modus operandi of the free and open source software community.

2 The data of transitions research

What are the data of transitions research? And where to procure them? To get a better idea of this, we will look into what kind of science transitions research would be, if it were one. Transitions research would be a *historical science*, in the sense Diamond (1997, Epilogue) used it. This is not to say that all transitions have now passed, nor that we can or should not study ongoing or future transitions. It means that most questions in transitions research are of the form: 'why did things unfold in this particular way?' or 'how did this situation come about?'. Though their subject matter is at first blush wildly different, this reveals transitions research to be of the same family as evolutionary biology, geology, cosmology, historical sociology – that is, the historical sciences which include history itself and, for example, historical sociology in as far as they are scientific.

This suggests that the data of transitions research ought to be descriptions of the transitioning system over time. Let us call such a description, in line with the literature, a *pathway*. For the purposes of our discussion we will distinguish two main approaches to describing pathways. The first approach describes a pathway as a discrete sequence or continuous progression of *system states*, where each system state is described, for example, in terms of a set of key variables. The second approach describes pathways as sequences of *events* that capture episodes in which a system changes from one state to another, in some qualitatively or quantitatively appreciable way. While the first approach focusses on describing the qualities of the system over time, the second approach focusses on describing the *changes* to that system over time.

In principle these approaches are equivalent in terms of the information they carry about a transition pathway. If the state of the system is known at *some* time on the pathway, one can reconstruct the system-state description over the *entire* pathway by cumulatively applying the changes each event describes. Conversely, an event sequence can be distilled from a system-state description by aggregating appropriate clusters of change into events. For the mathematically inclined, the former approach represents pathways with the system state, S, as a discrete or continuous function of time, that is, St or $S(t)$. The latter approach represents pathways in terms of a difference or differential equation of the system state as a function of time, like $\frac{dS(t)}{dt} = f(t)$ or $\Delta S_t = f(t)$. Clearly, if some initial – or, in fact, any – condition is known, the former can be completely reproduced with the latter by integrating or summing the equations. In other words, the representations are equivalent.

Our distinction parallels Abell's (2004; 2007) distinction between *states of the world* (system states in our language) and *actions* that transform elements in states of the world (events in our language). Like Abell (2007), we suggest that, although both system states and events can be considered together in analysis, most often our interest will be primarily in understanding what changes have led from one system state to another, without necessarily wanting to explicitly consider each intervening system state that lies on this path. Moreover, this seems more parsimonious, as an event description would only mention those aspects (i.e., key variables) that mattered in that event, whilst others that did not change or followed some uninteresting trend can be ignored. In fact, it allows one to forego the exhaustive identification of key variables altogether[4]. Finally, an orientation on events also puts the focus of analysis more explicitly on how changes occur (Van de Ven and Huber, 1990), which is a key question in transitions research. In summary, in most cases we would want to describe our pathways in terms of events.

3 Methodological considerations regarding events

At the heart of our methodological proposition is the event as the 'unit' of empirical data. Therefore, some words on what we consider an event to be and how to recognise a description of one seem appropriate. An event is a change in some state of affairs occurring within a bounded period of time. Casually speaking, things are not the same after an event. The definition allows events to be of any duration *and* instantaneous.

Our definition of event leaves open a host of methodologically relevant issues, which we will discuss shortly. Amongst these are: what matters about an event, that is, what should one record about an event? What about agency, i.e. who is responsible for the event? Can events be part of other events? And, moreover, which events matter? Or, similarly, how heavily does one event count relative to another? Especially these latter two point are slippery. Lest these issues be used as arguments against a data-driven, event-based analysis of the kind we propose, it should be noted that these issues are of course also faced by the narrative analyst. In narratives, however, the choices and rationales of the analyst are typically implicit, if not unconscious. At least in a data-driven, event-based analysis such choices are transparent and can be the object of scrutiny. Indeed, the importance of certain events may then be established as a *conclusion* of the analysis, rather than remaining an unquestionable assumption.

Thus, we propose to view transition pathways as sequences of events, though we do not exclude at all the possibility that events occur in parallel, so perhaps we should be speaking of bundles, or even networks of events instead, as the sequences may interact or cross over as it were. Such a view has two characteristics which we think are advantages, namely that it is

- rather similar to the way narrative case studies are usually presented, which provides opportunities for *meta analysis*, and
- there is no advance commitment to a choice of variables, leaving identification of key variables as a potential research finding.

With the choice for an event-based perspective made, we can now explore a number of methodological considerations. What is involved, practically, in doing data-driven transitions research from an event-based perspective. Where to find events? How to record them? And, importantly, what to do with them? We propose a methodological frame consisting of two main parts:

1 Cataloguing of events
2 Categorisation of events

The aim of the first is to establish, with as little interpretation or application of theory as possible, a database of events. Such a database could be fed from primary research done in a particular project or built from a corpus of pre-existing literature. Ideally, a shared and openly accessible, online database would emerge — a sort of transitions data commons[5]. We will refer to such a database as an *event catalogue*, or simply catalogue. What we have in mind is very close to the event data sets described by Van de Ven and Poole (1990). The second is where actual analysis occurs. Categorising events is using events as data points, the equivalent of interpreting measurements. The categorisation of event data can be part of a theory-building approach, where categories are gradually constructed and refined through systematic comparison of event data (Langley, 1999), but categorisation can also entail sorting events into pre-made categories, (e.g., 'entrepreneurial activity', 'scaling up'), or directly tallying them as corroborating or refuting some hypothesis. Though the term 'categorising' may sound somewhat simplistic, all acts of theorising on the basis of data are instances of categorising events.

In the following, we will outline our methodological considerations and suggestions for building event catalogues. These considerations are based on our personal experience as well as informed by literature. After that, in Section 4, we will discuss several approaches that make use of an event catalogue. Some of this is again reporting on our own experience, while some of it describes work by others or new avenues.

3.1 Cataloguing events

The events in the catalogue should be what we could call *statements of historical fact*. This means two things: Firstly, it means that the catalogue should only contain events that either no-one would dispute to have actually happened[6], or that could be checked to verify[7] whether they happened or not, for example by simply checking the sources cited. Secondly, it means that the events should contain as little interpretation as possible, that is, one should be able to accept the described event as a historical fact without having to accept some additional theory or to hold particular values. We realise that especially this latter criterion may lead to meta-analytic paralysis if taken at philosophical extremes. We do, however, maintain that in practice it is mostly straightforward to distinguish between *events proper* (i.e. statements of historical fact) and *interpreted events*. Van de Ven and Poole (1990) make a similar distinction between *incidents* and *events*, where the former is a bracketed, factual description of an occurrence of interest, and the latter is a theoretical construct that we might arrive at through a theory-informed interpretation of incidents.

We said that an event is a change in some state of affairs. That is, something ought to have happened for an event to have occurred. To jest, the plaque in Columbus, Ohio, USA (Wikipedia, 2018) with the text 'On this site in 1897 nothing happened.' does not commemorate an event. Note that periods of stasis can in this fashion still be informative as their lack of events can positively be noticed. We explicitly allowed events to be instants (e.g., New Year) as well as anything longer. Thus, a historical transition as a whole may count as an event. This brings us to another important aspect of events, which is that they can be *nested*. Similar to the idea that a set of basic actions can together constitute a more complex action (Danto, 1965), we suggest that a set of 'smaller' events can constitute a more complex event.

The ability to view events as composed of other events introduces a natural way of identifying levels of analysis. One may seek explanations on a given level of analysis, based on the nature of the question or on practical considerations, like the level at which data are available. Analysing events into their constituent 'sub' events in itself can be enlightening. For example, the approach outlined in this chapter largely revolves around attempts to increase our understanding of transitions (which are deeply complex events) by looking at the smaller events that compose them. Obviously, there are limits to the added utility of more and more finely grained events – we do not expect many events at the level of molecular interactions to be particularly interesting for the purposes of understanding the course of a transition. In practice, the available data sources impose a lower limit to the level of analysis and higher levels can be accessed by aggregating events into complex events.

To speak of an event proper, some temporal delineation is necessary, though in practice sources are often vague or ambiguous on this matter. For example, the statement '*E* happened in 1977' could mean anything from *E* occurring at some instant in that year to *E* taking that entire year. This may require the keen judgement of the analyst to interpret – or additional background checks. In addition to the temporal aspect, there are other aspects that can be part of a proper event. For example the spatial aspect (whatever happens, happens somewhere, though a spatially contiguous requirement seems too strong[8]). Depending on the purposes of the analysis, the precise relative timing of events may be of crucial importance, for example, if the priority of some action or other leads to a different conclusion as to the responsibility for some event. Or, it may matter very little and the approximate order in which the events took place is sufficient. Either way, again the data sources usually impose a lower limit to the precision of temporal delineations.

If we agree that history is shaped by human action, there will often be an agency aspect to an event as well. Indeed, Poole et al. (2000) define events as what entities do or what happens to them. In the context of transitions research, identifying the key agents in a sequences of events may well be the principal aim of the analysis. Thus we are interested in the aspect of agency of events. The notion of agency introduces some additional methodological issues, for two main reasons:

- Agency is bound up with the notion of causation. Causation is some thing one may wish to establish, based on analysis. Therefore, to avoid circularity, it should not be *assumed* in order to establish agency in relation to events.
- What entities one allows as agents is not exactly uncontroversial. While common sense (and the law) would restrict agency to adult humans who are *compos mentis*, some scholars would maintain that technological artifacts or trees are agents. This would mean that identifying agents related to events entails a theoretical commitment, making the event an *interpreted* event, whereas we want to avoid interpretation until the categorisation stages of analysis.

Nevertheless, in building a catalogue of events based on a corpus of source data, one most likely wants to catalogue the agents as well. What we propose is an approach to cataloguing events that is as agnostic about matters of causation and theory as possible. We propose that in addition to, but separately from, the event proper, one catalogues the agents *associated* with the event *according to the sources*. To avoid second-guessing who ultimately caused something, associating agents with events should be on the basis of sources describing them as if they were necessary conditions for the event – though sources obviously need not use this terminology. Where the agency of an event cannot be resolved from the sources, one could associate with the event an 'unknown' agent, or 'forces of nature', depending on what the case may be.

Compare in this regard the uncontroversial 'James Watt improved the Newcomen steam engine near the end of the 18th century', with the more problematic, perhaps even dubious 'near the end of the 18th century, James Watt helped bring about the Industrial Revolution'. This latter statement does not describe an event proper, it is a (rather heavily) *interpreted* event. It could however be the *conclusion* of the analysis, or it may be taken as a hypothesis to be tested. Note, however, that from a text fragment such as the latter an analyst *can* safely catalogue Watt's technological innovation as an event, and, moreover, safely catalogue Watt as an associated agent for this event. Distinguishing between various classes of agents can be a straightforward and informative exercise that would be part of the event categorisation side of the analysis.

Summarising this into an event-cataloguing procedure we yield the following:

- Collect from the raw research data (if primary research) or from the original sources (if relying on existing literature) those parts that describe proper events, statements of historical fact.

- Distill, from those fragments, the temporal aspect of each event. That is, for each event, determine the beginning and end point in time (taking these to coincide for instantaneous events).
- Describe the change in the state of affairs that each event effected. In other words, describe what happened and keep it matter-of-fact. This should, at most, amount to a rewording or synopsis of what the sources say, refraining from interpretation or judgement.
- Identify the actors associated with each event. Again, this does not mean an inquiry into who *ultimately* caused an event. Record who did what was done in the event in as far as the source reveals it[9].

We keep emphasising that the event catalogue as such should steer clear from adding a layer of interpretation. The catalogue should be as close to a structured data version of the original sources as possible. The main aim of the data-driven approach is to facilitate transparent, systematic, scientific enquiry — any interpretation, conclusion or inference should be based on the data in the catalogue so it can be traced back to the analyst and the sources used. Note that this requires that the events in the catalogue are appropriately referenced, so that they also may be scrutinised. One may wish to keep a companion catalogue of source fragments, for maximum transparency and ease of reference.

4 Approaches to transitions event analysis

In this section we will discuss a variety of approaches to transitions analysis based on an event catalogue or database of the sort we have been proposing. The section is divided into two parts, one part focussing on work that some of us have actually done (the *What We Have Done* sections) and the other part focussing on work that is in progress, done by others in different contexts, or speculative (the *What Could Be Done* sections).

Perhaps as a clarification in advance, the sections on *What We Have Done* do not present work that was based on the methodological considerations outlined earlier, nor do these considerations follow directly from the experiences we present there. What we aim to provide is examples of data-driven, event-analysis based work that we engaged with ourselves. The *What Could Be Done* parts that follow discuss approaches, techniques or technologies that we think can play an important role in actually performing data-driven event analyses. The selection we discuss is no doubt woefully incomplete but these are approaches that we are personally particularly interested in or already working with. For an overview of various other ways in which one might 'work' with event data, see for example Langley (1999).

4.1 What we have done (i) – event network analysis

One of us has been involved in the development of a methodology that makes use of the kind of event data sets that we envisioned in the previous section

(Spekkink, 2015; Spekkink and Boons, 2016; Boons et al., 2014; Spekkink, 2016). This methodology was inspired primarily by the methodological advances that came out of the Minnesota Innovation Research Programme (Van de Ven and Poole, 1990; Poole et al., 2000), by Abell's (1987; 1993) theory and method of Comparative Narratives, Heise's (1989) Event Structure Analysis (also see Griffin, 1993) and by Abbott's (1988; 2001) thinking on event-based methods.

A core idea underlying the methodology is that social processes can be usefully modelled as networks of events. The networked nature of the events arises from the observation that social processes often consist out of multiple 'streams' of events that sometimes occur in parallel, sometimes diverge into multiple streams, and sometimes converge onto a single stream (Van de Ven et al., 2008; Van de Ven, 1992). To capture this complexity, we make use of *event graphs*[10] in which social processes are visualised and analysed as directed a-cyclic graphs. In these graphs nodes represent events, and the edges represent relationships between events. This approach is thus especially useful for answering research questions about patterns of relationships between events. For example, Spekkink and Boons (2016) used this approach to examine how regional collaborations on sustainable industrial cluster development emerged at the intersection of collections of smaller projects that initially unfolded independent from each other. In this use case, the approach was instrumental in tracing the collaborations to their various building blocks.

The types of 'event catalogues' that we discussed earlier are a useful starting point for the application of the methodology, which originally also starts out with the creation of event data sets that capture relevant occurrences in the process of interest in the form of chronologically ordered, bracketed, qualitative descriptions. These data are then coded with qualitative data analysis software[11] to identify attributes of events and *relationships* between events. The attributes capture theoretically meaningful aspects of events, and can be used, for example, to classify events into different types, to identify actors involved in the events, to identify the locations where events occurred etc. The relationships between events capture how events are connected to each other into complex sequences. These relationships can be of various types. For example, we may define a relationship that captures how events *shape the conditions* under which later events occur (similar to Abell's (1987) paths of social determination), or we may define a relationship that captures how events occur *in response to* events that happened earlier (similar to Schatzki's (2016) chains of action). As in other forms of network analysis, offering a clear definition of the type of relationship under consideration is one of the main tasks in the construction of event graphs, and it will have important consequences for the type of event graph that is produced.

What types of analysis then, does the reconstruction of social processes in event graphs enable? Since event graphs are essentially a network representation of processes, one obvious candidate is the use of network analysis. For example, Spekkink and Boons (2016) used community detection algorithms to identify *modules* in the processes they investigated, and examined *paths* of relationships in the network as one step in the identification of 'emergent linkages' between events. In another example, Mu and Spekkink (2018) used the *outdegree* of events as a measure of their importance in bringing about new events. The use of network statistics possibly also paves the way for the development of statistical models used in other types of network analysis. For example, exponential random graph models Lusher et al. (2013) might be useful for examining the tendencies of different types of events to connect with each other, while also taking into account other characteristics of the events, such as their time and place of occurrence, and the actors involved in them. However, networks of events have some peculiarities, such as the temporal ordering of the nodes, and their a-cyclic nature, which have to be taken into consideration when specifying and interpreting models, for which there are currently few precedents.

We also strongly encourage to always make the step back from examination of abstract patterns in event graphs, derived through quantitative analysis, to an interpretation of the underlying qualitative data. Event graphs may be useful in bringing out patterns that otherwise remain 'buried' in the rich qualitative data, but a deeper understanding of these patterns typically requires going back to the qualitative descriptions of events to flesh out these abstract patterns.

4.1.1 Methodological reflections

Event network analysis has some commonalities with the data-driven approach introduced in this chapter, primarily with regards to how events are defined, and how they are recorded in data sets. However, there also important differences to highlight. In previous applications of event network analysis (e.g., Spekkink and Boons, 2016; Mu and Spekkink, 2018) decisions about what events to include in the data sets were already from the outset heavily influenced by the particular research questions and theoretical perspective of the analysts. Consequently, the resulting data sets are inevitably biased in what events are included and (perhaps more importantly) what events are excluded, or ignored. Even if the analysts are careful in recording *events proper* rather than *interpreted events* (see Section 3.1), their particular interpretations are implicit in the boundaries of the data set. The data-driven approach, by actively postponing interpretation, encourages the inclusion of a much wider range of events, making the data suitable as an empirical basis for tests of a wider range of theories.

In event network analysis, the particular interpretations of the analysts play an even stronger role in the stages that follow data collection. The fact that event attributes and relationships between events are identified through qualitative coding means that they capture, first and foremost, the analysts' interpretation of the data, rather than an objective account of how the process unfolded (cf. Heise, 1989; Griffin, 1993)[12]. However, this is generally true for any narrative account (including case descriptions) of social processes. The advantage of explicitly coding for attributes and relationships between events is that the choices made by the analyst in this regard are made transparent. Coding the event data in the ways described above makes them amenable to various types of quantitative (network) analysis. However, in performing these analyses one should always take into consideration that it is a subjective account of a social process that is being analysed, and not the social process 'an sich'.

The comparison with the data-driven approach reaches its limits here, because the qualitative coding of events is already a step beyond the cataloguing of events. Instead, most of the steps of event network analysis can be understood as a particular approach to *categorising events*, perhaps with the addition of *categorisation of relationships* between events, which is something we have not explicitly discussed as part of the data-driven approach (see Section 3). Ideally, the data collection stage of event network analysis is made redundant by the creation of event catalogues through a data-driven approach. It is replaced with the task of sampling relevant events from the wider catalogue. The added benefit of this approach is that the analyst has to justify not only decisions about what events to include, but also why other events in the catalogue were not included. Everything that follows this stage (the qualitative coding of events, the creation of event graphs, the summary of analysis in a narrative etc.) would still be imbued with the particular interpretations of the analysts (various checks of inter-coder reliability are available to compensate somewhat for this). However, since the catalogue of events is available to a wide audience, possibilities to scrutinise these interpretations improve, and comparisons with the interpretations of others (possibly based on other types of analysis) are easier to make as well.

4.2 What we have done (ii) – meta-analysis

Another one of us undertook meta-analysis of existing literature aimed at deriving 'lessons from the past' (Martínez Arranz, 2017). Meta-analysis is a versatile, resource-saving research tool that, in its simplest definition, relies on primary analyses to produce a secondary analysis. In other words, it re-uses and re-appraises the abundant good work carried out by a large number of scholars and institutions to elicit new insights (Borenstein et al., 2011). Given our considerations above, regarding positive parallels to other sciences like medicine and the large number of single-case transition studies, the use of meta-analysis in transitions research has been scarce to say the least: an inclusive search in Scopus for 'meta-analysis' and 'technolog* transition*' within social science, energy or environment journals yields exactly *two* directly relevant results: an article by Raven et al. (2016) and the above contribution by one of us. Wiseman et al. (2013) also appears but its evaluation of *future* transition strategies as published in government reports is only tangentially related to our work here, which deals with historical facts.

At a fundamental level, Raven et al. (2016) concur with us that large *N* studies are required to move the field of transitions forward. In their work they test a number of propositions against six low-carbon technology cases. The empirical section of the article is based on original research by the authors and, in that sense, it is a comparative analysis rather than a meta-analysis proper. Nevertheless it certainly is a big step forward from the usual focus on one or two transitions or technologies. A downside of their approach is that the original sources are not mentioned when the cases are used and compared, nor is the raw data available, e.g., as a table in an appendix or online. This is not to say we do not trust their findings (we do), rather, we and other researchers are now not in a position to verify those findings, nor to use their data to test other hypotheses. Making 'uninterpreted' data available is an integral part of the approach we advocate.

For its part, the meta-analytical work in Martínez Arranz (2017) more closely combined the two efforts that we have described in this article: on the one hand, there was a focus on extracting (and *abstracting*) elements from sources, although focusing on 'factors of change' rather than 'events'. These were identified for each of the 34 transitions analysed. On the other hand, there was categorisation by grouping these factors of change into more abstract categories that denote the motivations behind actions, for example: health & lifestyle, economic opportunity, economic necessity, ideals, etc. Despite the larger number of cases, issues remain around the validation of both the abstractions and the categories, which prompted some of our reflections here.

4.2.1 Methodological reflections

There has been criticism of meta-analysis when reduced to a mathematical/statistical tool, but the requirement for systematic and consistent analysis in the face of multifarious evidence only emerges reinforced after such criticism, even if with a different name (Stegenga, 2011). We are not married to the label of meta-analysis or to the statistical procedures used in medicine or other fields. Meta-analysis is merely one example of a data-driven tool that has important teachings for our event catalogue approach.

Amongst the well-known difficulties in meta-analysis are: how to select 'good' primary studies (including avoiding journals' publication bias towards 'success' and 'good stories'), how best to summarise findings (e.g., what metric is valid across all cases), whether meta-analysis can truly provide evidence for causality, and yet some others (Aguinis et al., 2011). Below are some of these methodological reflections adapted specifically to our endeavour in transitions research.

BIASES IN THE SELECTION AND CONTENTS OF SOURCES

Meta-analysis involves selecting a corpus of texts that *adequately* describe transitions. This is shared with any attempt at creating an event-catalogue. Biases in the selection of sources are therefore just as important as the procedures for extracting and analysing the 'events'. This has naturally been one of the earliest concerns of meta-analysis (Aguinis et al., 2011).

We concur with Lustick (1996) that any account has its own specific purposes (from confirming a theory to re-telling a story). Accounts of technological change that focus exclusively on quantifiable elements (as many economists tend to do (Fouquet, 2016) are by definition excluding certain explanatory events from our data set. By the same token, other accounts could be overloading it with contingent socio-cultural detail under the premise that some sociological construct *C* 'matters' (Healy, 2017). See, for instance, Geels (2011).

Accumulating 'events that made a difference' and that demonstrably took place during transitions should smooth out such *content bias* (e.g., too much discussion of raw material prices or of gender relations) in the aggregate. However, this potential bias may require us to be aware and record the theoretical background of inputs into the data set to avoid systematic bias (Lustick, 1996), such as the early MLP emphasis on heroic niche stories (Genus and Coles, 2008).

HOW TO ANALYSE ACROSS SOURCES

The flip side of allowing a wide enough input is the difficulty of comparability. We may ponder, firstly, are all socio-technical transitions from all sectors comparable? That is, how do we avoid *type bias*? Secondly, are *all* events recorded for *any* transition equivalent? How do we avoid *scale bias*? To pick one example, we can ponder whether events in the transition from hydropower to fossil-generated electricity in Mexico (Jano-Ito and Crawford-Brown, 2016) should count the same as that from alcohol to petroleum feedstocks in British chemicals (Bennett, 2012).

The first question may be easier to tackle as the number of sectors seems relatively manageable: chemicals, electricity generation, ground transport, sanitation, water supply, etc. Our database can incorporate a coding by sector, and each analysed independently or merged as needed. This can eventually help identify where and how transitions have common or different patterns across sectors in a much more reliable way. Martínez Arranz (2017) suggests that differences may be traced back to measurable differences in terms of incumbents. This leads to an answer to the second question, it should be possible to take recourse to relevant indicators, such as market share of new technologies and overall energy or emissions share of the sector concerned to ascertain what importance (weighting) to attach to each transition.

APPLYING LESSONS FROM THE PAST

An often overlooked aspect of meta-analysis with relevance for our event cataloguing efforts is that any 'lessons from the past' may not be applicable today. Firstly, key events may not be reproducible at will. Our database could contain entries for all three types of elements in Martínez Arranz's (2017) to differentiate between points of action for future policy-makers and the macro-trends that will condition them:

Indicators – reflect scientifically/systematically acquired information either about completely natural events (e.g., earthquakes), or human-related impacts (e.g., from the emissions of a particular factory to its oil supply figures). The transitions database may only be required to store rather local/not easily available data; for example, global CO_2 concentrations or oil production are widely and easily found for any given period.

Unintentional pressures – are fully purposeful activities not specifically targeting the socio-technical regimes under scrutiny (macro-political trends, e.g., liberalisation; macro-economic trends, e.g., information revolution, or trends in the perception of the natural world, e.g. environmentalism). These may affect regimes directly or through induced intentional pressures.

Intentional pressures – are deliberately initiated in order to induce change in or maintain the socio-technical regime under scrutiny. For example, the Arab oil embargo or the Soviet blockade of Berlin clearly targeted the energy regimes of the West. Similarly, as climate policy deliberately attempts to tackle climate change.

Secondly, the events that triggered transitions in the past may not have the same effects in these 'confusing times' (Castells, 2011). There seems to be a consensus that a new period started some time in the late 1980s. It has been variously called fifth Kondratiev wave (Freeman et al., 2001), 'second/late/liquid modernity' (Beck, 1986; Castells, 2000; Bauman, 2012). Regardless of its moniker and the exact start date, the characteristics of today's political economy have significant implications for technological transitions. We observe consolidated global supply chains, preoccupation with global ecological deterioration, a telecommunications revolution and an increasingly multipolar and ambiguous world order. An often asserted (and ignored) implication is that a Manhattan or Apollo Project for X or Y sustainable technologies cannot count on the concentration of manufacturing capacity, technological know-how, and single-mindedness of national will of the mid-20th century USA (Popp and Newell, 2012; Delina and Diesendorf, 2013; Mowery et al., 2010; Hargadon, 2010).

Considering that many of the events that may be included in our proposed catalogue will stem from periods where a completely different set of circumstances applied than in the present, the *interpretation of findings* should always be done with strict reference and contrast to the current context.

4.3 What could be done

4.3.1 Databases

Much in this chapter revolves around the notion of the event catalogue, which is in essence a database of events. What we mean by 'database' here more specifically is what is usually meant by it in the context of computing – the software implementation of a database or a database management system. Clearly, analysis of an event catalogue would be greatly empowered if it were accessible in the form of a database that can be queried programmatically. We preluded on this desirability in Section 3.1. The kind of data-driven approaches we are proposing benefit from databases in these (and likely other) ways:

Exportability / data-exchange – Once in electronic data format, events can be exported in various useful formats and, moreover, imported by other researchers to combine into larger data sets. Even a badly maintained database is better than a bunch of spreadsheets or publications with notes.

Scalability – Manually, scaling up an analysis from across a dozen to across a dozen million events is an insurmountable increase in work. If these data are in a database, the query will look basically the same. The increase in resolution that could be obtained by working with large volumes of data may well change the nature of the research altogether. Because of the previous characteristic of databases, building such large data sets is feasible.

Manipulability – This applies both to the data themselves and to the analytical tools one can apply to them. With regard to the former, this mostly refers to the ability to clean and organise data. Using script and similar tools for this greatly expedites these processes. With regard to the latter, a vast range of statistical and visualisation tools are available, often as free and open source software, that readily converse with databases. Moreover, writing one's own software to do so is fairly straightforward.

To further expand on the last point, numerous database management systems are available, in the free[13] and open source world the SQL-based offerings range from the simple, self-contained SQLite to the industrial strength PostgreSQL, amongst various others. So-called NoSQL options, which do not make use of relational database principles, abound also. Popular languages for data analysis, such as before R and Python (again free and open source) have built up lively communities of contributors developing and sharing massive amounts of software resources.

4.3.2 Ontologies

In computer science, the term 'ontology' has a precise meaning that is somewhat distinct from the philosophical usage. An ontology, in this sense, is a formal classification of the concepts and entities in some domain of discourse – in our case, transitions research. A database built on an ontology is referred to as a *knowledge base*. The difference between a 'normal' database on the one hand and an ontology or knowledge base on the other hand may not be intuitive to the uninitiated, but it is significant. The aid of an ontology would potentially allow a richer analysis than one just based on an event catalogue, but building an ontology for a large domain of discourse like transitions is far from trivial.

Ontologies provide, at the very least, a shareable vocabulary. Splendiani et al. (2014) remark that the use of ontologies has gained a 'critical dimension' in some areas of science (e.g., the life sciences) because of their 'convergence of disciplines', something which should also appeal to our field also. Ontologies are of course more than glorified dictionaries and they can be useful for transitions analysis in several ways: (1) as the frame of data- and knowledge bases (see e.g., Horrocks 2008 for a presentation of this), which then (2) enables improved meta-analyses; (3) as a basis for statistical and computer-based analyses; and (4) to support modelling exercises (e.g., van Dam and Lukszo (2006).

In other fields of research, ontologies and knowledge bases are successfully being used, notably bio-informatics and medicine, but also by the United Nations Development Programme, for example in their UNSPSC which codifies products and services (UNDP 2016; see e.g. Noy and McGuinness 2001). Closer to transitions studies, ontologies have been developed and used to represent and classify socio-technical systems, notably by van Dam in the context of the development of agent-based models (e.g., Keirstead and van Dam 2010; van Dam and Lukszo 2006; van der Sanden and van Dam 2010).

Referring back to our discussion of looking at transitions as changes in system states versus sequences of events – perhaps ontologies can play an important role in data-driven approaches based on the former, as they can provide a formal framework to capture what such a system state would encompass.

4.3.3 Extracting event data

One of the main tasks in cataloguing events is to extract event data from primary and/or secondary sources of data. This can simply be a matter of reading through the sources of data, and writing event descriptions whenever relevant events are found. This is how one of us has built event data sets in the past (see Section 4.1). This approach is easy enough to carry out, but it is also relatively sensitive to the analyst missing potentially important details, and to the analyst's particular understanding of what is important in the first place. Poole et al. (2000) suggest that the reliability of the process of recording event data can be improved by involving multiple researchers in the identification and description of events. However, this means that multiple researchers have to go through the same sources of data, which is an inefficient approach, and it may not even be feasible if the amounts of data to be processed are very large. In addition, if we wish to approach the construction of event data sets as a collective effort, it would be necessary to impose a clear set of standards on how events are to be identified and recorded, in order to ensure consistency in the way that event data are recorded.

It is therefore important to explore possibilities for the development of annotation schemes that can be used by researchers to explicitly identify events in primary and secondary sources of data, as well as relevant attributes of events (e.g., time, place, actors). Annotation schemes that capture most of the details of events that are important to us are currently already being developed in the field of text mining (e.g., Pustejovsky et al., 2011). Indeed, manually coding sources of data with such annotation schemes is still very time consuming. However, if many people are making contributions to shared data sets using the same annotation scheme, this becomes a feasible approach. A next step in this development would be to develop and apply tools that are able to assist in, or even fully automate the annotation of sources of data. Some work is being done in this area as well (Zhang et al., 2019), but this is still in an early stage, and there are plenty of difficulties to overcome, such as being able to distinguish between events that are relevant and events that are irrelevant, and being able to recognise events that are described across different sources as the same. In the short term, it is unlikely that we can rely on automated event extraction for the creation of event data sets. Nonetheless, we believe that these developments hold promise for the development of robust approaches to the creation of event data sets. It is worthwhile keeping an eye on these developments in our thinking about event extraction, to ensure the possibility for synergies 'down the road'.

5 Conclusion

We conclude this chapter briefly. We believe we have made a clear case for data-driven transitions research, in particularly in the form of event-based analysis. We have argued the desirability and possibility of data-driven transitions research and pointed out examples of other fields grappling with similar challenges that are successfully employing data-intensive methods. We discussed a range of methodological considerations about event-based analysis and provided examples of approaches we have worked with as well as tools and techniques we think may be fruitfully applied. As a final comment we pose the question: If not data-driven, then driven by what?

Notes

1 For example in the sense of Hedström's (2005) 'social mechanism', though the notion of mechanism is common in more recent philosophy of science, see, for example, Salmon (1990).

2 Nuance may appear a desideratum in a theory, but this is misguided. Nuance entails embracing exceptions, rather than precisely delineating the applicability of concepts and hypotheses. Healy (2017) calls this the 'nuance of the conceptual framework', describing it as the 'ever more extensive expansion of some theoretical system in a way that effectively closes it off from rebuttal or disconfirmation by anything in the world' and 'an evasion of the demand that a theory be refutable'.

3 In the Novum Organum, Bacon (1620) suggests a straightforward tabulation of empirical observations and to infer from their presence, absence and correlation in relation to a hypothesis of interest whether it should be accepted. Moreover, '[i]n forming our axioms from induction (…) we must observe, whether it confirm its own extent and generality by giving surety, as it were, in pointing out new particulars'. In other words, that our concepts and theories should not only accurately describe the phenomena they were inspired by, but also predict or explain other phenomena.

4 See also our observations about ontologies in Section 4.3.2.

5 This should preferably employ a Creative Commons share-alike, copyleft type licencing scheme (e.g., CC BY-SA https://creativecommons.org/licenses/by-sa/4.0/). Such a scheme would empower researchers to share their data for two reasons: (1) they can rest assured that their work would have to be acknowledged when used by others as a matter of law, and (2) any work based on it would have to be made openly available under the very same conditions. This would be in direct emulation of large collaborative, free and open source software projects such as the Linux kernel. Linus Torvalds, the project's initiator credits the role of the copyleft licencing (here the GPLv2) as follows at LinuxCon 2016: 'I really think the license has been one of the defining factors in the success of Linux because it enforced that you have to give back, which meant that the fragmentation has never been something that has been viable from a technical standpoint' https://www.zdnet.com/article/linus-torvaldss-love-hate-relationship-with-the-gpl/.

6 Unless perhaps as part of some philosophical debate. What we mean is a *common sense* understanding of what 'actually happened'.

7 Here again, we are not inviting a debate about the (im)possibility of verification in principle.

8 Where does a lunar eclipse happen? (Think about it.) Or more pertinent to this methodology, where did the Cuban Missile Crisis happen? In Washington? In Moscow? In the Caribbean sea, or in all of them somehow? Where does a cyber attack happen? Is the internet a place?

9 We should call this Mulisch' Rule of Guilt: *whatever is done, is done by those who did it.* In *De Aanslag* (1982), Mulisch describes how his protagonist gets stuck tracing who is ultimately responsible for the summary execution of his parents. Their execution was a Nazi retaliation in response to the assassination – with which they had nothing to do — of a local collaborating policeman. Was the resistance to blame for their execution because they assassinated the policeman? Were their neighbours to blame, who relocated the corpse to their doorstep? After three decades, the protagonist's back-then neighbour resolved the issue for him. His parents were killed by their killers, they were responsible for their deaths.

10 The methodology also makes use of other graph-theoretical concepts and visualisations to visualise networks of relationships that are enacted in social processes, and to visualise co-occurrences of, for example, attributes of events. For the sake of brevity, these are not discussed in detail here.

11 One of the authors has developed a dedicated software suite for this that will be released in the near future.

12 Of course, as with other forms of analysis that rely on qualitative coding, we may ask multiple coders to code the same data set, check for inter-coder reliability, and via this route attempt to achieve some degree of inter-subjectivity

13 'Free' as in 'freedom', not as in 'free beer' is the catchphrase often used by Richard M Stallman (RMS) and the Free Software Foundation to clarify that free software is not about price. See also https://www.gnu.org/philosophy/free-sw.html.en and https://en.wikipedia.org/wiki/Free_software_movement.

References

Abbott, Andrew (1988). Transcending General Linear Reality. *Sociological Theory 6* (2), 169–186.

Abbott, Andrew (2001). *Time Matters: On Theory and Method.* Chicago: University of Chicago Press.

Abell, Peter (1987). *The Syntax of Social Life: The Theory and Method of Comparative Narratives.* Oxford, Angleterre: Clarendon.

Abell, Peter (1993). Some aspects of narrative method. *The Journal of Mathematical Sociology 18* (2-3), 93–134.

Abell, Peter (2004, January). Narrative Explanation: An Alternative to Variable-Centered Explanation? *Annual Review of Sociology 30*, 287–310.

Abell, Peter (2007). Narratives, Bayesian Narratives and Narrative Actions. *Sociologica. Italian Journal of Sociology Online* (3), 1–21.

Aguinis, Herman, Charles A. Pierce, Frank A. Bosco, Dan R. Dalton, and Catherine M. Dalton (2011). Debunking Myths and Urban Legends About Meta-Analysis. *Organizational Research Methods 14* (2), 306–331.

Bacon, Francis (1620). *Novum Organum: Or True Suggestions for the Interpretation of Nature* (The Project Gutenberg EBook of Novum Organum, from the P.F. Collier & Son, New York, 1902 Edition. Release Date: June 15, 2014 [EBook #45988] ed.). Project Gutenberg.

Bauman, Zygmunt (2012). *Liquid Modernity*. Cambridge, England, UK and Malden, MA, USA: Polity Press.

Beck, Ulrich (1986). *Risikogesellschaft: Auf Dem Weg in Eine Andere Moderne*. Frankfurt am Main: Suhrkamp.

Bennett, S. J. (2012). Using past transitions to inform scenarios for the future of renewable raw materials in the UK. *Energy Policy 50*, 95–108.

Boons, Frank, Wouter Spekkink, and Wenting Jiao (2014). A Process Perspective on Industrial Symbiosis: Theory, Methodology, and Application. *Journal of Industrial Ecology 18* (3), 341–355.

Borenstein, Michael, Larry V Hedges, Julian PT Higgins, and Hannah R Rothstein (2011). *Introduction to Meta-Analysis*. Chichester, England, UK: John Wiley & Sons.

Castells, Manuel (2000). Materials for an exploratory theory of the network society. *The British Journal of Sociology 51* (1), 5–24.

Castells, Manuel (2011). *The Rise of the Network Society: The Information Age: Economy, Society, and Culture*, Volume 1. Chichester, England, UK: John Wiley & Sons.

Danto, Arthur C (1965). Basic Actions. *American Philosophical Quarterly 2* (2), 141–148.

Delina, L. L. and M. Diesendorf (2013). Is wartime mobilisation a suitable policy model for rapid national climate mitigation? *Energy Policy 58*, 371–380.

Diamond, Jared (1997). *Guns, Germs and Steel: A Short History of Everybody for the Last 13,000 Years* (2005 ed.). London: Vintage Books.

Fouquet, Roger (2016). Historical energy transitions: Speed, prices and system transformation. *Energy Research & Social Science 22*, 7–12.

Freeman, Christopher, Francisco Louçã, and Francisco Louçã (2001). *As Time Goes by: From the Industrial Revolutions to the Information Revolution*. Oxford University Press.

Geels, Frank W. (2011, June). The multi-level perspective on sustainability transitions: Responses to seven criticisms. *Environmental Innovation and Societal Transitions 1* (1), 24–40.

Genus, Audley and Anne-Marie Coles (2008). Rethinking the multi-level perspective of technological transitions. *Research Policy 37* (9), 1436–1445.

Griffin, Larry J. (1993). Narrative, Event-Structure Analysis, and Causal Interpretation in Historical Sociology. *American Journal of Sociology 98* (5), 1094–1133. bibtex: Griffin1993.

Hargadon, Andrew (2010). Technology policy and global warming: Why new innovation models are needed. *Research Policy 39* (8), 1024–1026.

Healy, Kieran (2017, June). Fuck Nuance. *Sociological Theory 35* (2), 118– 127.

Hedström, Peter (2005). *Dissecting the Social*. Cambridge, New York, Melbourne: Cambridge University Press.

Heise, David R. (1989). Modeling event structures. *The Journal of Mathematical Sociology 14* (2-3), 139–169.

Horrocks, Ian (2008). Ontologies and databases. *Presentation at the Semantic Days 4.*

Jano-Ito, M. A. and D. Crawford-Brown (2016). Socio-technical analysis of the electricity sector of Mexico: Its historical evolution and implications for a transition towards low-carbon development. *Renewable and Sustainable Energy Reviews 55*, 567–590.

Keirstead, James and Koen Haziël van Dam (2010, May). A comparison of two ontologies for agent-based modelling of energy systems. In *Proceedings of the First International Workshop on Agent Technologies for Energy Systems (ATES 2010)*, https://repository.tudelft.nl/islandora/object/uuid:6685b86e-6586-4680-9102-58fbb1efae62, pp. 21–28.

Langley, Ann (1999). Strategies for Theorizing from Process Data. *The Academy of Management Review 24* (4), 691–710.

Lusher, Dean, Johan Koskinen, and Gary Robins (Eds.) (2013). *Exponential Random Graph Models for Social Networks.* Cambridge: Cambridge University Press.

Lustick, Ian S. (1996). History, Historiography, and Political Science: Multiple Historical Records and the Problem of Selection Bias. *The American Political Science Review 90* (3), 605–618.

Martínez Arranz, Alfonso (2017, May). Lessons from the past for sustainability transitions? A meta-analysis of socio-technical studies. *Global Environmental Change 44*, 125–143.

Mowery, David C., Richard R. Nelson, and BR Martin (2010). Technology policy and global warming: Why new policy models are needed (or why putting new wine in old bottles won't work). Report, Nesta.

Mu, Rui and Wouter Spekkink (2018). A running start or a clean slate? How a history of cooperation affects the ability of cities to cooperate on environmental governance. *Sustainability (Switzerland) 10* (6).

Mulisch, Harry (1982). *De Aanslag.* Amsterdam: De Bezige Bij.

Noy, Natalya F and Deborah L McGuinness (2001). Ontology development 101: A guide to creating your first ontology. Technical report, http://protege.stanford.edu/publications/ontology_development/ontology101.pdf.

Poole, Marshall Scott, Andrew H. Van De Ven, Kevin J. Dooley, and Michael E. Holmes (2000). *Organizational Change and Innovation Processes: Theory and Methods for Research.* Oxford: Oxford University Press.

Popp, David and Richard Newell (2012). Where does energy R&D come from? Examining crowding out from energy R&D. *Energy Economics 34* (4), 980–991.

Pustejovsky, James, Robert Gaizauskas, and Beth Sundheim (2011). The TIMEBANK Corpus. In *Time and Event Recognition for Question Answering Systems Workshop*, Volume 30, pp. 647–656.

Raven, Rob, Florian Kern, Bram Verhees, and Adrian Smith (2016, March). Niche construction and empowerment through socio-political work. A meta-analysis of six low-carbon technology cases. *Environmental Innovation and Societal Transitions 18*, 164–180.

Salmon, Wesley C (1990). *Four Decades of Scientific Explanation.* Pittsburgh, PA, USA: University of Pittsburgh Press.

Schatzki, Theodore R. (2016). Keeping Track of Large Phenomena. *Geographische Zeitschrift 104*, 4–24.

Spekkink, Wouter (2015, July). Building capacity for sustainable regional industrial systems: An event sequence analysis of developments in the Sloe Area and Canal Zone. *Special Volume: Support your future today! Turn environmental challenges into opportunities. 98*, 133–144.

Spekkink, Wouter (2016). *Industrial Symbiosis as a Social Process: Developing Theory and Methods for the Longitudinal Investigation of Social Dynamics in the Emergence and Development of Industrial Symbiosis.* Ph. D. thesis, Erasmus University Rotterdam.

Spekkink, W. A. H. and F. A. A. Boons (2016, November). The Emergence of Collaborations. *Journal of Public Administration Research and Theory 26* (4), 613–630.

Splendiani, Andrea, Michele Donato, and Sorin Drăghici (2014). Ontologies for Bioinformatics. In N. Kasabov (Ed.), *Springer Handbook of Bio-/Neuroinformatics*, pp. 441–461. Berlin, Heidelberg: Springer Berlin Heidelberg.

Stegenga, Jacob (2011). Is meta-analysis the platinum standard of evidence? *Studies in History and Philosophy of Science Part C: Studies in History and Philosophy of Biological and Biomedical Sciences 42* (4), 497–507.

UNDP (2016, December). UNSPSC. http://www.unspsc.org/. GS1 US™ for the UN Development Programme (UNDP).

van Dam, Koen Haziël and Zofia Lukszo (2006, October). Modelling Energy and Transport Infrastructures as a Multi-Agent System using a Generic Ontology. In *Proceedings of the 2006 IEEE International Conference on Systems, Man, and Cybernetics (SMC)*, The Grand Hotel, Taipei, Taiwan, pp. 890–895.

Van de Ven, Andrew H. (1992). Suggestions for Studying Strategy Process: A Research Note. *Strategic Management Journal 13* (Special Issue: Strategy Process: Managing Corporate Self-Renewal), 169–191.

Van de Ven, Andrew H. and George P. Huber (1990, January). Longitudinal Field Research Methods for Studying Processes of Organizational Change. *Organization Science 1* (3), 213–219.

Van de Ven, Andrew H., Douglas E. Polley, Raghu Garud, and Sankaran Venkataraman (2008). *The Innovation Journey*. Oxford: Oxford University Press.

Van de Ven, Andrew H. and Marshall Scott Poole (1990, August). Methods for Studying Innovation Development in the Minnesota Innovation Research Program. *Organization Science 1* (3), 313–335. bibtex: Vande-VenPoole1990.

van der Sanden, Maarten C. A. and Koen Haziël van Dam (2010, November). Towards an ontology of consumer acceptance in socio-technical energy systems. In *Proceedings of the 3rd International Conference on Infrastructure Systems and Services: Next Generation Infrastructure Systems for Eco-Cities (INFRA)*, Shenzhen, China.

Wikipedia (2018, February). Clintonville, Columbus, Ohio. *Wikipedia, The Free Encyclopedia*. Page Version ID: 823516451.

Wiseman, John, Taegen Edwards, and Kate Luckins (2013). Post carbon pathways: A meta-analysis of 18 large-scale post carbon economy transition strategies. *Environmental Innovation and Societal Transitions 8*, 76–93.

Zhang, Hao, Frank Boons, and Riza Batista-navarro (2018). Whose Story Is It Anyway? Automatic Extraction of Accounts from News Articles. *Journal of Information Processing and Management 56* (5), 1837–1848.

13 Exploratory modelling of transitions

An emerging approach for coping with uncertainties in transitions research

Enayat A. Moallemi, Fjalar J. de Haan and Jonathan Köhler

1 Introduction

'Transitions research', as an emerging field (Markard, Raven, and Truffer 2012), analyses long-term transformative change, such as sustainability transitions in energy, transport and water systems. The analysis of transitions is challenged by complexity; that is, nonlinear and time-delayed interactions among agents and system components leading to emergent patterns and pathways (de Haan 2006). Transitions research is also challenged by prevalent uncertainties; that is, when the states of transitions and their surrounding environment are unknown because of our limited knowledge or because of the diversity of stakeholder perspectives and preferences. Such uncertainties can cause ambiguity in framing and conceptualising transitions and can create disagreements about the desirability and normative directions of outcomes.

Transitions modelling, pioneered in late 2000s in multiple articles by Timmermans, Haan, and Squazzoni (2008), Bergman et al. (2008), and Köhler et al. (2009), tried to introduce a variety of mathematical and computational modelling approaches to formalise and to cope with the complexity of transitions using computer simulation. However, the treatment of uncertainties has remained underdeveloped in transitions modelling as well as in the broader area of transitions research while being a significant topic from the early age of this field (Köhler et al. 2018; Rotmans 1998; Rotmans et al. 2001). The current chapter covers this gap and discusses how uncertainties could be managed in transitions research. The chapter focuses on the adoption of *exploratory modelling* for the treatment of uncertainties. Exploratory modelling is an emerging approach which uses a series of *computational experiments* to explore the implications of uncertainties manifested in varying assumptions and hypotheses (Bankes 1993; Bankes, Walker, and Kwakkel 2013). This chapter identifies sources of uncertainties in transitions. It investigates potential roles that exploratory modelling can have in advancing the use of models in transitions research for understanding, for policy insights and in stakeholder processes. The chapter then systematically analyses the strengths and limitations of exploratory modelling in coping with complexities of transitions, enabling systematic experimentation for theory development and advancing policy analysis of transitions.

What does the treatment of uncertainties in transitions research with exploratory modelling look like? Exploratory modelling analyses the implications of sets of input assumptions, including *scenarios*, *policies* and *hypotheses* about the mechanisms of transitions, to outcome variables using series of computational experiments. In the context of transitions research, scenarios are associated with the input data (i.e. exogenous factors, related boundary conditions and various contextual parameters such as population growth). A policy refers to input data, too, but those input data that are concerned with the values of alternative levers (e.g. tax rate) one can control and want to test using the model. A hypothesis is used in a broad sense as the mechanism (i.e. a certain conceptualisation or system model structure) one holds responsible for certain recurring observed patterns. Transitions models in an exploratory analysis operate like functions, called with a set of input assumptions and returning a set of outcome values. The models used in exploratory analyses do not necessarily need to be *new*. One can use existing models, such as the use of the MATISSE model (see Chapter 6 by Köhler in this volume) introduced initially by Bergman et al. (2008), Haxeltine et al. (2008), Schilperoord, Rotmans, and Bergman (2008), Köhler et al. (2009), and later in Moallemi and Köhler (2019) in sustainable mobility under uncertainty. This aspect of exploratory modelling is important as it can be used to get more out of the established models. Thousands to millions of simulations can be performed using models to investigate how transitions would behave in terms of the outcome variables if different input assumptions were correct. A range of statistical and data-mining techniques can be used afterwards to draw conclusions from generated simulations. Such conclusions, drawn based on ranges of (plausible and implausible) assumptions, are considered to be more reliable compared to conventional simulation modelling based on a limited number of best-estimate (most-likely) assumptions.

The rest of this chapter is structured as follows. Section 2 highlights the history of uncertainty discussions in transitions research and the sources of uncertainties in transitions. Section 3 defines potential roles for exploratory modelling in transitions research with examples from previous research. Section 4 shows how exploratory modelling can be applied to transitions research by presenting two examples from authors' previous work. Section 5 analyses strengths and limitations of exploratory modelling in informing transitions and sheds light on potential future research directions.

2 Transitions and uncertainties

The concept of uncertainty has had a prominent role in transitions research since its very inception. Notably, the 'bloodlines' of transitions research – coming from integrated assessment and its modelling exponents – have emphasised the importance of uncertainties continuously (Van Asselt and Rotmans 2002; Rotmans 1998). Uncertainty finds its way from integrated assessment into transitions research in particular through the work of Jan Rotmans (1998), who pioneered both fields. It is also important to observe the connection between the transitions, integrated assessment and what has become the 'deep uncertainty' fields of research, through the seminal publications by Walker et al. (2003) and Lempert, Popper, and Bankes (2003) and later in a dedicated edited volume by Marchau et al. (2019).

On the cusp of this transition, one finds a key publication by Rotmans (1998) on methods for integrated assessment mentioning 'uncertain*' 70 times, while the tail end of the article discusses 'transitions' as a concept that will become prominent in integrated assessment research. While this publication does not discuss uncertainty in the context of transitions directly, later articles explicitly do. An early transitions classic, Rotmans et al.'s (2001) 'more evolution than revolution' puts uncertainty at the heart of the nascent 'transition management' approach. Uncertainty had never lost this place in later works, as can be judged from the synthesising article by Loorbach (2010).

Interestingly, uncertainty has played a rather minor role in transitions *modelling* up until recently. For example, the benchmark-setting MATISSE modelling work of Haxeltine et al. (2008) and Köhler et al. (2009) do not directly mention uncertainty.[1] In what seems to be a prelude to exploratory modelling in transitions modelling, however, Schilperoord, Rotmans, and Bergman (2008, p. 292) say about the MATISSE model that it is 'capable of running large batches of simulation runs for producing the bandwidths for pathways, in order to shed light on the uncertainties that originate from non-linearities in agent behaviours'.

There does not seem to be literature providing a dedicated treatment of uncertainty in the context of transitions. This is a lacuna, not only for the modelling community, but definitely also for transitions research at large. Frameworks, such as those presented in Walker et al. (2003), however, are also applicable to transitions. Uncertainty is a lack of knowledge of transitions which can have various sources. An ad hoc, unsystematic categorisation of those sources could be:

- *Complexity*: Transitions are underpinned by non-linear mechanisms and many interacting entities or agents in complex systems, and much uncertainty about the state of such systems, as well as their temporal development, is due to this complexity.
- *Long time horizon*: Transitions involve medium and long time-scales, up to decades, which makes it difficult to rule out even small uncertainties at the present time as irrelevant for future developments.
- *Contingencies*: Transitions are historical processes in the sense that their developments are shaped by contingencies. Equivalently, they could unfold in different ways, depending on whatever happens to cross their path, be that a natural disaster, a charismatic leader or a miracle technology.
- *Multiple perspectives*: Because of uncertainty and complexity, multiple (epistemological, cultural, political) perspectives are possible. This in turn creates new uncertainties about policy choices. See Van Asselt and Rotmans (2002) for a treatment of perspectives in the context of uncertainty.

3 Potential roles for exploratory modelling in transitions research

Given the significance and various sources of uncertainties in transitions, exploratory modelling can be an approach to use for enhancing transitions research under uncertainty. Holtz et al. (2015) discuss the uses of simulation models in transitions research, based on the typology of Halbe et al. (2015), in three areas: for understanding transitions, providing case-specific policy insights, and facilitating stakeholder processes. We build on these three uses to articulate potential roles of exploratory modelling in transitions research.

3.1 Models for understanding

Models reflect hypotheses about elements and processes of real-world (socio-technical-environmental) systems. A transitions model can be used to test these hypotheses through implementing them with computer simulations and observing their results. Transitions models can be used to enhance the *understanding* of non-linear complex interactions over time and the identification of leverage points, among myriad of inter-related elements, which can stimulate or impede the transition (Köhler et al. 2018; Holtz et al. 2015). Exploratory modelling can serve the use of 'models for understanding' by enabling a *robust* understanding of transition dynamics, patterns and pathways. A robust understanding is less vulnerable to potential parametric changes and to misspecifications of model structures as it draws from diverse ranges of assumptions about elements and process of transitions. Bankes (1993) called this role of exploratory modelling a *data-driven* type of analysis where the model aims to explain regularities and patterns of datasets or a *model-driven* analysis where multiple models are tested to identify whether (and under what conditions) they can produce similar behaviour.

The role of exploratory modelling for understanding transitions can be used for generic (or case-specific) insight development, explaining a phenomenon of interest or to test and modify a set of hypotheses about transitions. Examples are when one generates potential transition pathways to a zero-emissions future to identify the most promising pathways among them (see, e.g. Pye et al. 2017) and when one investigates conditions for a seemingly unlikely renewable-dominated transition in energy sector to become viable in the future (see, e.g. Kwakkel and Yücel 2014; Moallemi et al. 2017). This role of exploratory modelling can also help to understand historical transitions and to refine proposed theories of transition dynamics, for example about lock-in effects. A few previous applications of exploratory modelling in transitions research have addressed the use of models for understanding. Among them are de Haan et al. (2016) who analyse the emergence of transitions pathways from a set of pre-specified transitions patterns, Guivarch, Rozenberg, and Schweizer (2016) who generate and explain the conditions for a diversity of scenarios which can share common outcomes (e.g. scenarios leading to high greenhouse gas emissions) and

Moallemi and Köhler (2019) who explain transition pathways in the mobility sector and conditions resulting in a transition to a public transport regime.

3.2 Models for policy insights

Models used for policy insights aim to generate advice for influencing a transition in a particular (normative) direction. Reliable policy insights require the development of ways to incorporate uncertainties in generating policy insights; uncertainties from model input misspecifications (randomness and natural variations of the system) and from a structurally incomplete knowledge (from, for example, technological developments and changes in the political and cultural landscape). Exploratory modelling can support the design of *robust* policies, which can remain effective under changing circumstances – rather than being relevant only under historical trends and predicted best-guessed scenarios (also see Chapter 10 by Malekpour in this volume). This role of exploratory modelling is what Bankes (1993) called a *question-driven* type of analysis which aims at specifying policy options for case-specific problems.

The use of exploratory modelling for policy insights can enhance the capability of transitions models in evidence-based policy making by generating a wide range of possibilities (e.g. policies and scenarios), experimenting with their consequences quantitatively and designing policies which rely on the obtained insights from experiments (Moallemi and Malekpour 2018). Exploratory modelling, in decision-making applications such as in the robust decision-making framework (Lempert, Popper, and Bankes 2003), can facilitate a forward-thinking approach in policy analysis of transitions by answering questions such as 'under what conditions could future targets be met?' (see, e.g. Moallemi et al. 2017) and 'under what circumstances, regardless of their likelihood, could policy interventions fail?' which is called scenario discovery (see, e.g. Lempert, Popper, and Bankes 2013, Bryant and Lempert 2010, Groves and Lempert 2007). Most of the previous applications of exploratory modelling in transitions research have been for the use of policy insights. Among them are Hamarat, Kwakkel, and Pruyt (2013), Kwakkel and Pruyt (2013), Kwakkel and Yücel (2014), Hamarat et al. (2014), Eker and van Daalen (2015), Pye, Sabio, and Strachan (2015), and Moallemi et al. (2017).

3.3 Models to facilitate stakeholder processes

Transitions emerge in a multi-stakeholder context with many (sometimes opposing) views and perspectives regarding how and in which direction a transition will unfold and how to manage the transition towards its normative direction (Loorbach 2007). Exploratory modelling can contribute to stakeholder processes in a range of ways, for example with smart games (Mayer 2009). Exploratory modelling can facilitate the participation of stakeholders to provide input on definitions and validity at early stages of the model building process. The computer-assisted approach of exploratory modelling can create an experimentation space for testing and observing the consequences of various assumptions (e.g. visions, policies, scenarios, values and perspectives) among groups of stakeholders, with potentially divergent views. Exploratory modelling – by sampling systematically over possibilities and by running batch simulations based on the samples – can help to test a variety of assumptions of stakeholders, testing what their consequences would be in case the assumptions are right (e.g. regarding the normative direction they want to pursue). This can enable transitions models to bridge various perspectives and areas of expertise and create room for dialogue: a common understanding and consensus about transitions, their normative directions and the ways they could be managed. Such collaborative platforms can promote negotiation, reflexivity and learning among stakeholders (Malekpour et al. 2017). Exploratory modelling can also facilitate stakeholder processes at *later* stages of the modelling process to contribute to policy analysis or the communication of resulted insights. The incorporation of – sometimes opposing and contradictory – views through exploratory thinking can help to resolve disagreements, create a shared understanding and increase the acceptability of final results among stakeholders (Malekpour, de Haan, and Brown 2016) (also see Chapter 11 by Halbe and Chapter 10 by Malekpour in this volume). Few previous studies have used exploratory modelling for enabling stakeholder processes in transitions research; amongst those few are Eker, van Daalen, and Thissen (2017) and Moallemi and Malekpour (2018).

4 Two examples from the exploratory analysis of transitions

This section reviews two examples to show what the previous section discussed as potential roles of exploratory modelling in transitions research, in practice. The first example is by Moallemi and Köhler (2019) where they use exploratory modelling to investigate the various transitions pathways that a pre-existing model (MATISSE, see Chapter 6 by Köhler in this volume) produces based on certain assumptions about niches–regime interactions and their societal support. These pathways are then statistically analysed. The transition of interest is from an internal combustion engines (ICEs) regime to a new public transport regime in the UK mobility sector from 2015 to 2065. The transitions model used in this research is the MATISSE model (Bergman et al. 2008). The MATISSE model formalises the multi-level perspective's niches, regime and landscape concepts of transitions (Geels 2002) using agent-based modelling to simulate how different mobility systems grow or decline over time in response to shift in consumer support. The model assumes agents (i.e. mobility systems and consumers) are distributed in a multi-dimensional practice space. The dimensions of this practice space represent features of mobility systems and consumer attitude (e.g. emissions and convenience of a mobility system). Each feature ranges along a spectrum (e.g. low emissions to high emissions). Consumer attitude regarding each feature is reflected by their position in the practice space along the axes of each dimension. Therefore, the shift in the location of consumers and mobility systems in the practice space over time reflects changes in mobility system features and consumer behaviours. The result of these changes can destabilise a mobility regime and let a new regime emerge from growing niches. A major uncertainty investigated in this exploratory analysis lies in behavioural changes of consumers in the practice space. While it is obvious that consumer behaviours are constantly changing, the direction and the extent of their change remain unknown. To address this challenge of uncertainty, the exploratory analysis of the MATISSE model implements: (1) transition pathways to sustainable mobility under the uncertainty of behavioural changes of consumers and (2) the sensitivity analysis of unfolding transition pathways to various dimensions of behavioural changes. Based on the exploratory analysis of the MATISSE model, the results show that a transition from ICE to public transport is most likely than any other transitions under supportive behavioural changes (regarding distance travelled using owned individual vehicle and the use of mechanised public mobility). There are also (marginal) chances for alternative mobility systems (e.g. car sharing and battery electric vehicle) to become dominant in limited future scenarios. The results of sensitivity analysis reveal that any forms of behavioural changes (within the specified ranges) could lead to a transition away from the ICE regime in the UK mobility sector, which implies a low likelihood for the regime to stay in power.

The second example is by de Haan et al. (2016) where they use exploratory modelling to analyse the uptake and phasing out of urban water management solutions – more precisely speaking, drainage solutions. The transition of interest is from a regime around centralised 'pipes underground' solutions (e.g. drainage sewers) to decentralised 'green-blue infrastructure' (e.g. biofilters and wetlands). The transitions model used is called the societal transitions module (STM), underpinned by the multi-pattern approach (de Haan and Rotmans 2011) as its theoretical framework. The theory understands transitions to be sequences of patterns, driven by conditions, such as needs and constraints. A pattern is a discrete chunk of change, or rather, an ideal-type of change. A limited number of patterns is recognised in the multi-pattern approach. As the name of the approach suggests, several patterns are possible. The theory is agnostic as to *which* pattern will occur given certain circumstances and only describes the outcome when a certain pattern eventuates in those circumstances. The STM also implements this agnosticism and at each point in time where multiple patterns *could* eventuate, the model splits of several branches, one for each pattern. Thus, starting from one initial state of the model, a tree structure off possible developments of the system is obtained – an ensemble of theoretically possible transition pathways. The number of branches grows as the number of patterns to the power of the number of time steps. This quickly becomes computationally unwieldy. To address this challenge, the exploratory analysis of STM implements (1) sampling, to prune the pathway tree before it grows, and (2) clustering of the obtained pathways. The results show that although each pathway is different in detail, a relatively small number (a handful) of clusters (e.g.

backlash or successful uptake) can often be obtained that contain *qualitatively quite similar pathways*. Importantly, it was found that for any[2] given scenario of needs and constraints, multiple *types* of pathways are possible – many roads to Rome – with greater or smaller numbers of pathways of each type (backlash, successful transition) depending on the particular scenario. The model was also tested on a historical scenario, which yielded a portfolio of theoretically allowed histories – one of which corresponding to the actual historical pathway.

5 An appraisal of exploratory modelling for transitions research: strengths, limitations and future research directions

This section evaluates the strengths and limitations of exploratory modelling in transitions research. We go beyond what has been done and concluded in previous research and want to discuss what other benefits (which are not necessarily tested so far) exploratory modelling can have for transitions research.

5.1 Enhancing capabilities of transitions models in coping with complexities

The contribution of exploratory modelling has been mostly discussed in dealing with uncertainties. However, exploratory modelling can also help to cope with the complexities of the real world (Bankes 2002; Bankes, Lempert, and Popper 2001). In complex systems when knowing which details matter is difficult and when models are intrinsically incomplete (Bankes, Lempert, and Popper 2001), the use of exploratory modelling can enable modellers to obtain valid conclusions from a series of *untestable* individual exploratory models. For example, the series of exploratory models can represent various framings of transitions (e.g. different system feedback loop structures and behavioural agent rules of interaction in a transition model). This allows modellers to draw valid conclusions not necessarily from individually correct and comprehensive models of transitions, but from an ensemble of alternative models which all together can represent an acceptable level of complexity that real-world transitions entail. An example is Moallemi et al. (2017) who used six different realisations of a single energy transitions model (varying in parametrisation and structure) to incorporate multiple framings of transition pathways into the analysis.

Exploratory modelling, however, faces limitations in dealing with complexities of transitions. Some classes of transitions models – in particular agent-based models with multiple agent classes and behavioural characteristics – can have substantial structural and parametric complexity. The search strategy of exploratory modelling with such an extensive size of model and dimensions of complexities can become computationally highly demanding and can result in prolonged and costly (sometimes impractical) analytical processes. The large size of generated data also becomes more and more difficult for humans to analyse and grasp. A potential solution to mitigate this limitation is to use multi-resolution models (or sometimes called fast simple models) of a highly complicated model where exploratory modelling screens and analyses interesting behaviours at a high level of abstraction (Davis 2005). Finer resolutions of models are then only used to limit the analysis to identified areas of interest.

5.2 Enabling systematic experimentation for theory development

Earlier in this chapter, the first role identified for modelling, following Halbe et al. (2015) and Holtz et al. (2015), was for understanding, which closely parallels the role that modelling has in the sciences (Bankes 2009). Understanding is the result of explanation and theories. With this in mind, there are two distinct ways in which modelling can contribute to understanding. The first is through assuming a theory or hypothesis and by explaining an observed situation or event by reproducing it. Here the theory is not being tested, but is taken for granted, and the aim is to explain a *particular* state of affairs or events. The second is through testing a candidate theory or hypothesis (or multiple candidate theories or hypotheses) and attempting to reproduce known states of affairs or events. Successful reproduction is an indicator that the candidate hypothesis was not all that bad. Here theory is being tested, while phenomena (states or events) serve as reference data, and the aim is to identify *generic* explanations for classes of phenomena.

The current state of the art of transitions research has no theory on offer that could be used in the first variant sketched above – by which we mean that they do not clearly state hypotheses or predictions that could be implemented or tested with models. This means there is even more reason to identify how the second variant can be explored systematically, for a more rigorous testing of hypotheses and theories. Exploratory modelling can play an important role in such systematic theory testing and theory development based on the variation of scenarios or hypotheses (as defined in Section 3). Depending on the 'location' of the uncertainties to consider,[3] there are two ways in which this can be approached:

- *Hypothesis fixed, vary scenarios*: For a given hypothesis, one scans over a, potentially large, set of scenarios. The output would likewise be a potentially large set of transition pathways or descriptions of an event. What conclusions to draw from such an output set and how are not entirely straightforward and depend on how the hypothesis explains and what it is supposed to explain. That is, all scenarios may give rise to the same output behaviour, or each scenario to a qualitatively distinct output or there may be a spectrum of smoothly varying outputs.
- *Scenario fixed, vary hypothesis*: In this approach, an ensemble of hypotheses is used against a single scenario. This single scenario may well be a stylised or ideal-typical one. An ensemble of hypotheses may consist of a discrete set of qualitatively different assumptions of the mechanisms of transitions at play. Another possibility is that the hypothesis is parameterised, enabling

a scan over a range of versions of a supposed mechanism. The conclusions being drawn from such an approach can vary. For example, it can be to ascertain which hypotheses were 'better' given the outputs they yielded against historical data.

In the first approach, the uncertainties explored are in the external scenarios, while in the second approach the uncertainties explored are in or amongst the hypotheses. The first approach was the one taken in Moallemi and Köhler (2019) (in most parts) for systematic experimentation of transitions pathways in the mobility sector. The article by de Haan et al. (2016) could be considered an example of the second approach, though their aim was not actually to ascertain which hypothesised mechanisms were 'better' but rather to find out whether certain qualitative developments occur *regardless* of the uncertainties surrounding those mechanisms.

While exploratory modelling can enable the systematic and tractable experimentation of assumptions, it can face limitations in the experimentation of models with large datasets. The generation and analysis of extensive data in exploratory modelling requires thousands to millions of samples to be taken, a similar large number of simulations to be run, and a resulting large volume of data to be analysed. This can impose excessive computational burdens, can cause challenges for visualisation, and can make it harder for humans to grasp (and therefore legitimise) the results (for stakeholders). A solution to address this limitation is to improve the speed and efficiency of the search process in exploratory modelling with a more systematic design of experiments, a goal-oriented design which can reduce the computational demand. See, for example, Islam and Pruyt (2016) for improving sampling, and Moallemi, Elsawah, and Ryan (2018b, 2018c) for more effective delineation of uncertainties in exploratory modelling.

5.3 Going beyond the standard setting of policy analysis in transitions modelling

The established approach to policy analysis with transitions models often relies on forecasting the impacts of (or analysing the sensitivity of) a set of pre-specified policies (identified, for example, by experts) on a set of known (best-estimate, or ideal-typical) scenarios. We call it a standard trade-off between a set of known policies over known (discrete or probabilistic) scenarios. One example is the evaluation of pre-specified policy mixes under pre-specified transition pathway scenarios to explore tipping points in niche growth and decline in Raven and Walrave (2018) and the sensitivity analysis of the critical factors of energy transitions under well-characterised probabilistic (triangular) distributions of future scenarios in Pye, Sabio, and Strachan (2015). While this approach has worked successfully in several previous analyses, it may not be effective under certain conditions. In conditions when numerous variations of policies are possible (e.g. a continuous range of carbon tax), relying on a limited set of pre-specified policies can simplify the actual diversity of (possible) policies that exist and can lead to sub-optimal policy making (Fischhoff and Davis 2014). Moreover, in conditions where the future is highly uncertain, the premature aggregation of assumptions to characterise scenarios can lead to policies whose effectiveness does not hold up under future contingencies (Bankes 1993; Walker, Rahman, and Cave 2001).

Exploratory modelling can broaden the currently established approach to policy analysis with transitions models, using techniques from other areas such as uncertainty analysis and scenario planning (Guivarch, Lempert, and Trutnevyte 2017; Trutnevyte et al. 2016; Pye et al. 2018) as well as operations research (Eker and Kwakkel 2018; Watson and Kasprzyk 2017; Moallemi, Elsawah, and Ryan 2018a). We explain these alternative approaches to policy analysis based on stakeholder knowledge about policies and scenarios in the matrix of Figure 13.1. The vertical axis of the matrix in Figure 13.1 represents different degrees of knowledge about possible policies to be tested with transitions models:

- The least uncertainty is a set of 'pre-specified' policies selected based on expert opinion (e.g. set of fixed-rate feed-in tariff and fixed-price emission trading scheme).
- A middle amount of uncertainty is a set of 'iterative' policies when an initial list of pre-specified policies can be still identified, but the initial list should be modified over time to respond to changing circumstances (e.g. when the permit price of an emission trading scheme needs to be updated over time to remain effective).
- The most uncertainty is when numerous possible policies within ranges exist, and candidate policies are selected through a 'search' over all possibilities (e.g. continuous ranges for feed-in tariff rate).

The horizontal axis represents different degrees of knowledge about scenarios (Walker et al. 2003):

- The least uncertainty reflects scenarios where one can identify critical uncertain factors of the future and can agree how they shape the future scenarios. This can include conditions where there is a complete certainty (i.e. a set of deterministic scenarios) or there are alternative futures that can be predicted well enough (i.e. known probabilistic futures).
- The middle amount of uncertainty reflects scenarios where there are a limited set of plausible futures and one can prioritise and extract reference scenarios. A scenario in this case provides a plausible description of what *can* happen rather than a prediction of what *will* happen in the future (Maier et al. 2016).
- The high uncertainty reflects scenarios where one can identify critical uncertain factors in the future, but the presence of future surprises and shocks prevents the ranking of uncertainties or the association of a probability

distribution to them (i.e. deep uncertainty) (Lempert, Popper, and Bankes 2003).

- The most uncertainty which is a total ignorance sitting at the end of the spectrum of knowledge about scenarios reflects 'unknown unknown' circumstances when the knowledge of both possible scenarios and probabilities of scenarios is contested (Stirling 2010).

The established trade-off and scenario analysis approaches in transitions model fall within an area bound by a least or middle amount of uncertainty about policies and a least or middle amount of uncertainty about future scenarios. This can involve designing (short-term) policies for well-defined models that can rely on aggregated information (e.g. from historical trends) to predict the futures and can use a simple sensitivity analysis to consider the impacts of potential (but small) variations around the input parameters. The traditional scenario analysis approaches are also relevant where policies are identified that can lead to desired outcomes in a few reference plausible scenarios. The traditional trade-off and scenario analysis approaches, however, do not work effectively in the rest of other conditions regarding our knowledge about policies and scenarios. Following the taxonomy of exploratory modelling from Herman et al. (2015) and Kwakkel and Haasnoot (2019), we summarise alternative approaches from exploratory modelling to policy analysis of transitions under different degrees of knowledge about policies and scenarios as follows (see Figure 13.1):

- *When there are deterministic or probabilistic future scenarios or we can imagine few plausible scenarios and when candidate policies are to be searched amongst a large number of policies*: One approach to policy analysis is 'sensitivity analysis' to investigate the sensitivity of transitions to a set of pre-specified scenarios, given that policies can vary within wide ranges in the future. The aim is to identify the scenario(s) that can shape the most favourable conditions for a transition, given the diversity of possible policies. An example is to assess the sensitivity of an energy transition to electricity systems scenarios given uncertainty around the state of fossil fuel price and financing costs (see, for example, Trutnevyte et al. 2015). Another approach is 'multi-objective optimisation' (Watson and Kasprzyk 2017). It searches over possible policies to find those good (i.e. Pareto optimal) policies which make trade-offs between the fulfilment of multiple objectives of transitions under pre-specified (few plausible) scenarios. An example is to find a good rate of carbon tax which maximises emissions reduction over a set of pre-specified socio-economic pathway scenarios. Policy analysis in this setting can be performed using evolutionary optimisation algorithms (Reed et al. 2013).
- *When there are many plausible (or unknown) futures and when there are limited number of pre-specified candidate policies to test:* One approach to policy analysis is (another type of) 'sensitivity analysis' where the aim is to understand policies that can leverage transitions the most under a variety of

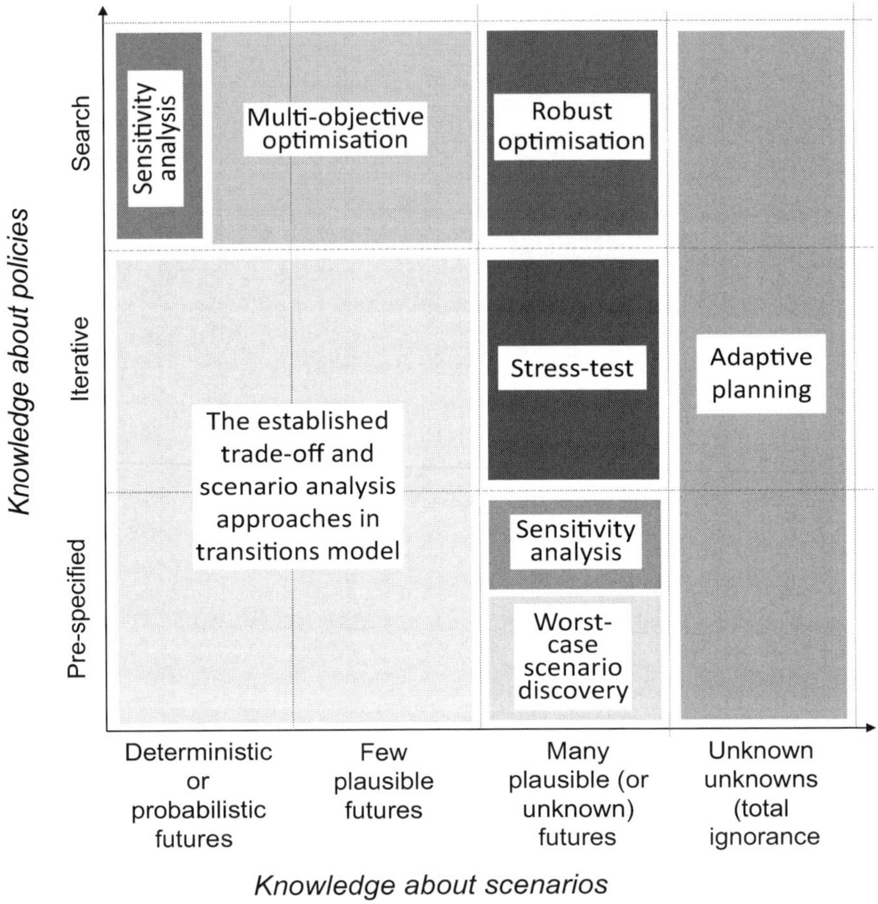

Figure 13.1 Alternative approaches which can be potentially applicable to the policy analysis of transitions, based on the level of knowledge about policies and scenarios (also see Moallemi et al., In Press 2019).

possible scenarios. An example is to identify critical uncertainties which have strong impact on delivering emissions reduction and which can leverage the decarbonisation process the most (see, e.g. Pye, Sabio, and Strachan 2015). Another approach to policy analysis is 'worst-case scenario discovery' (or perfect storm) where the aim is to find worst possible outcomes of a set of policies and scenarios leading to those outcomes (Halim, Kwakkel, and Tavasszy 2015). An example is to specify the worst level of emissions reduction and to identify the energy policies (i.e. future scenarios) which could lead to this worst outcome.

- *When there are many plausible (or unknown) futures and when policies are selected iteratively*: 'Stress-testing' is an approach to policy analysis where the aim is to identify when the initial set of policies should be adapted (and how) in order to remain effective over time (Lempert and Groves 2010). An example is to identify when a policy fails to deliver its intended impact on transitions (i.e. policy vulnerabilities) and how to be prepared to respond to failures by designing supportive, protective measures (see, e.g. Eker and van Daalen 2015; Hamarat, Kwakkel, and Pruyt 2013). Stress-testing can be performed using different methods and approaches such as scenario discovery (Bryant and Lempert 2010), adaptation tipping point analysis (Kwadijk et al. 2010), decision scaling (Brown et al. 2012), and Dynamic Adaptive Policy Pathways (Haasnoot et al. 2013).

- *When there are many plausible (or unknown) futures and when candidate policies are searched amongst a large number of policies*: An approach to policy analysis is 'robust optimisation' where the aim is to find robust policies which can fulfil multiple objectives over different possible future scenarios (rather than optimal policies in pre-specified reference scenarios). Robust optimisation can be performed with a mixed use of evolutionary optimisation algorithms (Reed et al. 2013) where candidate policies are generated by searching over possibilities and systematic sampling where an ensemble of random scenarios are sampled. Candidate policies, identified in each iteration of optimisation, are tested over sampled scenarios to identify the robust policy which can lead to satisfactory results under deep uncertainty (see, e.g. Hamarat et al. 2014).

While exploratory modelling can enhance the policy analysis of transitions under new (less or not at all investigated) conditions, it still faces limitations in analysing problems arising from 'unknown unknown' circumstances. This is rooted in the limited power of existing approaches and part of the very nature of the unknown unknown in understanding the unknown consequences of unknown futures (which is not limited to exploratory modelling). Previous research has suggested that while the surprises of unknown unknowns cannot be avoided, adaptive planning could be a way to at least proactively respond to them (Stirling 2010). Among these adaptive planning approaches are transition management (Loorbach 2010) and dynamic adaptive policy pathways (Haasnoot et al. 2013) that focus on monitoring, flexible commitments and the design of multiple pathways to be prepared for future contingencies.

Notes

1 Haxeltine et al. (2008) mention uncertainty once (p. 99) in a summary of one of the patterns by Geels and Schot (2007) Köhler et al. (2009) do not mention uncertainty, but they analyse transitions pathways under a limited number of runs. See Chapter 6 by Köhler in this volume for an explanation of the stochastic variables in the MATISSE model.

2 Within the limits of the model runs actually performed of course.

3 See Walker et al. (2003) for, amongst other things, a framework to locate uncertainties in the context of modelling.

References

Bankes, Steve. 1993. "Exploratory modeling for policy analysis." *Operations Research* 41 (3): 435–49.

Bankes, Steve. 2009. "Models as lab equipment: Science from computational experiments." *Computational and Mathematical Organization Theory* 15 (1): 8. http://doi.org/10.1007/s10588-008-9046-y.

Bankes, Steven C. 2002. "Tools and techniques for developing policies for complex and uncertain systems." *Proceedings of the National Academy of Sciences* 99 (suppl 3): 7263–6. http://doi.org/10.1073/pnas.092081399.

Bankes, Steven C, Robert J. Lempert, and Steven W. Popper. 2001. "Computer-assisted reasoning." *Computing in Science & Engineering* 3 (2): 71–7. http://doi.org/10.1109/5992.909006.

Bankes, Steven, Warren E. Walker, and Jan H. Kwakkel. 2013. "Exploratory modeling and analysis." In *Encyclopedia of Operations Research and Management Science*, edited by Saul I. Gass and Michael C. Fu, 532–7. New York: Springer.

Bergman, Noam, Alex Haxeltine, Lorraine Whitmarsh, Jonathan Köhler, Michel Schilperoord, and Jan Rotmans. 2008. "Modelling socio-technical transition patterns and pathways." *Journal of Artificial Societies and Social Simulation* 11 (3): 7.

Brown, Casey, Yonas Ghile, Mikaela Laverty, and Ke Li. 2012. "Decision scaling: Linking bottom-up vulnerability analysis with climate projections in the water sector." *Water Resources Research* 48 (9).

Bryant, Benjamin P., and Robert J. Lempert. 2010. "Thinking inside the box: A participatory, computer-assisted approach to scenario discovery." *Technological Forecasting and Social Change* 77 (1): 34–49.

Davis, Paul K. 2005. *Introduction to Multiresolution, Multiperspective Modeling (MRMPM) and Exploratory Analysis*. Santa Monica, CA: RAND Corporation.

de Haan, Fjalar J. 2006. "How emergence arises." *Ecological Complexity* 3 (4): 293–301. http://doi.org/10.1016/j.ecocom.2007.02.003.

de Haan, Fjalar J., Briony C. Rogers, Rebekah R. Brown, and Ana Deletic. 2016. "Many roads to Rome: The emergence of pathways from patterns of change through exploratory modelling of sustainability transitions." *Environmental Modelling & Software* 85: 279–92. http://doi.org/10.1016/j.envsoft.2016.05.019.

de Haan, Fjalar J., and Jan Rotmans. 2011. "Patterns in transitions: Understanding complex chains of change." *Technological Forecasting and Social Change* 78 (1): 90–102.

Eker, Sibel, and Els van Daalen. 2015. "A model-based analysis of biomethane production in the Netherlands and the effectiveness of the subsidization policy under uncertainty." *Energy Policy* 82: 178–96. http://doi.org/10.1016/j.enpol.2015.03.019.

Eker, Sibel, Els van Daalen, and Wil Thissen. 2017. "Incorporating stakeholder perspectives into model-based scenarios: Exploring the futures of the Dutch gas sector." *Futures* 93 (Supplement C): 27–43. http://doi.org/10.1016/j.futures.2017.08.002.

Eker, Sibel, and Jan H. Kwakkel. 2018. "Including robustness considerations in the search phase of many-objective robust decision making." *Environmental Modelling & Software* 105: 201–16. http://doi.org/10.1016/j.envsoft.2018.03.029.

Fischhoff, Baruch, and Alex L. Davis. 2014. "Communicating scientific uncertainty." *Proceedings of the National Academy of Sciences* 111 (Supplement 4): 13664–71.

Geels, Frank W. 2002. "Technological transitions as evolutionary reconfiguration processes: A multi-level perspective and a case-study." *Research Policy* 31 (8–9): 1257–74.

Geels, Frank W., and Johan Schot. 2007. "Typology of sociotechnical transition pathways." *Research Policy* 36 (3): 399–417.

Groves, David G., and Robert J. Lempert. 2007. "A new analytic method for finding policy-relevant scenarios." *Global Environmental Change* 17 (1): 73–85.

Guivarch, Céline, Robert Lempert, and Evelina Trutnevyte. 2017. "Scenario techniques for energy and environmental research: An overview of recent developments to broaden the capacity to deal with complexity and uncertainty." *Environmental Modelling & Software* 97: 201–10. http://doi.org/10.1016/j.envsoft.2017.07.017.

Guivarch, Céline, Julie Rozenberg, and Vanessa Schweizer. 2016. "The diversity of socio-economic pathways and CO2 emissions scenarios: Insights from the investigation of a scenarios database." *Environmental Modelling & Software* 80: 336–53. http://doi.org/10.1016/j.envsoft.2016.03.006.

Haasnoot, Marjolijn, Jan H. Kwakkel, Warren E. Walker, and Judith ter Maat. 2013. "Dynamic adaptive policy pathways: A method for crafting robust decisions for a deeply uncertain world." *Global Environmental Change* 23 (2): 485–98. http://doi.org/10.1016/j.gloenvcha.2012.12.006.

Halbe, Johannes. 2019. "Participatory modelling in sustainability transitions research." In *Modelling Transitions – Virtues, Vices, Visions of the Future*, edited by Enayat A. Moallemi and Fjalar de Haan. This volume: Routledge.

Halbe, Johannes, Dominik E. Reusser, Georg Holtz, Marjolijn Haasnoot, Annette Stosius, Wibke Avenhaus, and Jan H. Kwakkel. 2015. "Lessons for model use in transition research: A survey and comparison with other research areas." *Environmental Innovation and Societal Transitions* 15: 194–210. http://doi.org/10.1016/j.eist.2014.10.001.

Halim, Ronald A., Jan H. Kwakkel, and Lóránt A. Tavasszy. 2015. "A scenario discovery study of the impact of uncertainties in the global container transport system on European ports." *Futures* 81: 148–60. http://doi.org/10.1016/j.futures.2015.09.004.

Hamarat, Caner, Jan H. Kwakkel, and Erik Pruyt. 2013. "Adaptive robust design under deep uncertainty." *Technological Forecasting and Social Change* 80 (3): 408–18. http://doi.org/10.1016/j.techfore.2012.10.004.

Hamarat, Caner, Jan H. Kwakkel, Erik Pruyt, and Erwin T. Loonen. 2014. "An exploratory approach for adaptive policymaking by using multi-objective robust optimization." *Simulation Modelling Practice and Theory* 46: 25–39. http://doi.org/10.1016/j.simpat.2014.02.008.

Haxeltine, Alex, Lorraine Whitmarsh, Noam Bergman, Jan Rotmans, Michel Schilperoord, and Jonathan Kohler. 2008. "A conceptual framework for transition modelling." *International Journal of Innovation and Sustainable Development* 3 (1–2): 93–114.

Herman, Jonathan D., Patrick M. Reed, Harrison B. Zeff, and Gregory W. Characklis. 2015. "How should robustness be defined for water systems planning under change?" *Journal of Water Resources Planning and Management* 141 (10): 04015012.

Holtz, Georg, Floortje Alkemade, Fjalar de Haan, Jonathan Köhler, Evelina Trutnevyte, Tobias Luthe, Johannes Halbe, George Papachristos, Emile Chappin, Jan Kwakkel, and Sampsa Ruutu. 2015. "Prospects of modelling societal transitions: Position paper of an emerging community." *Environmental Innovation and Societal Transitions* 7: 41–58. http://doi.org/10.1016/j.eist.2015.05.006.

Islam, Tushith, and Erik Pruyt. 2016. "Scenario generation using adaptive sampling: The case of resource scarcity." *Environmental Modelling & Software* 79 (Supplement C): 285–99. http://doi.org/10.1016/j.envsoft.2015.09.014.

Köhler, Jonathan. 2019. "Modelling the multi-level perspective: The MATISSE agent-based model." In *Modelling Transitions – Virtues, Vices, Visions of the Future*, edited by E. Moallemi and F. de Haan. This volume: Routledge.

Köhler, Jonathan, Fjalar de Haan, Georg Holtz, Klaus Kubeczko, Enayat A. Moallemi, George Papachristos, and Emile Chappin. 2018. "Modelling sustainability transitions: An assessment of approaches and challenges." *Journal of Artificial Societies and Social Simulation* 21 (1): 8. http://jasss.soc.surrey.ac.uk/21/1/8.html.

Köhler, Jonathan, Lorraine Whitmarsh, Björn Nykvist, Michel Schilperoord, Noam Bergman, and Alex Haxeltine. 2009. "A transitions model for sustainable mobility." *Ecological Economics* 68 (12): 2985–95. http://doi.org/10.1016/j.ecolecon.2009.06.027.

Kwadijk, Jaap C.J., Marjolijn Haasnoot, Jan P.M. Mulder, Marco Hoogvliet, Ad Jeuken, Rob A.A. van der Krogt, Niels G.C. van Oostrom, Harry A. Schelfhout, Emiel H. van Velzen, and Harold van Waveren. 2010. "Using adaptation tipping points to prepare for climate change and sea level rise: A case study in the Netherlands." *Wiley Interdisciplinary Reviews: Climate Change* 1 (5): 729–40.

Kwakkel, Jan H., and Marjolijn Haasnoot. 2019. "Supporting DMDU: A taxonomy of approaches and tools." In *Decision Making Under Deep Uncertainty – From Theory to Practice*, edited by V.A.W.J. Marchau, W.E. Walker, P. Bloemen, and S. Popper. New York: Springer.

Kwakkel, Jan H., and Erik Pruyt. 2013. "Exploratory modeling and analysis, an approach for model-based foresight under deep uncertainty." *Technological Forecasting and Social Change* 80 (3): 419–31. http://doi.org/10.1016/j.techfore.2012.10.005.

Kwakkel, Jan H., and Gönenç Yücel. 2014. "An exploratory analysis of the Dutch electricity system in transition." *Journal of the Knowledge Economy* 5 (4): 670–85. http://doi.org/10.1007/s13132-012-0128-1.

Lempert, Robert J., and D.G. Groves. 2010. "Identifying and evaluating robust adaptive policy responses to climate change for water management agencies in the American west." *Technological Forecasting and Social* 77: 960.

Lempert, Robert J., N. Kalra, S. Peyraud, Z. Mao, S.B. Tan, D. Cira, and A. Lotsch. 2013. "Ensuring robust flood risk management in Ho Chi Minh City." In *Policy Research Working Paper 6465*, World Bank.

Lempert, Robert J., Steven W. Popper, and Steven C. Bankes. 2003. *Shaping the next one hundred years: New methods for quantitative, long-term policy analysis*. Santa Monica, CA: RAND Corporation.

Loorbach, Derk Albert. 2007. *Transition Management: New Mode of Governance for Sustainable Development*. (PhD Dutch Research Institute for Transitions (DRIFT)), Erasmus University.

Loorbach, Derk Albert. 2010. "Transition management for sustainable development: A prescriptive, complexity-based governance framework." *Governance* 23 (1): 161–83.

Maier, H.R., J.H.A. Guillaume, H. van Delden, G.A. Riddell, M. Haasnoot, and J.H. Kwakkel. 2016. "An uncertain future, deep uncertainty, scenarios, robustness and adaptation: How do they fit together?" *Environmental Modelling & Software* 81: 154–64. http://doi.org/10.1016/j.envsoft.2016.03.014.

Malekpour, Shirin. 2019. "Models as scenario tools for developing robust transformative plans." In *Modelling Transitions – Virtues, Vices, Visions of the Future*, edited by E. Moallemi and F. De Haan. This volume: Routledge.

Malekpour, Shirin, Rebekah R. Brown, Fjalar J. de Haan, and Tony H.F. Wong. 2017. "Preparing for disruptions: A diagnostic strategic planning intervention for sustainable development." *Cities* 63: 58–69. http://doi.org/10.1016/j.cities.2016.12.016.

Malekpour, Shirin, Fjalar J. de Haan, and Rebekah R. Brown. 2016. "A methodology to enable exploratory thinking in strategic planning." *Technological Forecasting and Social Change* 105: 192–202. http://doi.org/10.1016/j.techfore.2016.01.012.

Marchau, Vincent A.W.J., Warren E. Walker, Pieter J.T.M. Bloemen, and Steven W. Popper. 2019. *Decision Making Under Deep Uncertainty: From Theory to Practice*. New York: Springer.

Markard, Jochen, Rob Raven, and Bernhard Truffer. 2012. "Sustainability transitions: An emerging field of research and its prospects." *Research Policy* 41 (6): 955–67. http://doi.org/10.1016/j.respol.2012.02.013.

Mayer, Igor S., and J. Simulation. 2009. "The gaming of policy and the politics of gaming: A review." *Simulation & Gaming* 40 (6): 825–62.

Moallemi, Enayat A., Fjalar de Haan, Jan Kwakkel, and Lu Aye. 2017. "Narrative-informed exploratory analysis of energy transition pathways: A case study of India's electricity sector." *Energy Policy* 110: 271–87. http://doi.org/10.1016/j.enpol.2017.08.019.

Moallemi, Enayat A., Sondoss Elsawah, and Michael J. Ryan. 2018a. "Model-based multi-objective decision making under deep uncertainty from a multi-method design lens." *Simulation Modelling Practice & Theory* 84: 232–50. http://doi.org/10.1016/j.simpat.2018.02.009.

Moallemi, Enayat A., Sondoss Elsawah, and Michael J. Ryan. 2018b. "An agent-monitored framework for the output-oriented design of experiments in exploratory modelling." *Simulation Modelling Practice and Theory* 89: 48–63. http://doi.org/10.1016/j.simpat.2018.09.008.

Moallemi, Enayat A., Sondoss Elsawah, and Michael J. Ryan. 2018c. "Informing the delineation of input uncertainty space in exploratory modelling using a heuristic approach." *Insight* 21 (3): 27–33. http://doi.org/10.1002/inst.12209.

Moallemi, Enayat A., and Jonathan Köhler. 2019. "Coping with uncertainties of sustainability transitions using exploratory modelling: The case of the MATISSE model and the UK's mobility sector." *Environmental Innovation and Societal Transitions*. https://doi.org/10.1016/j.eist.2019.03.005

Moallemi, Enayat A., and Shirin Malekpour. 2018. "A participatory exploratory modelling approach for long-term planning in energy transitions." *Energy Research & Social Science* 35: 205–16. http://doi.org/10.1016/j.erss.2017.10.022.

Moallemi, Enayat A., Fateme Zare, Patrick M. Reed, Sondoss Elsawah, Michael J. Ryan, and Brett A. Bryan. 2019. "Structuring and evaluating decision support processes to enhance the robustness of complex human–natural systems." *Environmental Modelling & Software*: 104551. https://doi.org/10.1016/j.envsoft.2019.104551.

Pye, Steve, Francis G.N. Li, Arthur Petersen, Oliver Broad, Will McDowall, James Price, and Will Usher. 2018. "Assessing qualitative and quantitative dimensions of uncertainty in energy modelling for policy support in the United Kingdom." *Energy Research & Social Science* 46: 332–44. http://doi.org/10.1016/j.erss.2018.07.028.

Pye, Steve, Francis G.N. Li, James Price, and Birgit Fais. 2017. "Achieving net-zero emissions through the reframing of UK national targets in the post-Paris Agreement era." *Nature Energy* 2: 17024. http://doi.org/10.1038/nenergy.2017.24 www.nature.com/articles/nenergy201724#supplementary-information.

Pye, Steve, Nagore Sabio, and Neil Strachan. 2015. "An integrated systematic analysis of uncertainties in UK energy transition pathways." *Energy Policy* 87: 673–84. http://doi.org/10.1016/j.enpol.2014.12.031.

Raven, Rob, and Bob Walrave. 2018. "Overcoming transformational failures through policy mixes in the dynamics of technological innovation systems." *Technological Forecasting and Social Change*. http://doi.org/10.1016/j.techfore.2018.05.008.

Reed, Patrick M., David M. Hadka, Jon Herman, Joseph Kasprzyk, and Joshua B. Kollat. 2013. "Evolutionary multiobjective optimization in water resources: The past, present, and future." *Advances in Water Resources* 51: 438–56. http://doi.org/10.1016/j.advwatres.2012.01.005.

Rotmans, Jan. 1998. "Methods for IA: The challenges and opportunities ahead." *Environmental Modeling & Assessment* 3 (3): 55–179.

Rotmans, Jan, René Kemp, and Marjolein van Asselt. 2001. "More evolution than revolution: Transition management in public policy." *Foresight* 3 (1): 15–31. http://doi.org/10.1108/14636680110803003.

Schilperoord, Michel, Jan Rotmans, and Noam Bergman. 2008. "Modelling societal transitions with agent transformation." *Computational and Mathematical Organization Theory* 14 (4): 283–301. http://doi.org/10.1007/s10588-008-9036-0.

Stirling, Andy. 2010. "Keep it complex." *Nature* 468 (7327): 1029–31.

Timmermans, Jos, Hans de Haan, and Flaminio Squazzoni. 2008. "Computational and mathematical approaches to societal transitions." *Computational and Mathematical Organization Theory* 14 (4): 391–414. http://doi.org/10.1007/s10588-008-9035-1.

Trutnevyte, Evelina, Céline Guivarch, Robert Lempert, and Neil Strachan. 2016. "Reinvigorating the scenario technique to expand uncertainty consideration." *Climatic Change* 135 (3): 373–9. http://doi.org/10.1007/s10584-015-1585-x.

Trutnevyte, Evelina, Neil Strachan, Paul E. Dodds, Danny Pudjianto, and Goran Strbac. 2015. "Synergies and trade-offs between governance and costs in electricity system transition." *Energy Policy* 85: 170–81. http://doi.org/10.1016/j.enpol.2015.06.003.

Van Asselt, Marjolein B.A., and Jan Rotmans. 2002. "Uncertainty in integrated assessment modelling." *Climatic Change* 54 (1–2): 75–105.

Walker, Warren E., Poul Harremoës, Jan Rotmans, Jeroen P. van der Sluijs, Marjolein B.A. van Asselt, Peter Janssen, and Martin P. Krayer von Krauss. 2003. "Defining uncertainty: A conceptual basis for uncertainty management in model-based decision support." *Integrated assessment* 4 (1): 5–17.

Walker, Warren E., S. Adnan Rahman, and Jonathan Cave. 2001. "Adaptive policies, policy analysis, and policy-making." *European journal of operational Research* 128 (2): 282–9.

Watson, Abigail A., and Joseph R. Kasprzyk. 2017. "Incorporating deeply uncertain factors into the many objective search process." *Environmental Modelling & Software* 89: 159–71. http://doi.org/10.1016/j.envsoft.2016.12.001.

14 Epilogue

Quo vadis transitions modelling?

Fjalar J. de Haan and Enayat A. Moallemi

1 Introduction

In the concluding part of this volume on transitions modelling, it is appropriate to reflect on the state of affairs and the road ahead. Of course, these are the very subjects of this book so in some sense, if the book delivers on its title, we could end our conclusion right here. We have some reason to believe the book will be able to deliver in this manner. We were fortunate enough to have contributions from key figures in the transitions modelling community and we think their chapters can make this book the reference work we hope it will become. We will see whether it will live up to these lofty ambitions.

Therefore, we would like to seize the opportunity in this epilogue to discuss the state, status and potential aspirations of transitions modelling in a rather more broad sense. Whereas the book treats transitions modelling from a mostly technical, methodological point of view, here we will drag the discussion into an arena of much wider range, namely the relation of transitions modelling to the broader enterprise of transitions research, to other fields and the Grand Challenges of transitions. We will use this broader context, however, to formulate some practical suggestions that, we think, will help transitions modelling to do justice to its potential.

We do not claim to speak on behalf of the transitions modelling community in this epilogue, there are recent and good position papers that do that, both from a general perspective (Holtz et al., 2015) and from a more technical perspective (Köhler et al., 2018). However, we think most of our colleagues would agree that modelling could and should play a more prominent role in transitions research. Apart from a bias towards one's own research methods (clearly justified in this case), this feeling may stem from the belief that modelling is crucial in making research *matter* or it may stem from skepticism about the methodologies of mainstream transitions research.

We invite modellers and mainstream researchers of transitions alike, to consider the grand intellectual and societal challenges that our topic presents and reflect upon our suggestions for a way forward. These suggestions are not of the 'more of what we do, but better' variety. They present an agenda of diversification, engagement with methodologically cognate fields and self-assertion in the home field. We think modelling should *lead* the field of transitions research, not *follow* it.

2 The imperative for transitions modelling

Transitions pose two Grand Challenges, one intellectual and one societal. The latter, since a sustainable future requires transitions, the former because we do not understand them yet. Both these challenges are eminently worthy of our efforts and devotion. Both challenges necessitate – so we claim – the same sort of approach: a scientific one. This not necessarily to the exclusion of other modes of inquiry, but it should be the hard core of our research enterprise. There will be those who object that a scientific approach to transitions research is either inappropriate or impossible and some will argue both. We disagree thoroughly with such a position but we will not here elaborate, Chapter 4 of this volume has done this exhaustively.

Rather than defending our position in favour of a scientific approach and modelling, we suggest that our opponents justify the approaches they deem more appropriate. We are of the opinion that the current mainstream approach to transitions research is woefully inadequate, indeed inappropriate, for the task at hand. Some methodological introspection would be a healthy exercise for the field and we think the logical outcome of that should be a greater appreciation of modelling in its many guises.

The point is that the two Grand Challenges require *positive* knowledge, by which we mean knowledge that goes beyond ways of framing and phrasing phenomena of interest. We need to know how transitions work, what factors influence them in what way and how much. Moreover, we need to know in which circumstances they work that way and in which they do not. Finally, this knowledge should be tested empirically. If this sounds like a large ask, to us it seems a bare minimum. They are, after all, *Grand* Challenges.

To put these demands on transitions-knowledge in context, think about what most everyone expects of medical and engineering knowledge. Is that an unfair comparison to make? As a field, we are confident to advise government agencies on decisions affecting (if we are right!) society on the long term, involving large sums of public money. That transitions research is not meeting the intellectual Grand Challenge is one thing, but that we are making little headway in addressing the societal Grand Challenge should bother us. For many researchers in the transitions field, the possibility that their work might contribute to stimulate actual transitions towards sustainability is an important part of their motivation. It is a noble pursuit.

We will not discuss everything that is wrong with transitions research, for that, see, again Chapter 4 of this volume. Summarising, we observe that with respect to the Grand Challenges and the kind of knowledge they require, as a field, we have achieved little and the knowledge we claim to have is in fact uncertain or insufficiently verified to say the least. This calls into question both the *efficacy* of actions based on that knowledge and the *legitimacy* of any advice or undertaking of such actions. When we put our knowledge forward as a basis for decision making with society-wide, generations-long impacts, we had better have a good case to make about its reliability. At the moment, we say, there is no such case to be made.

We are not saying anything as trivial as 'more modelling will help'. We say that transitions research needs to become scientific in its methodological approach and standards. Modelling, especially in the broadly conceived meaning of the term we will urge in the next section, is central to such a scientific approach and, as a practice, embodies the scientific attitude. This is the imperative for transitions modelling and it is a formidable one. It implies a task the current transitions modelling community is likely not able to take on in full. This notwithstanding, it suggests an important role for the modelling community, that of the scientific and methodological conscience of transitions research.

3 Avenues of progress for modelling and transitions

Having expressed our rather grave concerns about the adequacy and legitimacy of mainstream transitions research, it is only just to take a critical look at transitions modelling itself. Not a critical look at what has been and is being done, that is the topic of the rest of this volume, in particular the Prologue and Chapter 0.3 by Köhler and Holtz. Here we want to discuss what needs to be done to move further on this ambitious path. A part of this is technical, about tools and methods, but another important part is about our modelling community itself and how it relates to transitions research more broadly and other fields.

Summarising in advance, we think we need improvement in the following areas:

- *Quantity* – of research and researchers active in transitions modelling. There are too few of us and the challenges are great.
- *Quality* – of the modelling work. As a benchmark, we should be able to survive the scrutiny of modellers in the most demanding fields.
- *Coordination* – of transitions modelling research, both amongst the community and with methodologically cognate fields.
- *Portfolio* – of tools and approaches. Very many useful tools and methods are not used or even known in our community.
- *Scope* – of the notion of modelling. Our conception of modelling is far too narrow, mostly constrained to computer simulation.

Perhaps it is appropriate to start with the last point, in fact, we will go through the list in more-or-less reverse order, looking at Scope and Portfolio first and then at Coordination, Quality and Quantity.

3.1 Scope and portfolio

For most people in the transitions modelling community, as well as in transitions research more broadly, 'modelling' seems to mean computer simulation. Despite the fairly comprehensive working definitions in the review articles by

Holtz et al. (2015) and Köhler et al. (2018), they discuss mainly simulation-oriented work and methods, which suggests this apparent bias is simply a reflection of the actual modelling practices. We think there is much to gain by drawing the conceptual boundary of modelling a fair bit broader – and then, of course, to actually employ the implied broader portfolio of methods.

Essentially, what we are suggesting is a shift towards formal modelling being at the core of every aspect of transitions research. In the following, we will attempt to give an idea of what we think that would entail. This will be based on the methods and approaches we have been exposed to, and therefore inevitably incomplete. To bring some system into it, we discuss four important 'areas' of transitions research where formalisation, in our view, is readily possible: *Representation of Systems* (3.1.1), *Modelling Transitions Processes* (3.1.2), *Hypotheses and Mechanisms* (3.1.3) and *Data, Analytics and Metrics* (3.1.4).

3.1.1 Representation of systems

Good formal representations of the systems affected by transitions are instrumental for progress in theory development and modelling alike. It even seems a *prerequisite* to those endeavours. The concepts used and proposed in the theoretical literature ('regime', 'niche', 'frontrunner' and friends) are usually quite loosely defined, if defining is the right term at all. This leaves plenty of opportunity for modellers to develop a variety of formal representations and test their merits and demerits. Assuming these concepts have some empirical basis and a proven utility in narratives, there is clearly value in pursuing this road to formalisation — if only because it enables systematic empirical testing. This road has been pursued (see e.g. Chapter 6 by Köhler on the Matisse model in this volume) and will no doubt continue to be.

There are several other ways, however, in which we should explore formal representation of the systems we are interested in. In fact, this would be a sub-genre of transitions modelling in its own right. In this sub-genre, dynamics is not the primary object of investigation. Instead, the formalisation focusses on representing the structure of the system of interest in such a way that it can be interrogated. (Clearly, this can be with the aim of investigating possibilities of dynamics.) The formal apparatus of network science and graph theory (see e.g. Newman, 2003) seems especially pertinent to such approaches. Structural approaches were already highlighted in Holtz et al. (2015, p. 54–55) and Chapter 2.4 by Rojas and de Haan is an example of a representational model in this volume.

Clearly, formal representations do not have to describe, or be able to describe, all aspects of a system affected by a transition. Formalisms that describe a particular aspect very well may even be preferrable. The networks and graph theoretical approach referred to above is good case in point, it would be very relevant to have good formal representations of the social networks of, say, social innovators. Similarly, representations of supply chain networks would capture an important aspect. From these examples, we also see that many approaches to formal representation provide transitions modelling with connections to other research fields, in the above cases Social Network Analysis (see e.g. Scott, 1991; Wasserman and Faust, 1994) and Supply Chain Networks (see e.g. Kim et al., 2015; Perera et al., 2018).

An important approach to formal representation is the use of *ontologies*. Here, 'ontology' is not used in its usual philosophical sense (though there is a relation), but in the computer science sense of the word: a formalism to represent a domain of discourse (see Noy and McGuinness, 2001, for a primer). Indeed, this suggests a much broader range of applications. Using the representation of systems in transitions research as an example, an ontology provides not only a formalism for the concepts that describe those systems but also how these concepts relate to each other. Ontologies are being used in various fields, including bio-informatics and medicine. They can also be the basis for dynamical modelling approaches, a good example of this in the context of transitions modelling is the work of van Dam and Lukszo (2006).

3.1.2 Modelling transitions processes

What constitutes a transitions process? What changes and are there patterns that recur? Like with formal representation of systems, we have a lack of formal representations of transitions processes.

A potentially interesting approach would be to capture transitions processes in a so-called *action language*. Action languages are formal languages to describe the effect of actions on the state of some system,[1] which means that they also entail a formalisation of the system *representation*. Such languages are commonly used in Artificial Intelligence in the area of planning and also in robotics. In particular, the language family PDDL[2] – Planning Domain Definition Language – appears readily applicable. (Haslum et al., 2019) is a great, recent, introduction to the PDDL language family and approach. Another action language of interest is GOLOG (Levesque et al., 1997), which is explicitly geared at modelling dynamic environments. Using action languages like PDDL and GOLOG would connect transitions modelling with the field of planning in Artificial Intelligence. Not only would this provide access to a host of tools and methods, perhaps more importantly, it could foster fruitful collaborations and the development of shared research questions.

Another way of formalising processes in transitions is to formalise their structure, that is, how they are composed of events, actions or other units of descriptions. This is the approach taken by Spekkink (2015) under the name *event sequence analysis* (see also Chapter 12 by de Haan, Martínez Arranz and Spekkink of this volume). Essentially, this approach captures the dynamics as a graph, that is, a network of related events. The graph approach of event sequence analysis, as introduced by Spekkink, has been applied directly on empirical case data. The advantage of such an approach is that cases become comparable in novel ways, namely on the basis of their event structure.

3.1.3 Hypotheses and mechanisms

Formalisation is the road to cumulative theory development. Thus, the exhortation is that our approach to theory development have a core of formal modelling – rather than modelling being optional or auxiliary. Theoretical innovations, be they concepts or dynamical hypotheses, should be proposed in a minimalist, formal – *formalisable* at least – and, if they are hypotheses, predictive form. The minimalism is to make sure we are not piling on 'nuance' (Healy, 2017) and confusing vagueness for generality (Russell, 1923). 'Predictive' means that hypotheses should be more than summaries of the data that suggested them, they should enable statements beyond what was already known – note that these statements need not be about the future. Predictive power may seem a steep demand to some but it really is a *minimal* requirement: if a piece of theory cannot make a prediction *it cannot be proven wrong* and is therefore empirically vacuous. Theoretical speculation (which we need) in minimalist, formal and predictive propositions, is a basis for robust, cumulative knowledge. In other words, theory development should be a progressive enterprise of model making and model testing.

There are many ways in which hypotheses about transitions could be formalised. They could be described as (1) sequences of actions or events – for which the aforementioned approaches of action languages and event sequence analysis could be employed, (2) statistical patterns ('signatures') in data – more on this in Section 3.1.4 or as (3) mechanisms, all of these together with the circumstances (e.g. necessary preconditions or state of the environment under which they occur). This is another one of those lists that is neither complete nor consisting of mutually exclusive options, but it gives an impression of what we are after.

Mechanisms, in particular should be close to the heart of many a modeller. They are after all the core components of simulation models. The usual interpretation of mechanism is a causal description of how antecedent circumstances lead to consequences. Here, we argue for modularity. Mechanisms should be modules that can be tested seperately. This does not only make sense from a software development point of view, but also scientifically. This is very much the approach argued in analytic sociology under the header of 'social mechanisms (e.g. Hedström, 2005, see also Chapter 5 of this volume by Bianchi and Squazzoni). We would, additionally, argue for a broad conception of 'mechanism' to also include non-causal hypotheses, e.g. covering-law style explanations.[3]

A genre of formal modelling we think is much under-utilised in transitions research is what we would call mathematical 'toy models'. These models are deliberately simplified and abstracted as much as possible, even at the expense of sometimes not being models of anything in 'real' life at all. The idea is that, nevertheless, important aspects of dynamics (or structure!) are captured, and, moreover, can be more properly understood without the noise introduced by reality. The advantage of toy models lies in their susceptibility to mathematical analysis, often even without recourse to simulation. And even if simulation is used, the outcomes can typically be understood directly in terms of the model's basic assumptions. Toy models are the bread and butter of theoretical physics and important advances in social science have also made use of them. Good examples are Watts and Strogatz's (1998) seminal work on small-world networks and Dodds et al.'s (2003) model of information exchange in organisational networks. John Holland also strongly advocated the use of toy models[4] (e.g. in Holland, 2006) for research on complex adaptive systems – arguably precisely the class of systems in which transitions occur.

3.1.4 Data, analytics and metrics

We believe that transitions research' relation with data has become severely crippled by the field's emphasis on narrative case studies. The result is an obsession with the particular, with the exceptions. From a scientific perspective, we should be interested in the general, the rules. Thus, we should be thinking of empirical data in terms of *data sets*, rather than case data. Clearly, we should be treating those data with formal methods as well. Nevermind the false dichotomy between qualitative and quantitative, if the ascendance of 'big data' teaches us anything it is that qualitative data are just as amenable to formal methods as quantitative data.

If we want to learn about transitions – in the plural – reproducing a case, or validating against one is not going to take us very far. Probably the very aim of reproducing a case is misguided – unless one is actually interested in the particulars of that case, rather than in the general transitions phenomenon it is an example of. We should first identify patterns in the case data, that is, recurring patterns across cases, and then explain those patterns with our models. Thus we need data *sets* and methods to work with them. Below we present a small list of things we should have data sets about, again a list non-exhaustive and possibly overlapping items:

- *Events* – A natural place to start would be to take the existing body of case studies and turn them in a large, structured data set. Though not quite as easy as that sounds, a methodological proposal is given in Chapter 12 of this volume by de Haan, Martínez Arranz and Spekkink. An immediate benefit from such an approach would be an ability to perform serious meta-analysis across cases.
- *Social networks* – Much is being said about the role of actors in transitions, including individuals. Constructing and analysing data sets of the social networks of actors in transitions seems a straightforward opportunity. Social Network Analysis is a long-established field with a host of ready-to-use tools, most of them available as free and open source software.
- *Organisations* – Like with individual actors and their social networks, so we should look in the networks of organisations involved in transitions and ideally cross-analyse them with social networks. A data-focus on organisations could also more simply look into the (changing) attributes

of organisations over time, including simple statistics regarding their numbers, sizes, assets and so on.

- *Infrastructure and technology* – Long recognised as a key factor in transitions, so an obvious candidate. Data sets could focus on cataloguing varieties in use and their change over time as well as registring more quantitative aspects like levels of adoption, installed capacities or capital in assets etc.
- *Laws, policies, grants, subsidies, patents, bibliometrics etc.* – Despite these being mostly qualitative entities, there is no reason why they should not be considered sources of data. Simply counting numbers of laws enacted of particular kinds across transitions and their relative timing for example would probably teach us a great deal. This is similar to what is done with patents in related fields.
- *Perceptions, values*, consumer demand and sentiment – Often considered an important factor in transitions but too often seen as the domain of narrative, qualitative research. With text analysis software and social media, these are now well within the scope of more data-driven approaches.

Where to obtain these data is of course a crucial question. Some of these data sets will have to be purposely and actively constructed. For some of the above, e.g. events, a fair amount of raw data are available but they are currently not in a readily usable form and making them so is serious work. For others, in fact most, even the raw data still needs to be collected. More positively though, the availability of data in machine-readable formats has never been greater. Social media provide a rich, new source of data about ongoing transitions, e.g. regarding consumer perceptions, communities of practice and initiatives by key individuals and organisations. More traditional data sources, like bureaux of statistics, government agencies and chambers of commerce now often make their data available for download or via APIs.

Since data is so important and since usable data sets are so hard to come by, we should be creating open, shared and cummulative data sets. Taking cues from the free and open software community and its off-shoot open culture, data sets can be published in, or as, public repositories. There are many licencing options one can choose from[5] to protect the authorship of the creators of the content (e.g. you) while giving other researchers maximal opportunity to use, improve and extend your data sets. Another source that is becoming increasingly tractable is raw text (e.g. from news sources) by means of text mining software.

Then, once we have the data, what to do with them? In short, we use it to test our hypotheses or to explore it as a source of inductive generalisation, to inspire new hypotheses.

Hypothesis testing is of obvious and paramount importance in a scientific approach to transitions research and it can take many forms. One of the most basic, yet powerful ways is statistical hypothesis testing. However, to our knowledge there is not a single instance of a hypothesis being tested statistically in the field of transitions research, the large amount of empirical work notwithstanding. Admittedly, our own work is no exception here but the point remains. Perhaps the lack of available data sets like discussed above is part of what causes this lacuna.

We already discussed hypotheses from the model-building perspective earlier and this is another key way of approaching hypothesis testing: with formal models – other than statistical models, that is, though statistics would nearly always enter at some stage of the process. Some models produce results that are readily compared against data. For more abstract models (e.g. the genre of toy models discussed earlier) there is more 'distance' to data. In such cases it is necessary to derive ideal types, abstractions, patterns etc. from the data. An example would be that a toy model predicts adoption to develop over time as a certain mathematical function. To test the model, one then needs to be able to say whether across cases, this behaviour is indeed observed.

As regards data for exploration, for coming up with new hypotheses, there are many approaches available. First of all, various well-established methods for descriptive statistics are at one's disposal, again usually as free and open source software. Defining meaningful new metrics (also for hypothesis testing) can also be important to see patterns in perhaps overwhelming amounts of data. Other ways of using large volumes of data fall in the general category of 'mining', e.g. text mining, and the same story about the availability of free and open source software applies. Likewise for the important area of data visualisation. Graphical representations, especially of high-dimensional data often provide insight not otherwise obtainable.

3.2 Coordination, Quality and Quantity

The combined output of the transitions modelling community is modest. Even if seen in proportion to transitions research at large, production is not very high in our opinion. This can be measured in publications, but also in project funding, number of modelling PhD theses and actual models developed. Likewise, we also feel that the quality of the extant transitions modelling leaves a fair bit to be desired. This is perhaps somewhat more difficult to substantiate, but a good proxy would be the publication outlets. While our narrative colleagues are by now more or less regularly publishing in high end journals like the Nature and Science franchises, we rarely, if at all, reach these heights.

Within the absolute constraints we face, there are things we can do to improve this situation. A starting point, on a more positive note, is that despite its modest size, there at least *is* a modelling community that identifies as a community. This fact alone has inspired activity, conference sessions, journal articles and this book. Strengthening this community and its work can therefore play an important role. This strengthening can be considered in a technical sense but also in a social sense. We will return to the latter in the conclusion, here we will discuss a few opportunities for strengthening transitions modelling research in the technical sense: *initiatives*, *allies* and *infrastructure*.

- *Initiatives* – Projects of all sorts will serve to focus efforts and combine forces. This book is an example. Ideally, we would set up multi-year, well-funded research projects and attract a cohort of graduate students. Not only will such projects increase output, the effort of acquiring the funds for them will force us to be critical about the quality and relevance of our work.
- *Allies* – We should be more strategic about our research allies. We should participate, as individual researchers and as a community, in the activities of *methodologically cognate* fields and collaborate with them in projects. By this we mean other modelling-oriented research communities, such as computational social science, network science research and various others. We should also broaden our audience and publish more in journals that reach a non-transitions audience that would be interested in our work from a technical perspective. We occasionally do such things of course, our point is that we should do it systematically and strategically. Again, such efforts would have the double benefit of increasing our output as well as raise our standards.
- *Infrastructure* – We need to make our efforts *cumulative* instead of parallel. Like we described above in the context of data sets, we need our work – models and data sets – to be open, shared and cumulative. Since most of this work is code or otherwise in digital formats, we are in the fortunate position that the infrastructure for this is already in place and typically at no cost. Publishing the code of our models in publicly accessible repository with free and open source licences should be the norm, and for many of us it already is of course. Likewise it should be with data sets. Sometimes, intellectual property issues impede open practices, e.g. universities or funders may insist source to remain closed, so it may be good to address this as early as the grant writing stage.

4 Conclusion

This book appears in a time when transitions research has become a well-established field. In the prologue, we reminisce that modelling has been a part of the transitions research portfolio since the very beginning. So it is not without some irk that we observe that the growth and maturation of the larger field is not proportionately followed by the modelling subfield. We hope the reader forgive us for not providing a decent quantification of this observation, but it seems safe to say that of the more than 500 transitions articles appearing in 2018 (Köhler et al., 2019), the number of modelling contributions would be in the order of a percent, if that. Psychologically, we are loath to blame this on a lack of effort from the side of the modelling community. But then what is to blame?

The obvious reason for the relatively small output of transitions modelling is the fact that the transitions modelling community is small. But this really only moves the question one level up. Why is the transitions modelling community so small? Is it that modelling does not suit transitions, or vice versa? We do not think so. In fact, many of the reasons usually offered for modelling being unsuitable for transitions research (complexity, uncertainty, humans involved) are, to us, reasons to rely more on modelling rather than on the intrinsically ambiguous and unrigorous practice of interpretative storytelling typically suggested in its stead. Historically speaking, the development towards a small modelling niche was not at all obvious or inevitable. Early transitions research drew heavily on integrated assessment and complexity, both of which emphasise the importance of modelling.

We feel that the status of modelling in the broader transitions community and the nature of mainstream transitions research play a considerable role. While the transitions modelling community has generally advocated methodological diversity and positioned modelling as a *complementary* approach (see Holtz et al., 2015; Köhler et al., 2018), this generous and inclusive attitude is not reciprocated. More strongly put, modelling as an approach to transitions research is tolerated, at best. Qualitative, narrative approaches are considered naturally superior and modelling is heeded caution.

Consider, for example McDowall and Geels (2017): 'modellers should be cautious about over-promising, despite the attractions of policy influence and status that often come with confident projections. A core risk with apparently crisp quantitative tools (…)'. This is the world on its head. We leave it as an exercise to the reader to find such a 'confident projection', issued in a transitions modelling article without a raft of caveats, especially in the rare case where such an article dare speak directly to a policy audience. Contrast this with Geels et al.'s (2017) confidence when they open the dedicated 'Policy Implications' section of their Science article (taking up a good 15% or so of the text) with: 'General policy implications for accelerated low-carbon transitions can be derived from the above lessons'. In case the reader thinks this article presents rigorously tested scientific hypotheses from which those 'implications' derive – it is a Science article, after all – then the reader is wrong.

Geels et al. give their article the subtitle '[a]ccelerating innovation is as important as climate policy'. As this was published in the 'Policy Forum' section of Science, there is at least the suggestion that policy makers should consider their recommendations and re-allocate funds from climate policy to support innovation policy. If that is the case, then we should not be satisfied with a review of historical case studies fit in a template dynamic. Incidentally, the article claims that the 'sociotechnical' style of narrative case study research is the more appropriate mode for those aspects 'which are difficult to model but crucial in real-world transitions'. If they are difficult to model, then *a fortiori*, we should not be relying on the intrinsically ambiguous and error prone approach of case story analysis – let alone base sweeping policy recommendations on them. This is about the future of society and public money. We need models and data to frame and test hypotheses.

This brings us back to the Grand Challenges of transitions. The above examples should make transitions modellers and the broader transitions community acutely aware of the importance of transitions research, and moreover, of the importance of approaching it scientifically. We therefore see an important role for modelling and modellers in transitions research. Firstly, *modelling* is central in making transitions research a scientific endeavour and beginning to address the Grand Challenges. This has already been discussed in this epilogue and in Chapter 4. Secondly, and because of the first point, *modellers* are – or rather, could rise to the challenge and become – *the scientific and methodological conscience of transitions research.*

This exaltation is justified, again in two ways, (1) by virtue of the relative superiority of modelling as a set of methods in the pursuit of scientific transitions knowledge. And, (2) by virtue of modelling transgressing the boundaries of transitions research. Modelling methods are not wedded to their applications and shared across diverse research areas. Consequently, much higher standards apply. This also works in a social fashion. Modellers discuss models with modellers from radically different fields, while transitions case narratives are only interesting to other transitions researchers. Therefore, modelling is instrumental in avoiding a self-referential transitions 'bubble'.

Another away of phrasing this is that modelling should lead and not follow. Our attitude needs to change. Modelling should assert itself and not accept a subservient, mere complementary approach. Clearly we are not suggesting there is no place at all for non-modelling approaches in transitions research, methodological pluralism is still precious. We suggest that modelling can provide the methodological benchmark for all of transitions research to at least comply with scientific standards. This will no doubt also require modellers to raise their standards. Nevertheless, modelling should lead. Transitions are too important to leave to the storytellers.

Notes

1 See e.g. https://en.wikipedia.org/wiki/Action_language.

2 The PDDL family was inspired by the language for the *ST*anford *R*esearch *I* nstitute *P*roblem *S*olver (STRIPS) and others. It was developed to facilitate the International Planning Competition of 1998/2000 where algorithms are pitted against each other. It was envisioned that a common formal language would support progress in the field. See https://en.wikipedia.org/wiki/Planning_Domain_Definition_Language.

3 Not all scientific theory is in mechanistic form. In fact, the most fundamental theories of physics notoriously are not. For example, Newton's law of universal gravitation (or Einstein's General Relativity for that matter) does not provide mechanisms for gravity. It describes, very accurately, how gravitation behaves but not in a cause-and-effect manner. Strongly put, the equations do not justify any statement like: 'the sun causes the movements of the planets in their orbits'. Causation remains a philosophically problematic concept and we do not want to wed ourselves or anyone else to causal hypotheses only.

4 He calls them 'exploratory models' and they are not to be confused with the exploratory modelling discussed in this book, they are really very different things – and his emphasis is slightly different. He draws an analogy to the thought experiments of theoretical physics.

5 In fact, there are so many options that finding a suitable licence may be a bit of a quest. For content as such, typically one will want one of the Creative Commons licences (see https://creativecommons.org/share-your-work/, for example the one Wikipedia uses. For actual databases, there is the possibility of licencing the database separate from its content (see https://opendatacommons.org/), like OpenStreetMap does.

References

Dodds, Peter Sheridan, Duncan J. Watts, and Charles F. Sabel (2003, October). Information exchange and the robustness of organizational networks. *Proceedings of the National Academy of Sciences 100* (21), 12516.

Geels, Frank W., Benjamin K. Sovacool, Tim Schwanen, and Steve Sorrell (2017, September). Sociotechnical transitions for deep decarbonization. *Science 357* (6357), 1242.

Haslum, Patrik, Nir Lipovetzky, Daniele Magazzeni, and Christian Muise (2019). *An Introduction to the Planning Domain Definition Language.* Series Lectures on Artificial Intelligence and Machine Learning. San Rafael, CA, USA: Morgan & Claypool Publishers.

Healy, Kieran (2017, June). Fuck Nuance. *Sociological Theory 35* (2), 118–127.

Hedström, Peter (2005). *Dissecting the Social.* Cambridge, New York, Melbourne: Cambridge University Press.

Holland, John H. (2006, March). Studying Complex Adaptive Systems. *Journal of Systems Science and Complexity 19* (1), 1–8.

Holtz, Georg, Floortje Alkemade, Fjalar de Haan, Jonathan Köhler, Evelina Trutnevyte, Tobias Luthe, Johannes Halbe, George Papachristos, Emile Chappin, Jan Kwakkel, and Sampsa Ruutu (2015, December). Prospects of modelling societal transitions: Position paper of an emerging community. *Environmental Innovation and Societal Transitions 17*, 41–58.

Kim, Yusoon, Yi-Su Chen, and Kevin Linderman (2015, January). Supply network disruption and resilience: A network structural perspective. *Journal of Operations Management 33-34* (1), 43–59.

Köhler, Jonathan, Fjalar de Haan, Georg Holtz, Klaus Kubeczko, Enayat Moallemi, George Papachristos, and Emile Chappin (2018, January). Modelling Sustainability Transitions: An Assessment of Approaches and Challenges. *Journal of Artificial Societies and Social Simulation 21* (1), 8.

Köhler, Jonathan, Frank W. Geels, Florian Kern, Jochen Markard, Elsie Onsongo, Anna Wieczorek, Floortje Alkemade, Flor Avelino, Anna Bergek, Frank Boons, Lea Fünfschilling, David Hess, Georg Holtz, Sampsa Hyysalo, Kirsten Jenkins, Paula Kivimaa, Mari Martiskainen, Andrew McMeekin, Marie Susan Mühlemeier, Bjorn Nykvist, Bonno Pel, Rob Raven, Harald Rohracher, Björn Sandén, Johan Schot, Benjamin Sovacool, Bruno Turnheim, Dan Welch, and Peter Wells (2019). An agenda for sustainability transitions research: State of the art and future directions. *Environmental Innovation and Societal Transitions 31*, 1–32.

Levesque, Hector J., Raymond Reiter, Yves Lespérance, Fangzhen Lin, and Richard B. Scherl (1997). GOLOG: A logic programming language for dynamic domains. *The Journal of Logic Programming 31* (1), 59–83.

McDowall, Will and Frank W. Geels (2017, March). Ten challenges for computer models in transitions research: Commentary on Holtz et al. *Environmental Innovation and Societal Transitions 22*, 41–49.

Newman, M. E. J. (2003, January). The Structure and Function of Complex Networks. *SIAM Review 45*, 167–256.

Noy, Natalya F and Deborah L McGuinness (2001). Ontology development 101: A guide to creating your first ontology. Technical Report http://protege.stanford.edu/publications/ontology_development/ontology101.pdf.

Perera, Supun S., Michael G. H. Bell, Mahendrarajah Piraveenan, Dharshana Kasthurirathna, and Mamata Parhi (2018). Topological Structure of Manufacturing Industry Supply Chain Networks. *Complexity 2018*, 23.

Russell, Bertrand (1923, June). Vagueness. *Australasian Journal of Psychology and Philosophy 1* (2), 84–92.

Scott, J. (1991). *Social Network Analysis: A Handbook* (Second edition ed.).London, Thousand Oakes, New Delhi: SAGE Publications.

Spekkink, Wouter (2015, July). Building capacity for sustainable regional industrial systems: An event sequence analysis of developments in the Sloe Area and Canal Zone. *Special Volume: Support your future today! Turn environmental challenges into opportunities. 98*, 133–144.

van Dam, Koen Haziël and Zofia Lukszo (2006, October). Modelling Energy and Transport Infrastructures as a Multi-Agent System using a Generic Ontology. In *Proceedings of the 2006 IEEE International Conference on Systems, Man, and Cybernetics (SMC)*, The Grand Hotel, Taipei, Taiwan, pp. 890–895.

Wasserman, Stanley and Katherine Faust (1994). *Social Network Analysis: Methods and Applications*. Structural Analysis in the Social Sciences. Cambridge: Cambridge University Press.

Watts, Duncan J. and Steven H. Strogatz (1998, June). Collective dynamics of 'small-world' networks. *Nature 393* (6684), 440–442.

Index

Note: Page numbers in italic indicate a figure and page numbers in bold indicate a table on the corresponding page.